CISM COURSES AND LECTURES

The series presents lecture notes, monographs, edited works and proceedings in the field of Mechanics, Engineering, Computer Science and Applied Mathematics.
Purpose of the series in to make known in the international scientific and technical community results obtained in some of the activities organized by CISM, the International Centre for Mechanical Sciences.

INTERNATIONAL CENTRE FOR MECHANICAL SCIENCES

COURSES AND LECTURES - No. 356

CRACK AND CONTACT PROBLEMS FOR VISCOELASTIC BODIES

EDITED BY

G.A.C. GRAHAM
SIMON FRASER UNIVERSITY
AND
J.R. WALTON
TEXAS A & M UNIVERSITY

SPRINGER-VERLAG WIEN GMBH

Le spese di stampa di questo volume sono in parte coperte da
contributi del Consiglio Nazionale delle Ricerche.

This volume contains 28 illustrations

Originally published by Springer-Verlag Wien New York in 1995

In order to make this volume available as economically and as rapidly as possible the authors' typescripts have been reproduced in their original forms. This method unfortunately has its typographical limitations but it is hoped that they in no way distract the reader.

ISBN 978-3-211-82686-7 ISBN 978-3-7091-2694-3 (eBook)
DOI 10.1007/978-3-7091-2694-3

PREFACE

These Lecture Notes contain an account of the material presented at the Advanced School on Crack and Contact Problems for Viscoelastic Bodies held at the International Centre for Mechanical Sciences in Udine (Italy), September 5-9, 1994.

The main emphasis of these Lecture Notes is on constructing solutions to specific problems; however properties of the equations of viscoelasticity that provide the theoretical underpinnings for constructing such solutions are also covered.

Particular attention is paid to the solution of crack and contact problems. This work is of interest in the context of polymer fracture, modelling of material behaviour, rebound testing of polymers and the phenomenon of hysteretic friction.

The most general methods of solution presuppose that inertial effects may be neglected. The evolution of stress singularities at notches and at tips of cracks meeting at the interface between dissimilar viscoelastic media is traced as a function of viscoelastic material properties; with application to special cases, and extension to anisotropy.

For crack problems and contact problems where the contact area varies with time the source of the main difficulty is that the regions over which different types of boundary conditions are prescribed generally vary with time and may not be known apriori. A fundamental decomposition of the hereditary integrals of linear viscoelasticity leads to the solution of a wide variety of non-inertial problems.

Crack propagation in inhomogeneous and viscoelastic media in the presence of inertial effects is also discussed, from different points of view. Both constant speed and accelerating cracks are treated. Also, modelling and thermodynamic restrictions on viscoelastic behaviour are studied. Existence and uniqueness results and wave solutions for linear viscoelasticity are presented. This material complements that described in previous paragraphs.

An aim of the Lecture Notes is to contribute towards bringing linear viscoelasctic stress analysis to the same level of development as linear elastic stress analysis.

It is a pleasure to record here our thanks to the officers of the CISM, and in particular to Professor Giovanni Bianchi, Professor Sandor Kaliszky and Professor Carlo Tasso for inviting us to give the lectures and agreeing to publish them; and for ensuring that our stay in Udine was so very enjoyable and rewarding.

G.A.C. Graham

J.R. Walton

CONTENTS

Contents

STRESS SINGULARITIES IN VISCOELASTIC MEDIA AND RELATED PROBLEMS

C. Atkinson
Imperial College, London, UK

ABSTRACT

Stress singularities at sharp notch and crack tips in viscoelastic media are considered for both homogeneous and inhomogeneous media (particularly bimaterials). In Lecture 1 various invariants for elastic media are briefly reviewed and extended to the viscoelastic case. Some useful asymptotic results for both Laplace and Mellin transforms are also briefly reviewed. In Lecture 2 near crack tip behaviour is derived and a problem of a crack in an inhomogeneous viscoelastic strip treated by means of one of the invariants derived in Lecture 1. Dual variational principles are also derived for use in this problem. In Lecture 3 the problem of a crack meeting a viscoelastic bimaterial interface is considered and it is shown how the nature of the stress singularity evolves with time depending on the relaxation moduli of the bimaterial components. A partial interpretation of these results is given in terms of the force on a dislocation interacting with the bimaterial interface. In Lecture 4 the time evolution of the singular stress field at a wedge or notch tip in plane strain is discussed. Finally in Lecture 5 we consider the problem of interpreting a crack extent and location from measurements on the inside of a hole from which the crack is initiated.

LECTURE 1

1. INVARIANTS FOR ELASTIC AND VISCOELASTIC MEDIA

1.1 INTRODUCTION

Here we set up the fundamental equations of viscoelasticity required. A derivation is also given of the energy momentum tensor for a continuum system. Various invariant integrals are then deduced in terms of this tensor with some applications discussed in more detail in later lectures. Invariants based on ideas related to a reciprocal theorem are also discussed. Finally some useful asymptotic results for the Laplace and Mellin transforms are presented.

1.2 THE EQUATIONS OF VISCOELASTICITY

For an isotropic medium the viscoelastic constitutive equations can be written in terms of the deviatoric components of stress s_{ij} and strain e_{ij} as

$$s_{ij} = \int_{-\infty}^{t} G_1(t-\tau)\,\frac{de_{ij}(\tau)}{d\tau}\,d\tau \tag{1.1}$$

and

$$\sigma_{kk} = \int_{-\infty}^{t} G_2(t-\tau)\,\frac{d\epsilon_{kk}(\tau)}{d\tau}\,d\tau \tag{1.2}$$

with

$$\begin{aligned} s_{ij} &= \sigma_{ij} - \tfrac{1}{3}\delta_{ij}\sigma_{kk}\ , \qquad s_{ii} = 0 \\ e_{ij} &= \epsilon_{ij} - \tfrac{1}{3}\delta_{ij}\epsilon_{kk}\ , \qquad e_{ii} = 0 \end{aligned} \tag{1.3}$$

where the infinitesimal strain tensor ϵ_{ij} is defined as

$$\epsilon_{ij} = \tfrac{1}{2}[u_{i,j} + u_{j,i}]\ , \tag{1.4}$$

u_i are the cartesian components of displacement and a comma denotes partial differentiation.

In the above equations the relaxation functions $G_1(t)$ and $G_2(t)$ specify the material behaviour.

A creep integral form of the stress-strain relations can be written

$$e_{ij} = \int_{-\infty}^{t} J_1(t-\tau)\,\frac{ds_{ij}(\tau)}{d\tau}\,d\tau \qquad (1.5)$$

$$\epsilon_{kk} = \int_{-\infty}^{t} J_2(t-\tau)\,\frac{d\sigma_{kk}(\tau)}{d\tau}\,d\tau \qquad (1.6)$$

where $J_1(t)$ and $J_2(t)$ are two independent isotropic creep functions. From the above equations it is apparent that $G_1(t)$ and $J_1(t)$ are the relaxation and creep functions appropriate to states of shear while $G_2(t)$ and $J_2(t)$ are defined relative to states of dilation.

The creep and relaxation functions are not independent and can easily be defined in terms of their Laplace transforms. Thus with the definition

$$\bar{f}(s_1) = \int_0^{\infty} e^{-s_1 t} f(t)dt \qquad (1.7)$$

the Laplace transform of the above equations gives

$$\bar{s}_{ij} = s_1\bar{G}_1\bar{e}_{ij} \quad , \quad \bar{\sigma}_{kk} = s_1\bar{G}_2\bar{\epsilon}_{kk}$$

and

$$\bar{e}_{ij} = s_1\bar{J}_1\bar{s}_{ij} \quad , \quad \bar{\epsilon}_{kk} = s_1\bar{J}_2\bar{\sigma}_{kk} \qquad (1.8)$$

For consistency, these equations require

$$\bar{J}_\alpha = (s_1^2\bar{G}_\alpha)^{-1} \qquad \alpha = 1,2 \qquad (1.9)$$

where we have assumed that all functions are such that their Laplace transforms exist. Another useful result is the relation between the above equations and the differential form of the constitutive equations such as used in spring and dashpot models. For example if we have a version of equation (1.1) as

$$H_1\left(\frac{d}{dt}\right) s_{ij} = H_2\left(\frac{d}{dt}\right) e_{ij} \qquad (1.10)$$

where H_1 and H_2 are polynomial functions of the operator $\frac{d}{dt}$, then applying the Laplace transform gives

$$H_1(s_1)\bar{s}_{ij} = H_2(s_1)\bar{e}_{ij} \quad . \qquad (1.11)$$

Hence, by comparing with equations (1.8) we have

$$s_1\bar{G}_1 \equiv H_2(s_1)/H_1(s_1), \qquad s_1\bar{J}_1 \equiv \frac{H_1(s_1)}{H_2(s_1)} \tag{1.12}$$

when an equivalence exists. For the examples we consider later we shall use simple model relaxation functions for which either formulation is suitable. However, the methods presented later will be applicable to the most general viscoelastic relaxation functions.

In addition to the above equations we have the equation of motion, which in cartesian coordinates (x_1,x_2,x_3) is

$$\frac{\partial\sigma_{ij}}{\partial x_j} = \rho\,\frac{\partial^2 u_i}{\partial t^2} \qquad i,j = 1,2,3 \tag{1.13}$$

and there is a summation over repeated indices.

1.3 A 'PSEUDO' ENERGY MOMENTUM TENSOR

If it is assumed that all fields are zero prior to $t = 0$ then taking the Laplace transform of (1.13) gives

$$\frac{\partial\bar{\sigma}_{ij}}{\partial x_j} = \rho s_1^2\bar{u}_i \qquad i,j = 1,2,3 \tag{1.14}$$

Equations (1.8), (1.4) (Laplace transformed) and (1.14) can be deduced from the conditions of stationarity of the functional $\int L\,dV$ where L is an appropriate function of $\bar{u}_i$ and $\bar{u}_{i,j}$. Thus defining

$$L \equiv L(\bar{u}_i,\bar{u}_{i,j},x_j) \tag{1.15}$$

such that

$$\bar{\sigma}_{ij} = \frac{\partial L}{\partial\bar{u}_{i,j}} \tag{1.16}$$

and the Euler-Lagrange equations for stationarity of $\int L\,dV$ as

$$\frac{\partial}{\partial x_j}\left(\frac{\partial L}{\partial \bar{u}_{i,j}}\right) - \frac{\partial L}{\partial \bar{u}_i} = 0 \qquad (1.17)$$

one defines

$$L = \frac{1}{2}\bar{\sigma}_{ij}\bar{\epsilon}_{ij} + \frac{1}{2}\rho s_1{}^2\bar{u}_i\bar{u}_i \quad . \qquad (1.18)$$

We are allowing all material constants to be inhomogeneous here so they can vary with x_i. It can be readily checked that (1.16) gives

$$\bar{\sigma}_{ij} = s_1\bar{G}_1\bar{\epsilon}_{ij} + \frac{s_1}{3}\delta_{ij}(\bar{G}_2-\bar{G}_1)\bar{\epsilon}_{kk} \qquad (1.19)$$
$$\bar{\epsilon}_{ij} = \frac{1}{2}(\bar{u}_{i,j}+\bar{u}_{j,i})$$

and (1.17) gives (1.14). Once L has been defined we can deduce certain invariants in terms of the 'pseudo' energy momentum tensor $P_{\ell j}$ defined as

$$P_{\ell j} = L\delta_{\ell j} - \frac{\partial L}{\partial \bar{u}_{i,j}}\bar{u}_{i,\ell} \quad . \qquad (1.20)$$

Now

$$P_{\ell j,j} = \frac{\partial L}{\partial x_\ell} - \frac{\partial L}{\partial \bar{u}_{i,j}}\bar{u}_{i,\ell j} - \bar{u}_{i,\ell}\frac{\partial}{\partial x_j}\left(\frac{\partial L}{\partial \bar{u}_{i,j}}\right) \qquad (1.21)$$

and

$$\frac{\partial L}{\partial x_\ell} = \frac{\partial L}{\partial \bar{u}_i}\bar{u}_{i,\ell} + \frac{\partial L}{\partial \bar{u}_{i,j}}\bar{u}_{i,j\ell} + \left(\frac{\partial L}{\partial x_\ell}\right)_{exp} \qquad (1.22)$$

where $\left(\frac{\partial L}{\partial x_\ell}\right)_{exp}$ means differentiation of L partially with respect to x_ℓ keeping $\bar{u}_i$ and $\bar{u}_{i,j}$ constant. This allows L to depend explicitly on x_j and hence the material to be spatially inhomogeneous. Combining equations (1.21) and (1.22) and using (1.17) leaves

$$P_{\ell j,j} = \left(\frac{\partial L}{\partial x_\ell}\right)_{exp} \qquad (1.23)$$

which in turn means that the integrals

$$F_\ell = \int_S P_{\ell j} dS_j \qquad \ell = 1,2,3 \quad \text{(summation over j)} \tag{1.24}$$

will be zero when taken around closed surfaces enclosing no singularities provided L does not depend explicitly on x_ℓ. ($dS_j = n_j dS$ where n_j are components of the unit outward normal). This follows by an application of the divergence theorem since $\int_S P_{\ell j} dS_j = \int_V P_{\ell j,j} dV$ where the closed surface S encloses the volume V. There are other invariants that can be deduced from the properties of $P_{\ell j}$ under special circumstances we consider only one other here. Noting that the repeated index summation convention is continually used here, then

$$\begin{aligned}(x_\ell P_{\ell j})_{,j} &= \delta_{\ell j} P_{\ell j} + x_\ell P_{\ell j,j} \\ &= P_{jj} + x_\ell P_{\ell j,j}\end{aligned}$$

From (1.23) and (1.20) this equals zero provided the material is homogeneous (no explicit dependence on x_ℓ in the material properties) there are *no* inertia terms ($\rho = 0$) and conditions of plane strain exist so j takes values 1 and 2 only, hence $\delta_{jj} = 2$. Then the integral

$$M = \int_S x_\ell P_{\ell j} dS_j \tag{1.25}$$

is invariant where S is now a cylinder with axis in the x_3 direction and there are no variations in the displacement fields in the x_3 direction so the integral can be viewed as a line integral in the (x_1,x_2) plane j and ℓ taking values 1 and 2 only. Later we will also consider a similar plane version of (1.24) in our applications.

1.4 A RECIPROCAL THEOREM

This theorem can be written as

$$\int_S \bar{\sigma}_{ij}\hat{u}_i dS_j = \int_S \hat{\sigma}_{ij}\bar{u}_i dS_j \tag{1.26}$$

where the hat functions $\hat{u}_i$ etc. satisfy the same field and constitutive equations as $\bar{\sigma}_{ij}$ but different boundary conditions. Thus

$$\hat{\sigma}_{ij,j} = \rho s^2 \hat{u}_i \quad \text{and} \quad \hat{\sigma}_{ij} = \bar{c}_{ijk\ell}\, \hat{u}_{k,\ell} \tag{1.27}$$

where we have written the Laplace transformed constitutive equation as if the material were anisotropic. The tensor of Laplace transformed viscoelastic coefficients satisfies the symmetry relations $\bar{c}_{ijh\ell} = \bar{c}_{jih\ell} = \bar{c}_{ij\ell h}$. The result (1.26) then follows by an application of the divergence theorem. By appropriate selection of the functions $\hat{u}_i$ the relation (1.26) can be used to relate near tip singular behaviour of the actual stress and displacement fields $\bar{u}_i$ and $\bar{\sigma}_{ij}$ to far field behaviour. Examples of this are to be given later.

1.5 ASYMPTOTICS OF LAPLACE AND MELLIN TRANSFORMS

We have already defined the Laplace transform in equation (1.7) as

$$\bar{f}(s_1) = \int_0^\infty e^{-s_1 t} f(t)dt \tag{1.7}$$

If we assume that f(t) is locally integrable on $[0,\infty)$ and that as $t \to \infty$

$$f(t) = O(e^{at}) \tag{1.28}$$

for some real number a, then the integral defining $\bar{f}(s_1)$ exists and $\bar{f}(s_1)$ will be analytic in the complex s_1 half plane $R\ell\ s_1 > a$.

If we further suppose that as

$$t \to 0+ \qquad f(t) \sim \sum_{m=0}^{\infty} c_m t^{a_m} \tag{1.29}$$

with $\mathrm{Re}(a_m)$ increasing monotonically to $+\infty$ as $m \to \infty$ and $\mathrm{Re}(a_0) > -1$, then it can be shown that

$$\bar{f}(s_1) \sim \sum_{m=0}^{\infty} c_m \int_0^{\infty} e^{-s_1 t}\, t^{a_m} dt$$

$$= \sum_{m=0}^{\infty} \frac{c_m \Gamma(a_m+1)}{s_1^{a_{m+1}}} \tag{1.30}$$

Thus we can relate the asymptotic expansion of $f(t)$ as $t \to 0+$ with the asymptotic expansion of the Laplace transform $\bar{f}(s_1)$ as $|s_1| \to \infty$. Results which relate the behaviour of $\bar{f}(s_1)$ as s_1 tends to zero with that of $f(t)$ as t tends to infinity are less well defined. Expanding the exponential in (1.7) in powers of s_1t and integrating term by term shows that an expansion of $\bar{f}(s_1)$ in powers of s_1 has coefficients which are the moments of $f(t)$ (*i.e.* $\frac{(-s_1)^N}{N!} \int_0^{\infty} t^n f(t)dt$) assuming that $f(t)$ is such that these integrals exist. This will be so if $f(t)$ decays exponentially as t tends to infinity. However, to extract such exponential decay from the behaviour of $\bar{f}(s_1)$ we would be better served by detecting the poles of $\bar{f}(s_1)$ and/or the left hand limit of its region of analyticity. This follows from the well known inversion formula for the Laplace Transform:-

$$f(t) = \frac{1}{2\pi i} \int_{c-i\infty}^{ci+\infty} \bar{f}(s_1) e^{s_1 t} ds_1 \tag{1.31}$$

where $c = a$ when $f(t) = o(e^{at})$ as $t \to +\infty$ as in equation (1.28).

1.5.1 The Mellin Transform

This transform which will be useful when we consider certain wedge problems later is defined as

$$\bar{\bar{\phi}}(z,\theta,s_1) = \int_0^\infty \bar{\phi}(r,\theta,s_1)r^{z-1}dr \tag{1.32}$$

where we have assumed that this transform (an integral over r a polar coordinate) is taken over a function which has already been Laplace transformed.

It will usually be clear (or explicitly stated) which transform is being used when the overbar notation is used later on.

The Mellin transform of a function $\bar{\phi}(r,s_1)$ converges absolutely and is holomorphic in the strip

$$a < \mathrm{Re}(z) < \beta \tag{1.33}$$

where

$$\alpha = \inf\{\alpha^* \mid f=O(r^{-\alpha^*}) \quad \text{as} \quad r \to 0+\}$$
$$\beta = \sup\{\beta^* \mid f=O(r^{-\beta^*}) \quad \text{as} \quad r \to +\infty\} \tag{1.34}$$

Thus when the Mellin transform of a function $\bar{\phi}$ converges absolutely it does so in a vertical strip of the complex z-plane whose boundaries are determined by the asymptotic behaviour of $\bar{\phi}$ in the limits $r \to 0+$ and $r \to +\infty$. The inversion formula for the Mellin transform (analogous to that for the *2-sided* Laplace transform and the Fourier transform) is

$$\bar{\phi}(r,\theta,s_1) = \frac{1}{2\pi i}\int_{c_1-i\infty}^{c_1+i\infty} r^{-z}\bar{\bar{\phi}}(z,\theta,s_1)dz \tag{1.35}$$

with $\alpha < c_1 < \beta$.

For the problems we consider later such as the singular stress fields at notch tips key information for r small can be obtained by closing contour integrals like (1.35) into the left hand half of the complex z-plane. Similarly information as $r \to \infty$ can be obtained by closing the integral into the right hand half of the complex z-plane. The residues at the poles of $\bar{\bar{\phi}}$ in these regions of the complex z-plane will be particularly important.

There are more complicated problems for which asymptotic behaviour at a finite value of r might be required *e.g.* at r = 1. This can be obtained by the use of theorems such as that which leads to equation (1.30) for the Laplace transform. We shall not require them here but see Atkinson (1975) for some applications which do.

LECTURE 2

2. STRESS SINGULARITIES AT VISCOELASTIC CRACK AND NOTCH TIPS (1) ANTI-PLANE STRAIN

2.1 INTRODUCTION

The well-known square root singular stress field at an elastic crack tip in a homogeneous medium is rederived as an eigenvalue problem and shown to apply also for anti-plane strain deformation in viscoelastic media. Later these results are generalised to singular fields at wedge shaped notches using Mellin and Laplace transform methods. The application of some of the invariants of lecture 1 to these problems will also be considered.

2.2 ANTI-PLANE STRAIN DEFORMATION

For this mode of deformation there is only one non-zero component of displacement u_3 which acts in the x_3 direction but is a function of x_1 and x_2. The only non-zero components of stress are σ_{13} and σ_{23} and the equilibrium equations (2.13) reduce to

$$\frac{\partial \sigma_{13}}{\partial x_1} + \frac{\partial \sigma_{23}}{\partial x_2} = \rho \frac{\partial^2 u_3}{\partial t^2} \tag{2.1}$$

From equations (1.1) and (1.5) we also have the constitutive equations in the relaxation, or creep forms,

$$\sigma_{13} = \int_{-\infty}^{t} G_1(t-\tau) \frac{d\sigma_{13}(\tau)}{d\tau} d\tau \tag{2.2}$$

$$e_{13} = \int_{-\infty}^{t} J_1(t-\tau) \frac{d\sigma_{13}(\tau)}{d\tau} d\tau \tag{2.3}$$

$i = 1$ or 2

with $$s_1^2 \bar{J}_1(s_1)\bar{G}_1(s_1) = 1 \tag{2.4}$$

and $$e_{13} = \frac{1}{2} \frac{\partial u_3}{\partial x_1} \tag{2.5}$$

s_1 being the Laplace transform parameter. If all fields are zero for $t < 0$ then the Laplace transform of equations (2.1), (2.2) and (2.3) give

$$\frac{\partial\bar{\sigma}_{13}}{\partial x_1} + \frac{\partial\bar{\sigma}_{23}}{\partial x_2} = \rho s_1^2 \bar{u}_3 \tag{2.6}$$

and

$$\bar{\sigma}_{13} = \frac{s_1}{2} G_1(s_1)\partial u_3/\partial x_1 \quad . \tag{2.7}$$

2.2 Anti-Plane Crack Problems

We consider the problem shown in Figure 1, a viscoelastic strip of width 2h has a central stress-free semi-infinite crack on $x_2 = 0$, $-\infty < x_1 < 0$ and on the sides $x_2 = \pm h$ a time dependent displacement is applied. The boundary conditions are thus

$$\begin{array}{llll} \text{on} & x_2 = \pm h & u_3 = \pm f(t)H(t) & \text{for all } x_1 \\ \text{and on} & x_2 = 0 & -\infty < x_1 < 0 & \sigma_{23} = 0 \end{array} \tag{2.8}$$

The strip may be inhomogeneous so the viscoelastic moduli may depend on x_2 but this property is assumed symmetric with respect to $\pm x_2$. If convenient the problem (2.8) can thus be replaced by

$$\left.\begin{array}{ll} \text{on } x_2 = +h & u_3 = f(t)H(t) \\ \text{on } x_2 = 0 & -\infty < x_1 < 0,\ \sigma_{23} = 0 \\ \text{on } x_2 = 0 & x_1 > 0 \quad u_3 = 0 \end{array}\right\} \tag{2.9}$$

$$H(t) = 1 \quad t > 0 \text{ is the Heaviside step function,}$$
$$= 0 \quad t < 0$$

and all displacements are zero for $t < 0$.

Applying the Laplace transform (1.7) to these boundary conditions and combining (2.6) and (2.7) gives

$$\frac{\partial}{\partial x_i}\left(\frac{s_1}{2}\bar{G}_1(s_1,x_2)\frac{\partial\bar{u}_3}{\partial x_i}\right) = \rho s_1^2\bar{u}_3, \quad i = 1,2 \tag{2.10}$$

with

$$\text{on } x_2 = h \qquad \bar{u}_3 = \bar{f}(s)$$

$$\text{on } x_2 = 0 \quad \begin{cases} \bar{\sigma}_{23} = 0 & -\infty < x_1 < 0 \\ \bar{u}_3 = 0 & x_1 > 0 \end{cases} \tag{2.11}$$

2.2.1 Near-Crack Tip Behaviour on Eigensolutions

Before proceeding with the solution of this problem it is useful to study the behaviour at the crack tip. Since close to the crack tip $x_2 = 0$ it can be seen by a Taylor expansion about $x_2 = 0$ that the singular near crack tip field is dominated by the solutions of the equation

$$\frac{\partial}{\partial x_i}\left(\left(s_1 \bar{G}_1(s_1,0)\right) \frac{\partial \bar{u}_3}{\partial x_i}\right) = 0 \tag{2.12}$$

Solutions of this equation which satisfy the boundary conditions (2.11) are given by

$$\bar{u}_3 = Ar^{\lambda}\sin(\lambda\theta) \tag{2.13}$$

where $\cos\lambda\pi = 0$ to satisfy condition (2.11).

Thus $\lambda = \pm \dfrac{(2N+1)}{2}$ and the least singular solution compatible with an integrable energy density is that with $\lambda = \frac{1}{2}$. Thus

$$\bar{u}_3 = Ar^{\frac{1}{2}}\sin(\theta/2)$$

$$\bar{\sigma}_{23} = \frac{As_1\bar{G}_1(s_1)}{4} r^{-\frac{1}{2}}\cos(\theta/2)$$

$$\bar{\sigma}_{13} = \frac{As_1\bar{G}_1(s)}{4} r^{-\frac{1}{2}}\sin(\theta/2) \tag{2.14}$$

are the leading terms in the near crack tip stress and displacement fields, and A is a function of s_1 to be determined from the full solution of the problem. A is independent of the crack tip polar coordinates (r,θ). We see also that the form of this near crack tip field the power of r and the function θ are *independent* of the viscoelasticity of the medium.

The effect of the viscoelasticity is seen only through the factor A (related to the stress-intensity factor). It should be noted that the above result is not true if the crack is moving and the inertia terms are taken into account, the power $r^{\frac{1}{2}}$ is still independent of the medium but the angular dependence is not. It is also clear from (2.14) that r and θ factors remain the same in the real time expressions of stress and displacement. For plane strain the power of r is as above but only when Poisson's ratio is constant is the real time θ dependence independent on time.

2.2.2 A 'Simple' Solution Using An Integral Invariant

An integral invariant for equation (2.10) can be deduced by the method of section (1.3). We deduce the Lagrangian

$$L = \frac{1}{4} s_1 \bar{G}_1(s_1,x_2) \frac{\partial \bar{u}_3}{\partial x_i} \frac{\partial \bar{u}_3}{\partial x_i} + \frac{1}{2}\rho s^2 \bar{u}_3^{\,2} \tag{2.15}$$

(summation over x_i for i = 1,2, of course).

Then from (1.20)

$$P_{\ell j} = L\delta_{\ell j} - \sigma_{j3}\bar{u}_{3,\ell} \; . \tag{2.16}$$

We now consider

$$F_1 = \int_S P_{1j} dS_j \tag{2.17}$$

with j = 1,2 and the integral following the arrowed line of Figure 1 *i.e.* a rectangle from $(-\infty,\pm h)$ to $(+\infty,\pm h)$.

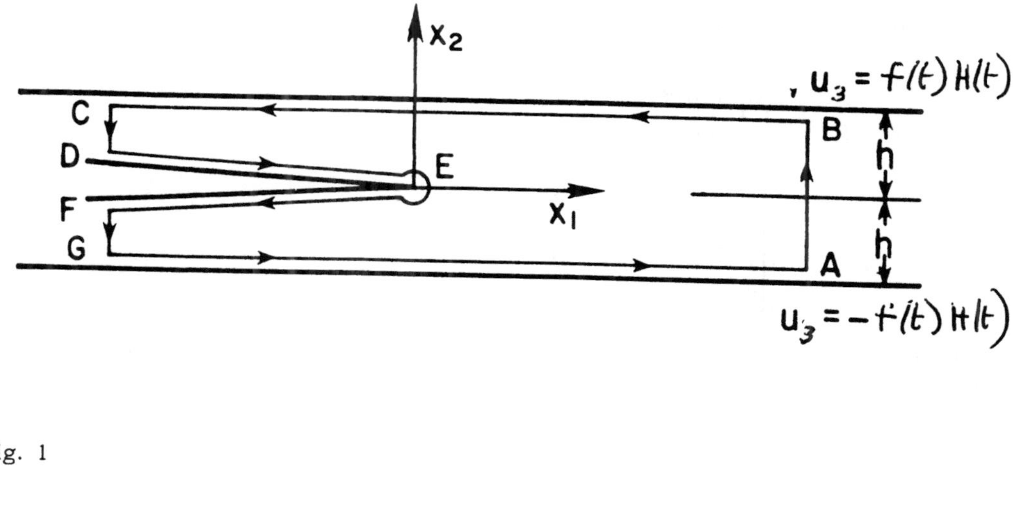

Fig. 1

On $x_2 = \pm h$, $\bar{u}_{3,1} = 0$ because of the boundary conditions (2.8). Since ds_2 is proportional to dx_1 on $x_2 = \pm h$ it follows that the contribution to the integral (2.17) is zero from these parts of the contour. Furthermore on the crack $x_2 = 0$, $x_1 < 0$, $\bar{\sigma}_{23} = 0$ (from the stress free boundary condition) so $P_{12} = 0$ here also. The only non-zero contributions to (2.17) come from $x_1 = \pm\infty$ $-h < x_2 < h$. For these limiting values of x, P_{11} tends to L provided $\bar{u}_{3,1}$ tends to zero as $x_1 \to \pm\infty$. To find L in this limit, we must solve the equilibrium equation which reduces to

$$\frac{d}{dx_2}\left(\frac{s_1}{2} G_1(s_1,x_2) \frac{d\bar{u}_3}{dx_2}\right) = \rho s_1^2 \bar{u}_3 \tag{2.18}$$

with

$$L = \frac{1}{4} s_1 \bar{G}_1(s_1,x_2)\bar{u}_{3,2}^2 + \frac{1}{2}\rho s_1^2 \bar{u}_3^2 \tag{2.19}$$

along the ligaments $-h < x_2 < h$, $x_1 = +\infty$, with boundary conditions $\bar{u}_3(\pm h) = \pm\bar{f}(s_1)$ and $0 < x_2 < h$, $-h < x_2 < 0$, $x_1 = -\infty$ with the additional conditions $\bar{u}_{3,2} = 0$ on x_2.

Finally one needs to calculate the contribution to F_1 from the integral around the crack tip at $x_1 = 0, x_2 = 0$. The only non-zero contribution comes from the expressions in (2.14) since they make $P_{\ell j} = O(1/r)$.

Integration of (2.17) then results in

$$F_1 = \frac{s_1 \bar{G}_1(s_1,0)\pi A^2}{8} \tag{2.20}$$

$$= \frac{1}{2}\left[\int_0^h L(+\infty,x_2)dx_2 - \int_0^h L(-\infty,x_2)dx_2\right]$$

where for the right hand side we have used the symmetry in the problem about $x_2 = 0$.

In general it is not possible to solve equation (2.18) exactly when the functions $\overline{G}_1(s_1,x_2)$ and $\rho(x_2)$ are not constants. Asymptotic methods can be used to get some information when s_1 is either very small or large. For intermediate values of s_1 bounds on the integrals appearing in (2.20) can be obtained by variational methods as follows.

2.2.3 Dual Variational Principles in Transform Space

We derive here some dual variational principles for the one-dimensional problem defined by equation (2.18) with various boundary conditions. The same methods can be applied in principle to more space dimensions (see Atkinson and Appleby (1994) for some applications to coupled thermo-viscoelasticity).

Beginning with the Lagrangian defined in (2.19) we define a conjugate variable U by

$$U = \frac{\partial L}{\partial \bar{u}_{3,2}} = \tfrac{1}{2} s_1 G_1(s_1,x_2)\bar{u}_{3,2} \quad . \tag{2.21}$$

Now define a Hamiltonian $H(x_2,\bar{u}_3,U)$ by means of

$$H(x_2,\bar{u}_3,U) = U\bar{u}_{3,2} - L(x_2,\bar{u}_3,\bar{u}_{3,2}) \quad . \tag{2.22}$$

Using (2.21) and (2.19), (2.22) can be written

$$H(x_2,\bar{u}_3,U) = \frac{U^2}{s_1\overline{G}_1(s_1,x_2)} - \tfrac{1}{2}\rho s_1^2\bar{u}_3^2 \quad . \tag{2.23}$$

Consider now the integral

$$I(x_2,\bar{u}_3,U) = \int_0^h (U\bar{u}_{3,2} - H(x_2,\bar{u}_3,U))dx_2 - \Big[U(\bar{u}_3 - \bar{u}_3^B)\Big]_0^h \tag{2.24}$$

where $\bar{u}_3^B$ are specified boundary values of $\bar{u}_3$ at $x_2 = 0$ and $x_2 = h$. If we vary I about the functions $\bar{u}_3$ and U assumed independent in the usual manner of the calculus of variations it can be shown that for I to be stationery at $\bar{u}_3$, U the following equations must be satisfied:-

$$\bar{u}_{3,2} = \frac{\partial H}{\partial U}, \qquad -U_{,2} = \frac{\partial H}{\partial \bar{u}_3} \tag{2.25}$$

and the boundary conditions $\bar{u}_3 = \bar{u}_3^B$ on $x_2 = 0$ and h must be satisfied. Equations (2.25) are just the usual Euler-Lagrange equations for a functional depending on *two independent* variables $\bar{u}_3$ and U. Furthermore, the first of equations (2.25) serves to define U as in equation (2.21) and the second of equations (2.25) is then just the equation of motion (2.18).

Let us use the following notation

$$\tilde{u}_3 \text{ and } U \tag{2.26}$$

are the exact solution of our problem with

$$\tilde{u}_3 = \bar{u}_3^B \quad \text{on the boundaries } x_2 = 0 \text{ and } h \tag{2.27}$$

Thus $\tilde{u}_3$ and $\tilde{U}$ satisfy equations (2.25) when u_3 and U are replaced by them. also by integrating the first term in the integral of (2.24) by parts we can rewrite

$$I(x_2,\bar{u}_3,U) = \int_0^h (-\bar{u}_3 U_{,2} - H(x_2,\bar{u}_3,U))dx_2 + \left[U\bar{u}_3^B)\right]_0^h . \tag{2.2\ \ }$$

If we now use the second of equations (2.25) to specify $\bar{u}_3$ in terms of U we have

$$\rho s_1^2 \bar{u}_3 = U_{,2} ,$$

$$F_1(U) = I(x_2, \frac{U_{,2}}{\rho s_1^2}, U)$$

$$= \int_0^h \left[-\frac{U_{,2}^2}{2\rho s_1^2} - \frac{U^2}{s_1 \bar{G}_1(s_1,x_2)} \right] dx_2$$

$$+ \left[U\bar{u}_3^B) \right]_0^h . \qquad (2.28)$$

If we now consider variaticns $U = \tilde{U} + \epsilon\eta(x_2)$ in the usual manner of the calculus of variations we find

$$F_1(U) - F_1(\tilde{U}) = -\epsilon^2 \int_0^h \left\{ \frac{[\eta^1(x_2)]^2}{2\rho s_1^2} + \frac{\eta^2(x_2)}{s_1 \bar{G}_1(s_1,x_2)} \right\} dx_2 . \qquad (2.29)$$

The test functions U should be continuous with piecewise continuous derivatives to first order. There is no requirement that $\eta(x_2)$ should be zero at $x_2 = 0$ and $x_2 = h$.

Thus from (2.29) it follows that

$$F_1(U) - F_1(\tilde{U}) \le 0 \qquad (2.30)$$

provided $\rho(x_2) > 0$ and $\bar{G}_1(s_1,x_2) > 0$ for all $0 \le x_2 \le h$.

Similar arguments can be applied to the expressions (2.24) with U replaced by (2.21) (which is the same as solving equation (2.25) for U) and varying the function $\bar{u}_3$ subject to the boundary conditions $\bar{u}_3 = \bar{u}_3^B$.

Thus defining

$$F_2(\bar{u}_3) = \int_0^h L(x_2,\bar{u}_3,\bar{u}_{3,2})dx_2 \qquad (2.31)$$

where the test functions $\bar{u}_3$ satisfy $\bar{u}_3 = \bar{u}_3^B$ at $x_2 = 0$ and h. The test functions $\bar{u}_3$ are to be continuous with continuous derivatives up to second order. One then obtains

$$F_1(U) \le F_1(\tilde{U}) = \int_0^h L(x_2,\tilde{u}_3,\tilde{u}_{3,2})dx_2$$

$$= F_2(\tilde{u}_3) \le F_2(\bar{u}_3) \tag{2.32}$$

with

$$\bar{u}_3^{\,B} = \bar{f}(s) \quad \text{on} \quad x_2 = h \tag{2.33}$$

$$= 0 \quad \text{on} \quad x_2 = 0$$

then the above formulation serves to give a method for the integral $\int_0^h L(+\infty,x_2)dx_2$ of equation (2.20). This can be done by trying various test functions U and $\bar{u}_3$ in (2.32) to calculate the bounds as is usual with the calculus of variations.

The procedure is the same for estimating $\int_0^h L(+\infty,x_2)dx_2$ except that we replace the integral I of (2.24) by

$$I_1(x_2\bar{u}_3,U) = \int_0^h [U\bar{u}_{3,2}-H(x_2,\bar{u}_3,U)]dx_2$$

$$- \{u(\bar{u}_3-\bar{u}_3^{\,B})\}_h$$

since the trial functions U have U = o at $x_2 = 0$ from the stress free boundary conditions on the crack; the subscript h means evaluated at $x_2 = h$.

LECTURE 3

ABSTRACT

The problem of a semi-infinite crack meeting a plane welded interface between two dissimilar viscoelastic materials under antiplane strain (mode 3 deformation) is considered. We investigate the manner in which two different viscoelastic solids affect the time evolution of the singular stress field.

To this end, integral transform methods are used to establish that the singularity behaviour is governed by an eigenvalue equation, in Laplace transform space identical in form to the corresponding elastic problem provided the effects of inertia are neglected. Explicit solutions to the eigenequation are available enabling the stress concentrations to be expressed as contour integrals within an appropriately cut complex z-plane using analytic continuation. Numerical inversion around appropriate paths is necessary due to the complicated nature of the integrands and the real time behaviour of the stress fields is compared to long and short time asymptotic expansions (relating the properties of the bimaterial) for distances near and far from the crack tip. A discussion of the results is made for intermediate times where there is a transition between these extreme elastic limits. The solution is also given for a screw dislocation interacting with a bimaterial interface and this simple solution is used to explain the qualitative behaviour of the crack tip stress field by means of an analogy between the crack and a dislocation pile up.

3.1 INTRODUCTION

There is a large literature on problems of stress concentrations and associated stress singularities at notch tips, corner and cracks in elastic bodies, plates, bimaterials, etc.; some of this work is reviewed in Atkinson (1977) and England (1971). For example, the problem of a crack meeting an interface between two different elastic media under longitudinal shear deformation was first considered by Williams (1959) who by means of an eigenvalue analysis showed that the stress at the crack tip was proportional to $r^{-\lambda}$, where r is the distance from the crack tip and λ is a constant lying between zero and one depending on the ratio of the shear moduli of the bimaterial. The case of a crack approaching the interface or just passing through has been considered by Atkinson (1975) who showed that the stress intensity factor (the stress is proportional to $r^{-0.5}$ for the crack tip lying in either medium, r again measured from the crack tip) is a function of the (small) distance ε the crack is from (or just through) the interface. For ε small enough this function is ε raised to a power depending on the coefficient λ. Similar ideas can be applied to other elastic stress singularities. It is our purpose here to extend results of this kind to the situation where the bimaterial is viscoelastic.

As will be seen in the analysis section the application of a correspondence principle (*i.e.* noting that the Laplace transform of the viscoelastic equations is identical to the Laplace transform of the elastic equations provided the constant elastic moduli are replaced by corresponding functions of s_1) gives an eigenvalue equation for the Laplace transformed problem which has the same form as the elastic problem when the effects of inertia are neglected. However, to elucidate the real time evolution of the singular crack tip field is by no means straightforward. We need to monitor the appropriate branch of the solution of the eigenvalue equation as a function of s_1 and hence effect the Laplace transform inversion. For the problem considered here, the mode -3 crack meeting an interface, the solution of the eigenvalue equation can be effected explicitly and hence analytic continuation through the complex s_1 plane performed. This allows various integration paths to be taken to complete the Laplace inversion. This is considered in detail in Atkinson and Bourne (1989) where the characteristics of the complex integrands are discussed. A complete solution of the problem for the real time evolution of the singular crack tip stress field under various internal loadings can thus be effected. Some results are presented later for a unit load acting on the crack for $0 \le r < 1$ where r is the distance along the crack front, and various material combinations where the viscoelastic media are represented either as a linear solid or an Achenbach-Chao solid (*cf.* Christensen (1971) and Achenbach and Chao (1962)). The results are combined with asymptotic results (valid for any linear viscoelastic solids) for large and short times. The short time results follow by use of an appropriate Tauberian Theorem.

A detailed discussion of the results is deferred to section 3.5; however we also derive the solution to the problem of a screw dislocation interacting with a bimaterial interface. This is given in section 3.4. Explicit real time solutions can be obtained for this problem and it is interesting to compare the nature of the time evolution of the force on the dislocation with that of the singular crack tip stress field. In each case a rationalisation of the results can be obtained. This being motivated by the interpretation of a crack as a pile up of virtual dislocations so a single dislocation is merely a rather crude approximation to the crack.

3.2 Formulation of the Problem

We consider a semi-infinite crack lying perpendicular to a plane welded interface between two dissimilar viscoelastic media. The crack lies on the plane $x_2 = 0$, extends from $x_1 = -\infty$ to $x_1 = 0$ and is of infinite extent in the x_3-direction (normal to the page), and the interface is taken to be the line $x_1 = 0$. The typical situation is illustrated in Fig. 2 (here we take the angle $\alpha = 0$). For the purpose of our analysis, however, we neglect the inertia terms on the right hand side of equation of motion, thereby working with

$$\frac{\partial \sigma_{ij}^{(m)}}{\partial x_j} = 0 \quad . \tag{3.1}$$

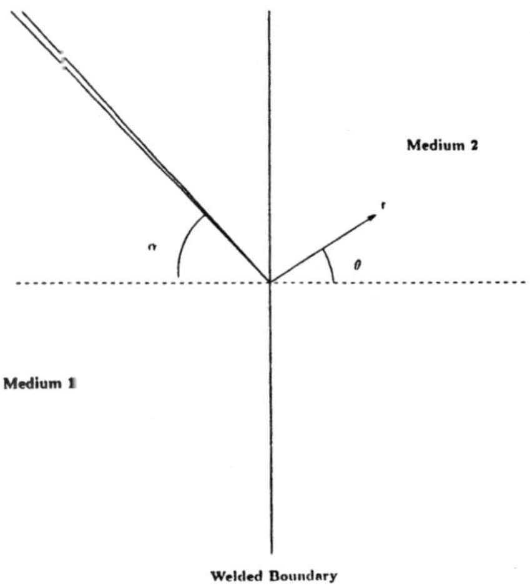

Fig. 2

The purpose of our analysis is to study how the singular terms of the stress field evolve with time due to the time dependence of the loading and the constitutive behaviour of the material and these singular terms are controlled to a large extent by the highest spatial derivative terms on the left hand side of (3.1). Thus, neglecting the inertia terms is not a major omission in the calculations of the singular stress field provided the crack remains stationery.

Since we are concerned with the case of anti-plane strain, the only non-zero component of displacement is $u_3^{(m)}(x_1,x_2,t)$ while the only non-vanishing stresses are $\sigma_{13}^{(m)}(x_1,x_2,t)$ and $\sigma_{23}^{(m)}(x_1,x_2,t)$ where the superscript (m) refers to the region (the crack lies in region 1).

The stress-strain constitutive equations are as given in Lecture 1 *i.e.*

$$\sigma_{i3}^{(m)} = \int_{-\infty}^{t} G_1^{(m)}(t-\tau)\,\frac{de_{i3}^{(m)}}{d\tau}\,d\tau \tag{3.2}$$

$$e_{i3}^{(m)} = \frac{1}{2}\left(\frac{\partial u_3^{(m)}}{\partial x_i}\right) \qquad i = 1,2 \tag{3.3}$$

or using the differential operator notation they are simply

$$H_1^{(m)}\left(\frac{d}{dt}\right)\sigma_{i3}^{(m)} = J_1^{(m)}\left(\frac{d}{dt}\right)e_{i3}^{(m)} \qquad i = 1,2 \quad . \tag{3.4}$$

Likewise, the set of three equations (3.1) reduce for anti-plane strain to the single equation

$$\frac{\partial\sigma_{13}^{(m)}}{\partial x_1} + \frac{\partial\sigma_{23}^{(m)}}{\partial x_1} = 0 \quad . \tag{3.5}$$

Based on the geometry of the problem it proves convenient to introduce an (r,θ) coordinate system based at the crack tip and to replace the displacement component $u_3^{(m)}(x_1,x_2,t)$ by $\phi^{(m)}(r,\theta,t)$. To proceed, define a Laplace transform of $\phi^{(m)}(r,\theta,t)$ over t as

$$\bar{\phi}^{(m)}(r,\theta,s_1) = \int_0^\infty \phi^{(m)}(r,\theta,t)e^{-s_1 t}\,dt \tag{3.6}$$

and denote by

$$\bar{\bar{\phi}}^{(m)}(z,\theta,s_1) = \int_0^\infty \bar{\phi}^{(m)}(r,\theta,s_1)r^{z-1}dr \tag{3.7}$$

its subsequent Mellin transform over the radial coordinate r. Applying these in turn to (3.5) gives

$$\frac{\partial^2\bar{\bar{\phi}}^{(m)}}{\partial\theta^2} + z^2\,\bar{\bar{\phi}}^{(m)} = 0 \tag{3.8}$$

Regularity conditions on the displacement as $r \to 0$ and $r \to \infty$ establish that (3.8) exists in the strip $-1 < d_1 < \mathrm{Re}(z) < d_2 < 1$, where d_1 and d_2 are to be determined. Elementary solutions to (3.8) for the transformed displacement fields are taken in the form

$$\bar{\bar{\phi}}^{(1)}(z,\theta,s_1) = A_1(z)\sin z(\theta-\tfrac{\pi}{2}) + B_1(z)\cos z(\theta-\tfrac{\pi}{2}) \tag{3.9}$$

$$\bar{\bar{\phi}}^{(2)}(z,\theta,s_1) = A_2(z)\sin(z\theta) \tag{3.10}$$

where the functions $A_1(z)$, $B_1(z)$ and $A_2(z)$ must be established from the remaining boundary conditions as stated below:

Continuity of stress and displacement at the interface:

$$\left.\begin{array}{l} \sigma_{\theta z}^{(1)} = \sigma_{\theta z}^{(2)} \\ \phi^{(1)} = \phi^{(2)} \end{array}\right\} \quad \text{on} \quad \theta = \frac{\pi}{2} \tag{3.11}$$

Symmetry:

$$\phi^{(2)} = 0 \qquad \text{on} \quad \theta = 0 \tag{3.12}$$

Specified internal crack loading:

$$\sigma_{\theta z}^{(1)} = -f(r)H(t) \qquad \text{on} \quad \theta = \pi \tag{3.13}$$

where H(t) is the Heaviside unit step function. Applying the Laplace and Mellin transforms in turn to the above conditions we obtain

$$\left.\begin{array}{l} \mu_1(s_1)\dfrac{\partial\bar{\bar{\phi}}^{(1)}}{\partial\theta} = \mu_2(s_1)\dfrac{\partial\bar{\bar{\phi}}^{(2)}}{\partial\theta} \\ \bar{\bar{\phi}}^{(1)} = \bar{\bar{\phi}}^{(2)} \end{array}\right\} \quad \text{on} \quad \theta = \frac{\pi}{2} \tag{3.14}$$

$$\bar{\bar{\phi}}^{(2)} = 0 \qquad \text{on} \quad \theta = 0 \tag{3.15}$$

and

$$\mu_1(s_1)\frac{\partial\bar{\bar{\phi}}^{(1)}}{\partial\theta} = \frac{f(z)}{s_1} \qquad \text{on } \theta = \pi \tag{3.16}$$

where

$$f(z) = \int_0^\infty f(r)r^z dr \tag{3.17}$$

and

$$\mu_m(s_1) = \frac{s_1}{2}\,\bar{G}_1^{(m)}(s_1) = \bar{J}_1^{(m)}(s_1)/2\bar{H}_1^{(m)}(s_1) \tag{3.18}$$

define the 'complex moduli'. Using expressions (3.9) and (3.10) it is then quite straightforward to establish that

$$A_1(z) = \frac{\mu_2(s_1)}{\mu_1(s_1)} \cos\left(\frac{z\pi}{2}\right) A_2(z) \tag{3.19}$$

$$B_1(z) = \sin\left(\frac{z\pi}{2}\right) A_2(z) \tag{3.20}$$

where

$$A_2(z) = \frac{2f(z)}{s_1 z[\mu_1(s_1)+\mu_2(s_1)]\left[\cos(z\pi) + \left(\frac{\mu_2(s_1)-\mu_1(s_1)}{\mu_2(s_1)+\mu_1(s_1)}\right)\right]} . \tag{3.21}$$

In particular,

$$\bar{\bar{\phi}}^{(2)}(z,\theta,s_1) = \frac{2f(z)\sin(z\theta)}{s_1 z[\mu_1(s_1)+\mu_2(s_1)](\cos(z\pi)-b)} \tag{3.22}$$

where

$$b = \frac{\mu_1(s_1) - \mu_2(s_1)}{\mu_1(s_1) + \mu_2(s_1)} . \tag{3.23}$$

The real time behaviour of the displacement and stress fields will follow on inverting the above transforms, and each is characterised by solutions to the eigenequation

$$\cos(z\pi) = b \tag{3.24}$$

Explicit solutions are available to (3.24) and one particular enumeration establishes the roots as

$$z = \pm \frac{i}{\pi} \log(b+(b^2-1)^{\frac{1}{2}}) + 2n \qquad n = 0,\pm1,\pm2,\ldots \tag{3.25}$$

where the choice of branch of $(b^2-1)^{\frac{1}{2}}$ is made so that the root to be taken has positive imaginary part and where it is understood that the principal branch of the logarithm has been taken. For small r, the Laplace transformed components of displacement and stress are found using the inverse Mellin transform, closing the contour in the left hand plane and evaluating the residue of the integrands at the first pole

$$z_1 = \frac{i}{\pi} \log(b+(b^2-1)^{\frac{1}{2}}) \quad . \tag{3.26}$$

Here the principal branch of the logarithm has been taken and $(b^2-1)^{\frac{1}{2}}$ denotes that root having positive imaginary part. The results are

$$\bar{\phi}^{(2)}(r,\theta,s_1) = \frac{-2f(z_1)\sin(z_1\theta)r^{-z_1}}{\pi s_1 z_1[\mu_1(s_1)+\mu_2(s_1)]\sin(z_1\pi)} \tag{3.27}$$

and

$$\bar{\sigma}_{\theta z}^{(2)}(r,\theta,s_1) = \frac{-2f(z_1)\mu_2(s_1)\cos(z_1\theta)r^{-(z_1+1)}}{\pi s_1[\mu_1(s_1)+\mu_2(s_1)]\sin(z_1\pi)} \tag{3.28}$$

as $r \to 0$.

Similar expressions are obtained for large r, with z_1 replaced by $-z_1$.

A Laplace inversion of expressions (3.27) and (3.28) will then give the displacements and stress fields in real time, *e.g.*

$$\sigma_{\theta z}^{(2)}(r,\theta,t) = \frac{1}{2\pi i}\int_{c-i\infty}^{c+i\infty} \sigma_{\theta z}^{(2)}(r,\theta,s_1)e^{s_1 t}\,ds_1 \quad . \tag{3.29}$$

For a brief account of the numerical method employed to invert the above contour integrals we refer the reader to the Appendix of Atkinson and Bourne (1989), where particular viscoelastic materials are treated.

3.3 Asymptotic Results

To support the numerical results given in section 3.5 one can deduce long and short time asymptotic expressions associated with the displacement and singular stress fields for the near and far crack tip behaviour. The results (see Atkinson and Bourne (1989)) are valid for any linearly viscoelastic solids and thus demonstrate how the solutions depend on the material properties of the two media and the effect of specific internal loadings applied to the crack. The small time results are obtained by considering the behaviour of the integrands for large s_1 [*i.e.* expand everything in inverse powers of s_1] and applying appropriate Tauberian Theorems, whereas for large times we expect the main contribution to the integrals to come from the residue of the pole at the origin. For reasons of space we omit the details here but note that such asymptotic results will lead to the approximations shown later in the results section and give rise to formulae which although complicated apply to any viscoelastic solid. For small times, for example, the singular stress field looks like the short time elastic one but multiplied by a polynomial in t (the time) with coefficients which depend on logr and θ. In this way we can get some idea as to how the singular fields might evolve. If time permits we shall try to derive some of these results, if not see the paper by Atkinson and Bourne (1989). Some numerical results will be presented later. However, to understand these it is worthwhile considering a simpler problem that of a dislocation interacting with a welded boundary.

3.4 DISLOCATIONS INTERACTING WITH A WELDED BOUNDARY BETWEEN TWO DISSIMILAR VISCOELASTIC MEDIA

We suppose a screw dislocation to occupy the position (-a,0) and to have an associated Burgers vector of magnitude b_0 with the welded interface between the two media at $x_1 = 0$. Our interest, here, lies in establishing explicit expressions for the time evolution of the force on the dislocation which we assume to be nucleated at t = 0.

We proceed as in section 3.2 by applying a Laplace transform to both the constitutive equations and equation of motion to conclude that $\bar{u}_3^{(m)}(x_1,x_2,s_1)$ (the only non-zero component of displacement in the anti-plane strain case) is a harmonic function of x_1 and x_2. Accordingly, the displacement fields in regions 1 and 2 can be written

$$\frac{2\pi s_1}{b_0}\bar{u}_3^{(1)} = \tan^{-1}\left(\frac{x_2}{a+x_1}\right) + K(s_1)\tan^{-1}\left(\frac{x_2}{a-x_1}\right) \tag{3.30}$$

and

$$\frac{2\pi s_1}{b_0}\bar{u}_3^{(2)} = (K(s_1)+1)\tan^{-1}\left(\frac{x_2}{a+x_1}\right) \tag{3.31}$$

where $K(s_1)$ is to be determined.

To establish the function $K(s_1)$ we appeal to conditions of continuity of displacement and stress at the interface, which, when transformed give rise to the requirements

$$\left.\begin{aligned} &\bar{u}_3^{(1)} = \bar{u}_3^{(2)} && (3.32)\\ &\mu_1(s_1)\frac{\partial\bar{u}_3^{(1)}}{\partial x_1} = \mu_2(s_1)\frac{\partial\bar{u}_3^{(2)}}{\partial x_1} && (3.33)\end{aligned}\right\} \text{ on } x_1 = 0$$

Equations (3.30) and (3.31) automatically satisfy (3.32) and condition (3.33) is met provided

$$K(s_1) = \frac{\mu_1(s_1)-\mu_2(s_1)}{\mu_1(s_1)+\mu_2(s_1)} . \qquad (3.34)$$

The force on the dislocation is given by the formula

$$F(t) = b_0\left[\sigma_{23}^{(1)}-(\text{self-stress})\right]_{x_1=-a,x_2=0} , \qquad (3.35)$$

the Laplace transform of which can be shown to be

$$\bar{F}(s_1) = \frac{b_0^2}{4\pi a}\left[\frac{\mu_1(s_1)K(s_1)}{s_1}\right] . \qquad (3.36)$$

Inverting (3.36) will then establish an expression for the force in real time, once a particular viscoelastic model has been assumed.

3.4 Some Numerical Results

For illustration we choose one example from the paper of Atkinson & Bourne (1989) in which medium one is visocelastic while medium two is elastic. The viscoelastic medium is represented by the standard linear solid written in the differential operator form

$$\left(\frac{d}{dt} + \beta_1\right)\sigma_{13}^{(1)} = 2\mu_{10}\left(\frac{d}{dt} + \alpha_1\right)e_{13}^{(1)} . \qquad (3.37)$$

For this representation the relaxation function $G_1^{(1)}(t)$ is given as

$$G_1^{(1)}(t) = 2\mu_{10}\left(\frac{\alpha_1}{\beta_1} - \left(\frac{\alpha-\beta_1}{\beta_1}\right)e^{-\beta_1 t}\right) . \qquad (3.38)$$

Medium 2 is elastic so

$$\sigma_{13}^{(2)} = 2\mu_{20}e_{13}^{(2)} \qquad (3.39)$$

and

$$G_1^{(2)}(t) = 2\mu_{20} \tag{3.40}$$

with the above expressions

$$\bar{\mu}_1(s_1) = \mu_{10}\left(\frac{s_1+\alpha_1}{s_1+\beta_1}\right), \quad \bar{\mu}_2(s_1) = \mu_{20} \quad . \tag{3.41}$$

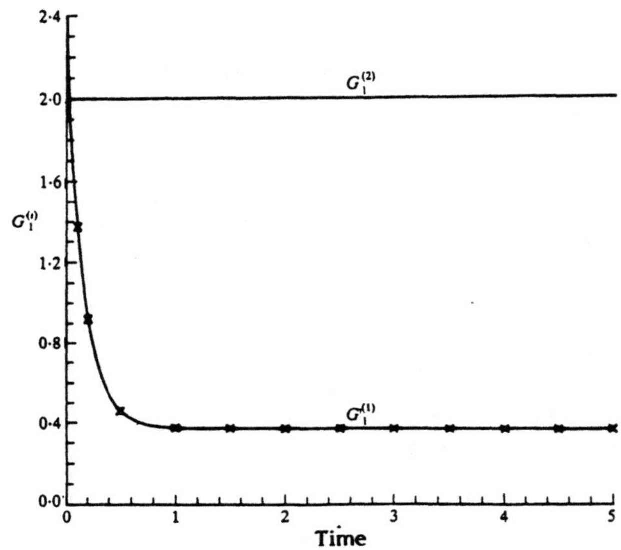

Fig. 3 Relaxation functions for media 1 and 2 with parameters $\mu_{20} = 1$, $\mu_{10} = 1.1$, $\alpha_1 = 1$, $\beta_1 = 6$.

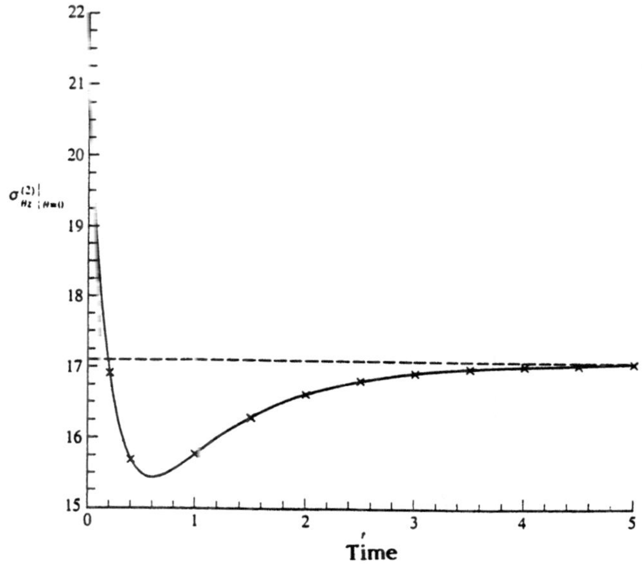

Fig.4 Time evolution of the singular crack tip stress field on $\theta = 0$ ($r = 0.001$) for the loading case (crack in material 1) and its comparison with small- and large-time asymptotic analysis (----).

Figure 3 shows the relaxation functions for a model problem with $\mu_{20} = 1$, $\mu_{10} = 1.1$, $\alpha_1 = 1, \beta_1 = \epsilon$. As can now be seen in Fig. 3 the viscoelastic solid relaxes from a value which is greater than that of medium 2 for short times to one much less for large times. The results shown in Fig. 4 give the evolution of the singular crack tip stress field evaluated at $r = 0.001$ as a function of time together with the asymptotic solutions. The nature of the time evolution for short times requires the asymptotic solution including terms proportional to t. The pure elastic singularity arguments (that the long and short time results for the stress singularity should be that for a composite elastic medium with appropriate long and short time elastic moduli) would just give two points on the graph $t = 0$ and $t = \infty$. It is assumed the crack is loaded internally by a specified internal load. Figure 4 shows a distinct minimum in the stress field as a function of time which cannot be anticipated from the asymptotic results. To try to understand the character of this time evolution curve it is useful to have an independent argument. Note however that a variety of bimaterial cases are considered in the forementioned paper including elastic/viscoelastic and viscoelastic/viscoelastic combinations.

Since a crack can be considered as a distribution of dislocations maybe the characteristics of the force on a single dislocation appropriately interpreted can be used to help understand the crack/bimaterial interaction results derived above.

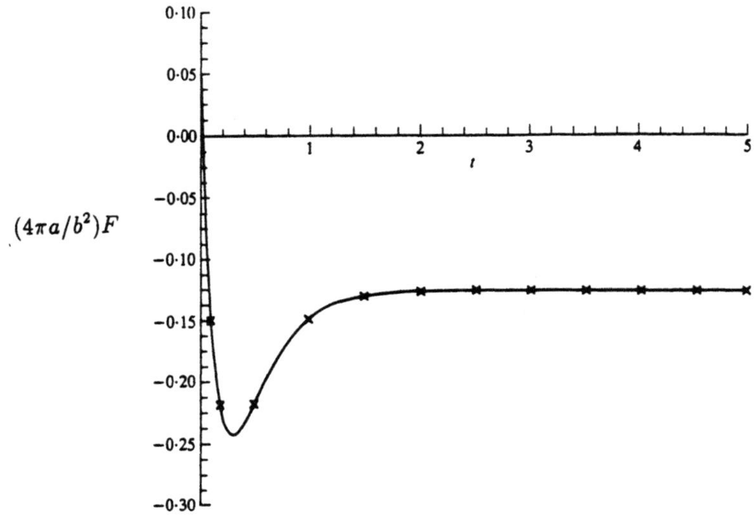

Fig. 5 A plot of $(4\pi a/b^2)F$ against t for the dislocation interaction problem in a bimaterial

Because we are concerned here with mode 3 deformation (the only nonzero displacement being in the 3 direction) we suppose that a screw dislocation occupies the position (-a,0) and to have an associated Burgers vector of magnitude b_0 with the welded interface between the two media at $x_1 = 0$ as before. The force on this dislocation is as derived in section (3.4) (equation (3.36)). Inverting this expression then establishes an expression for the force in real time once a particular viscoelastic model has been assumed. For simple viscoelastic models, such as the standard linear solid, an explicit inversion can be accomplished. Figure 5 shows a plot of $(4\pi a/b^2)F$ against t for the dislocation interaction problem in a bimaterial with properties given above in Equations (3.37) - (3.40) and in Fig. 3. Note that the analogy between the force on the single dislocation and the crack is not exact since the dislocation has been assumed to have a displacement jump which is fixed for all time whereas the crack opening displacement varies with time. Furthermore the crack behaviour is somewhat different depending on whether the loading is applied internally or remotely. Nevertheless, the dislocation solution is interesting in its own right and as the figures show, illustrates the influence of the boundary very clearly. Also, as far as the field far from a crack is concerned the field of a semi-infinite crack will be similar for large r to that of a dislocation with, in this case, some time dependent Burgers vector. This idea can have some useful consequences as we shall see later. See Atkinson and Bourne (1989) for more detailed arguments and a number of other results. Note that the ideas of this section can be applied to plane strain problems in a similar way.

4. PLANE STRAIN STRESS SINGULARITIES AT CORNERS

The main idea of Section 3 was that the eigenvalue equation for stress-singularities in viscoelastic media may have the same form in the Laplace transform domain as that in an elastic medium, but with the major difference that the parameters in the eigenvalue equations are not constants but depend on s_1 the Laplace transform variable through the Laplace transform of the viscoelastic relaxation moduli. This leads to results for the field near a notch or crack tip evolving with time in quite a complicated way. Here we consider the solution near a sharp wedge (or notch) as shown in Fig. 6 and look at the influence of a non-constant Poisson ratio on the singular field near a notch tip.

Figure 6 represents an infinite plate or re-entrant corner of included angle 2α, $\pi < 2\alpha < 2\pi$ with the corner tip based at the origin of an (r,θ) coordinate system. Fixed normal displacements are symmetrically distributed (with respect to the wedge axis $\theta = 0$) along the radial edges of the sector. A particular example considered by Bourne and Atkinson (1990) uses polar coordinates with displacements $u_r(r,\theta,t)$, $u_\theta(r,\theta,t)$ and boundary conditions

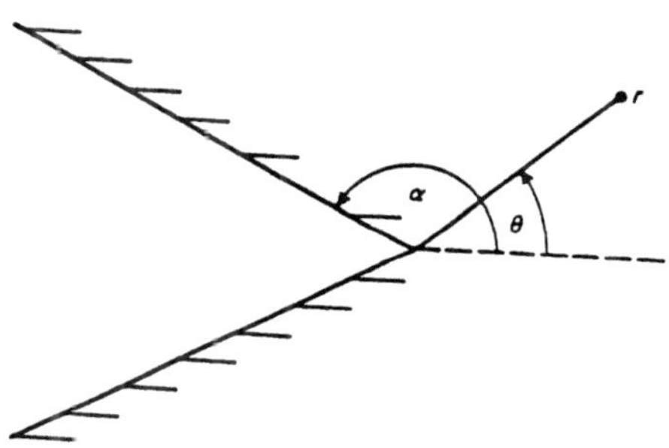

Fig. 6 An infinite plate or reentrant corner of included angle 2α, $\pi < 2\alpha < 2\pi$ with the corner tip based at the origin of an (r,θ) co-ordinate system

$$u_\theta(r,\alpha,t) = - u_\theta(r,-\alpha,t) = F(r)H(t), \quad 0 \le r < \infty \tag{4.1}$$

$$u_r(r,\alpha,t) = u_r(r,-\alpha,t) = 0, \quad 0 \le r < \infty \tag{4.2}$$

where F(r) denotes some arbitrary prescribed loading and H(t) is the Heaviside step function. More general loadings could be assumed without much difficulty. It is assumed that the displacements tend to zero in the body as r tends to infinity and the function F(r) is such that its Mellin transform exists. The following transforms are defined:

$$\bar{u}(r,\theta,s_1) = \int_0^\infty u(r,\theta,t_1)e^{-s_1t}\,dt \quad , \tag{4.3}$$

$$\bar{\bar{u}}(z,\theta,s_1) = \int_0^\infty \bar{u}(r,\theta,s_1)r^z dr \tag{4.4}$$

for the components of displacement and

$$\bar{\bar{\sigma}}(z,\theta,s_1) = \int_0^\infty \bar{\sigma}(r,\theta,s_1)r^{z+1}\,dr \tag{4.5}$$

for the components of stress. Taking the Laplace and Mellin transforms of the equilibrium equations and boundary conditions the transforms $\bar{u}_r(z,\theta,s_1)$ and $\bar{u}_\theta(z,\theta,s_1)$ are obtained as well as the corresponding expressions for the transformed stresses. Applying the inverse Mellin transform to a typical stress component $\bar{\sigma}$ gives

$$\bar{\sigma}(r,\theta,s_1) = \frac{1}{2\pi i}\int_{\gamma-i\infty}^{\gamma+i\infty} \bar{\bar{\sigma}}(z,\theta,s_1)r^{-z-2}dz \quad , \tag{4.6}$$

where γ is real and $-2 < \gamma < -1$ follows from conditions of regularity. We omit the details (see Bourne and Atkinson 1990) but a major feature in the solution is that the poles of the function $\bar{\bar{\sigma}}$ occur at the zeros of the eigenvalue expression

$$(z+1)\sin 2\alpha = \bar{\kappa}(s_1)\sin 2(z+1)\alpha \tag{4.7}$$

where $\bar{\kappa}(s_1) = 3-4s_1\bar{\nu}(s_1)$ for viscoelastic plane strain with $\bar{\nu}(s_1)$ being the time transform of Poissons ratio so $\bar{\nu}(s_1) = \nu_0/s_1$ for an elastic medium. The zeros of the Equation (4.7) move about as s_1 varies for the viscoelastic case and a careful treatment is necessary to analyse the situation in real time. In terms of the Laplace transforms of the relaxation functions $\bar{G}_1(s_1)$ and $\bar{G}_2(s_1)$ we may define the time transform of the viscoelastic Poissons ratio

$\nu(t)$ as

$$\bar{\nu}(s_1) = \frac{1}{s_1}\left(\frac{\bar{G}_2(s_1) - \bar{G}_1(s_1)}{\bar{G}_1(s_1) + 2\bar{G}_2(s_1)}\right) \tag{4.8}$$

the inversion of which gives its real time evolution. See for example Fig. 7 and Figs. 8, 9 for the time evolution of the singular corner tip stress components $\sigma_{\theta\theta}$ and σ_{rr} along the symmetry line $\theta = 0$ for a sector of included angle $2\alpha = \frac{3\pi}{2}$. Note that the shear stress is identically zero along this line. These results are for the standard linear solid where

$$\bar{G}_1(s_1) = \frac{2\mu_{10}}{s_1}\left(\frac{s_1+\alpha_1}{s_1+\beta_1}\right), \quad \bar{G}_2(s_1) = \frac{2\mu_{20}}{s_1}\left(\frac{s_1+\alpha_2}{s_1+\beta_2}\right) \tag{4.9}$$

for a loading such that

$$F(r) = \begin{cases} 1 \text{ for } 0 < r < 1 \\ 0 \; r > 1 \end{cases} \tag{4.10}$$

Note further that the stress field is plotted for $r = 0.001$ so that the singular near field is the dominant contribution to the stress

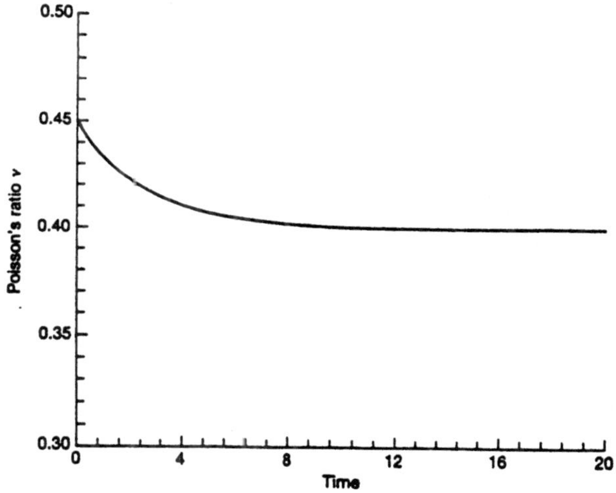

Fig. 7 Poisson's ratio versus time curve based on linear solid with parameters $\mu_{10} = 0.2$, $\alpha_1 = 0.29$, $\beta_1 = 0.4$, $\mu_{20} = 2.9$ $\alpha_2 = 0.28$, $\beta_2 = 0.8$

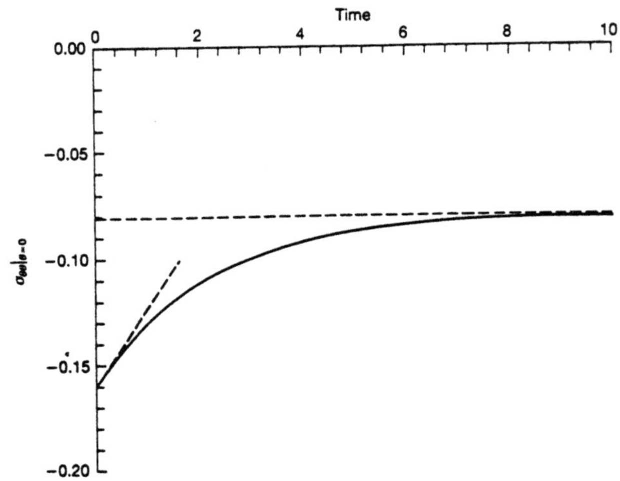

Fig 8 Time evolution of the near notch-tip stress component $\sigma_{\theta\theta}$ at $r = 0.001$, $\theta = 0$ (unit loading case) for a corner of semi-included angle $x = \frac{3\pi}{4}$ and its comparison with short- and long-time asymptotic analysis (-----). Material properties as in Fig.7.

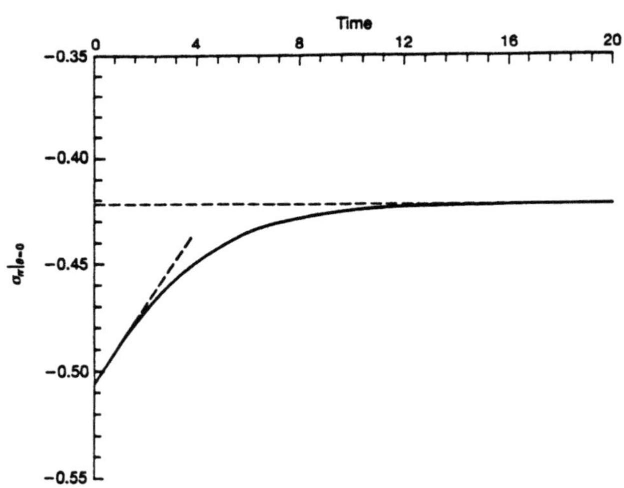

Fig 9 Time evolution of the near notch-tip stress component σ_{rr} at $r = 0.001$, $\theta = 0$ (unit loading case) for a corner of semi-included angle $x = \frac{3\pi}{4}$ and its comparison with short- and long-time asymptotic analysis (-----). Material properties as in Fig.7.

Furthermore, the results are compared with asymptotic expansions for small and large times. In the small time results (limit $s_1 \to \infty$) the inclined straight line (linear in t) follows from a detailed expansion away from the limit solution which follows from the first most singular eigensolution of Equation (4.7) with $\bar{\kappa}(s_1)$ replaced by the limit as $s_1 \to \infty$ $\bar{\kappa}(s_1)$. These limiting solutions simply convey the singular stress solution associated with a material with its instantaneous (short time) elastic moduli (glassy) and that associated with the long time (rubbery) moduli. In between times as seen in Section 3, a complicated evolution between the two limiting solutions is attained.

Various results are presented in Bourne and Atkinson (1990) in particular the angular variation of the near corner tip stress fields for different times. The similarity or variation between the curves for different times may be illuminated by comparison with analysis of the form of the stress field for the associated elastic problems corresponding to elastic material with moduli equal to the short and long time moduli.

For the elastic problem the singular field at the corner tip may be represented by

$$\sigma_{ij} = KF_{ij}(\nu,\theta)r^{-\lambda} \tag{4.11}$$

where r is measured from the corner tip and λ lies between 0 and 1 and is the solution of the relevant eigenequation. The angular variation within the plate is dependent solely on the Poisson ratio ν. Close to the corner tip the stress field is fully characterised by the coefficient K of the $r^{-\lambda}$ term and is dependent on the applied load and material constants. For viscoelastic materials one might hope that the singular stress field may be similarly characterised; however it is clear from the asymptotic results presented in Bourne and Atkinson (1990) that the angular variation will evolve with time also.

Nevertheless, if the dominant contribution to the time dependence of the problem is due to the loading rather than the geometry of the notch, then this time dependence may be absorbed into the parameter K, while the angular variation is slowly varying. In this respect the angular variation of stress components for the short/long time elastic solution to that for the viscoelastic response should not be dissimilar. This is borne out by the results presented in Bourne and Atkinson (1990). Of course, for the case considered there the important variations are in the Poisson's ratio ν which is fairly small for the examples considered.

4.1 Mixed Boundary Conditions on the Wedge Faces

A natural generalisation of the above analysis is to investigate the situation when mixed boundary conditions are applied to the wedge faces. This has been done by Atkinson and Bourne (1990). The geometry is shown in Fig. 6 and the boundary conditions now take the form

$$u_r = u_\theta = 0, \quad \text{on} \quad \theta = -\alpha \tag{4.12}$$

$$\sigma_{\theta\theta} = f(r)H(t), \quad \sigma_{r\theta} = 0 \quad \theta = \alpha \tag{4.13}$$

where again f(r) denotes some applied load whose Mellin transform exists and H(t) is the Heaviside unit step function and again more general loads could be considered without difficulty. This problem could be thought of as debonding on one side of a rigid wedge in a visocelastic medium. Applying in turn the Laplace and Mellin transforms to the above conditions and to the field equations we find that a typical double transformed stress field $\bar{\bar{\sigma}}(z,\theta,s_1)$ required in Equation (4.6) for the inversion integral to give $\bar{\sigma}(r,\theta,s_1)$ has in its denominator a factor determined by the zeros of the eigenequation

$$\sin^2 z_0 = K_1(s_1) - K_2(s_1)z_0^{\ 2} \tag{4.14}$$

where

$$K_1(s_1) = 4(1-s_1\bar{\nu}(s_1))^2/(3-4s_1\bar{\nu}(s_1)), \; K_2(s_1) = \left(\frac{\sin 2\alpha}{2\alpha}\right)^2 /(3-4s_1\bar{\nu}(s_1)) \tag{4.15}$$

and $z_0 = 2(z+1)\alpha$, the above equations holding for plane strain. For the near notch tip stress field the important zeros lie in the strip $-2 < \mathcal{R}(z) < -1$. Note also that the dual solutions for the method outlined in Section 1.4 could be obtained here also from the poles lying in $-1 < \mathcal{R}(z) < 0$ and elsewhere.

Although explicit solutions of the above eigenequation are not available, the Mellin inversion process may be applied for small r (and similarly for large r) by closing the contour in the left hand half-plane and picking up the relevant contributions from the roots $z_j(s_1)$, j = 1,2,3 each of whose real part lies between -2 and -1. Note that in this problem there is more than one solution of the eigenequation which contributes to the singular near crack tip stress field. The real time behaviour of the stress and displacement field follows from inverting the respective Laplace transforms.

As before, for large and small times the behaviour of the corner tip stress field is precisely that of the corresponding elastic problem based on the appropriate short time (glassy) moduli and long time (rubbery) moduli. Close to the corner tip the singular field consists of the sum of three terms each proportional to $r^{-\lambda}$ (r is the distance from the notch tip). The exponents λ each lie between 0 and 1 and are related to solutions of Equation (4.14) *via.* $\lambda = z+2$. The relevant values are tabulated below (Table 1), for the $\nu(t)$ variation given in Figure 11, for both short and long times.

Table 1

	z_1	z_2	z_3
Short time	-1.28179	-1.41275	-1.82176
Long time	-1.29437	-1.39538	-1.83884

For intermediate times the applied load and time dependent properties of the notch material influence the field and a transition between the instantaneous and long time elastic limit occurs. It should be noted that the leading term of the eigenexpansion will dominate the singular field as the corner tip is approached. In this respect each of the most singular fields arising from the leading singularity, given in column 1 of Table 1, should display similar trends as $r \to 0$ discounting rare cases where the coefficient of the leading singularity is zero. This means in magnitude one might expect that the dominant singular stress terms will decrease monotonically between their appropriate elastic limits (*i.e.* from short to long time). However, as an elastic eigenvalue analysis of all three singularities shows this is limited to very small regions confined to the corner tip (typically $r < 10^{-12}$ where $r = 1$ is the typical loading length). For more realistic values of r (*e.g.* $r = 0.001$ shown in Fig. 10) all three singular terms shown in Table 1 come into play and as can be seen from the table even a crude interpretation based on these short and long time results shows the possibility of a complicated interplay between them since z_1 and z_3 decrease algebraically while z_2 increases algebraically between short and long times. This is further complicated by the presence of angular variations of the coefficients in each case (the exponent is, of course, independent of the angle θ). As an illustration of this interplay table 2 shows the instantaneous and long time limits associated with each of the three singular terms in the eigenexpansion of

$$\sigma(r,\theta,t) = \sum_{j=1}^{3} \sigma_j(r,\theta,t) \tag{4.16}$$

where $\sigma_j(r,\theta,t) = K_j(t)F_j(\nu,r,\theta,t)r^{-(z_j(t)+2)}$.

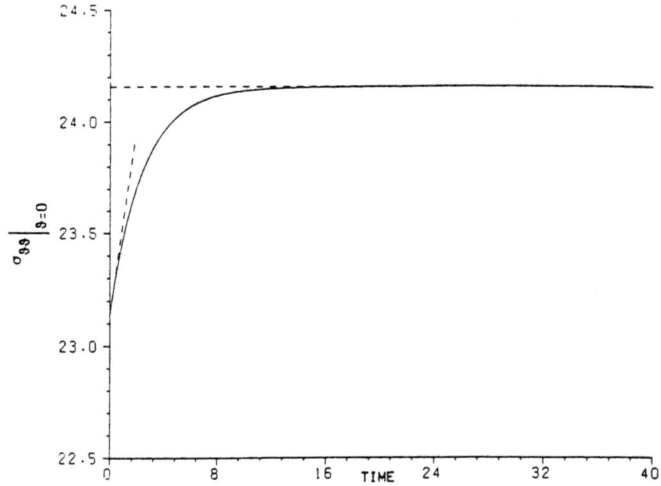

Fig. 10 Time evolution of the near notch-tip stress component $\sigma_{\theta\theta}$ at $r = 0.001$, $\theta = 0$ (unit loading case) for a corner of semi-included angle $x = \frac{3\pi}{4}$ and its comparison with small- and long-time asymptotic analyses.

Table 2

The instantaneous and long time limits associated with each of the three singular terms in the eigenexpansion (4.16) for $\sigma_{\theta\theta}$ on $\theta = 0$, $r = 0.001$

	Short time (t=0)	**Long time (t=∞)**
$\sigma_1(r,0,t)$	21.12702	26.08082
$\sigma_2(r,0,t)$	0.48207	-3.61271
$\sigma_3(r,0,t)$	1.53063	1.68825

The cases we show in Table 2 and Fig. 10 are for $\sigma = \sigma_{\theta\theta}$. Here the coefficient K_j depends on the applied load and material properties of the medium and F_j is the angular variation at the notch tip. Note that at intermediate times F_j may also depend on r, but for very small and very large times F_j is independent of r and $s_j(t)$ tends to an appropriate constant value. We refer to Atkinson and Bourne (1990) for detailed examples of these results and for attempted rationalisations of the angular variations.

The above example shows that it is important to consider all three stress singularities and discuss the regions in which any one might dominate. For the current problem, there are other interesting possibilities. For exmaple if Poisson's ratio varies with time as in Fig. 11 we see that the most singular exponents z_1 and z_2 are real, but evolve to complex long time values (see Table 3).

Table 3

	z_1	z_2	z_3
Short time	-1.3333	-1.35190	-1.85974
Long time	-1.33809 + 0.10110i	-1.33809 - 0.10110i	-1.95350

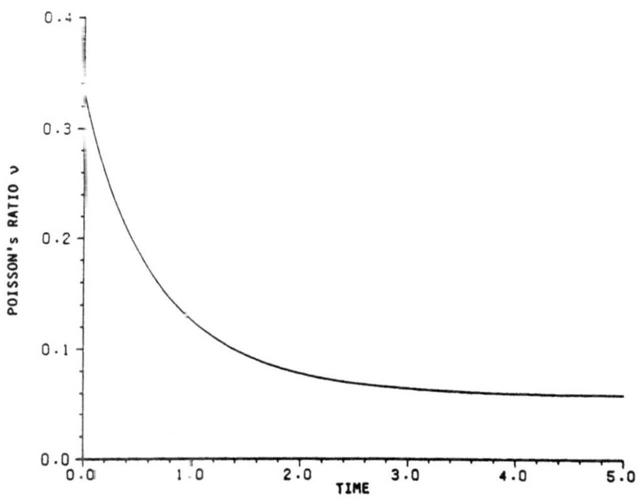

Fig. 11 Poisson's ratio r(t) versus time based on the linear solid with parameters $\mu_{20} = 0.5$, $\alpha_1 = 2$, $\beta_1 = 3$, $\mu_{20} = 2$, $\alpha_2 = 1$, $\beta_2 = 5$.

For this case for large times a quite different response should be seen around the notch tip where the singularity behaviour has the oscillatory form

$$r^{-0.6691} \begin{pmatrix} \cos \\ \sin \end{pmatrix} (0.1011 \log r) \tag{4.17}$$

4.2 Cracks Meeting Interfaces for Anisotropic Viscoelastic Bimaterials

In the previous section isotropic viscoelastic stress singularity problems have been analysed using Mellin transforms and by isolating the solutions of the eigenequations as functions of s_1 the Laplace transform parameter. The Laplace inversion of these functions was then performed numerically and the complete near tip results (*i.e.* the singular stress terms) compared with various asymptotic results obtained independently. Also in Section 3 a rationalisation of some crack bimaterial results was made by comparison with the force on a dislocation (of fixed Burgers vector). For the case of anisotropic viscoelastic bimaterials Eftaxiopoulos and Atkinson (1991) have considered a crack at an angle to the interface. Since the crack is assumed semi-infinite it is possible to effect a formal solution in the Laplace/Mellin transformed domains for the most general anisotropy (*i.e.* one for which the three modes of displacement u_1,u_2 and u_3 are coupled even though each depends only on the plane coordinates x_1 and x_2.

Because of the complexity of the algebra both transforms (Laplace and Mellin) are inverted numerically to get the real time results. Asymptotic results for small and large times are also presented which compare favourably with the numerical results. The solution for a single dislocation is also given and the force on it produced by its interaction with the welded boundary is determined. This force is used to interpret the crack tip stress reuslts in favourable cases. A number of examples are considered in Eftaxiopoulos and Atkinson (1991).

4.3 The Dual Function Method and Reciprocity (A Simple Example)

To illustrate this method we consider a simple example of mode 3 deformation of a wedge. The problems considered earlier in this section could also be treated by this method but the details are much more complicated. We neglect inertia so the key equations are as in section 2 *i.e.*

$$\frac{\partial \sigma_{13}}{\partial x_1} + \frac{\partial \sigma_{23}}{\partial x_2} = 0 \ . \tag{4.18}$$

If we assume the medium is viscoelastic but homogeneous and with zero stress and displacement prior to time t = 0 then the Laplace transformed equation of motion become:-

$$\frac{\partial^2 \bar{u}_3}{\partial x_i \partial x_i} = 0 \tag{4.19}$$

We consider a problem in which

$$\bar{u}_3 = 0 \quad \text{on} \quad \theta = 0 \quad t \geq 0$$

$$\left.\begin{aligned} \sigma_{\theta 3} &= H(t) \quad & 0 < r < 1 \\ &= 0 & r > 1 \end{aligned}\right\} \text{on } \theta = \alpha \tag{4.20}$$

so

$$\bar{u}_3(r,\theta,s_1) = 0 \quad \text{on} \quad \theta = 0 \quad \text{for all } r \geq 0$$

$$\left.\begin{aligned} s_1\bar{G}_1(s_1)\frac{\partial u_3}{\partial\theta} &= \frac{r}{s_1} \quad & 0 < r < 1 \\ &= 0 & r > 1 \end{aligned}\right\} \quad \text{on} \quad \theta = \alpha \tag{4.21}$$

The Mellin transform of equation (4.17) gives

$$\bar{\bar{u}}_3 = A(z)\sin(z+1)\theta \tag{4.22}$$

where $\bar{\bar{f}}(z) = \int_0^\infty \bar{f}(r)r^z dr$ is the definition of the Mellin transform used here.

Expression (4.22) satisfies $(4.21)_1$ and for $(4.21)_2$ we require

$$s_1\bar{G}_1(s_1)\frac{\partial\bar{\bar{u}}_3}{\partial\theta} = \frac{1}{s_1(z+2)} \quad \text{on} \quad \theta = \alpha \tag{4.23}$$

Thus

$$s_1\bar{G}_1(s_1)\bar{\bar{u}}_3 = \frac{\sin(z+1)\theta}{s_1(z+2)(z+1)\cos(z+1)\alpha} . \tag{4.24}$$

Suppose we take $\alpha > \pi/2$ and recall that the Mellin inversion theorem gives

$$\bar{u}_3(r,\theta,s_1) = \frac{1}{2\pi i}\int_{c-i\infty}^{c+i\infty} \bar{\bar{u}}_3(z,\theta,s_1)r^{-z-1}dz \tag{4.25}$$

where $-1 - \dfrac{\pi}{2\alpha} < c < -1$ to ensure that $\bar{u}_3$ is finite as r tends to zero. Closing the integral (4.25) in the left hand half of the complex z-plane we pick up a pole at $z+1 = -\pi/2\alpha$, so the leading behaviour as r tends to zero is

$$s_1\bar{G}_1(s_1)\bar{u}_3 \sim \frac{r^{\pi/2\alpha}\sin(\pi\theta/2\alpha)}{s_1\left(\frac{\pi}{2}\right)\left(1-\frac{\pi}{2\alpha}\right)} \quad . \tag{4.26}$$

However, suppose we evaluate the pole at z+1 = $\pi/2\alpha$ and don't worry about any constants in the solution. We then get

$$\hat{\bar{u}}_3 = r^{-\pi/2\alpha}\sin(\pi\theta/2\alpha) \quad . \tag{4.27}$$

This satisfies the conditions

$$\hat{\bar{u}}_3 = 0 \quad \text{on} \quad \theta = 0 \quad \text{for all } r$$

$$\frac{\partial\hat{\bar{u}}_3}{\partial\theta} = 0 \quad \text{on} \quad \theta = \alpha \quad \text{for all } r \quad . \tag{4.28}$$

If we now follow the method of section (1.4) we have from the reciprocal theorem (or Green's theorem in this case)

$$\int_C \bar{u}_3 \frac{\partial\hat{\bar{u}}_3}{\partial n}\, dS = \int_C \hat{\bar{u}}_3 \frac{\partial\bar{u}_3}{\partial n}\, dS \tag{4.29}$$

where C is some contour such as shown in the Figure and n is the outward drawn normal.

Suppose further, that instead of the stress boundary condition (4.20), the straight sides of the wedge at $\theta = \alpha$ is stress free and loads are applied at a remote boundary

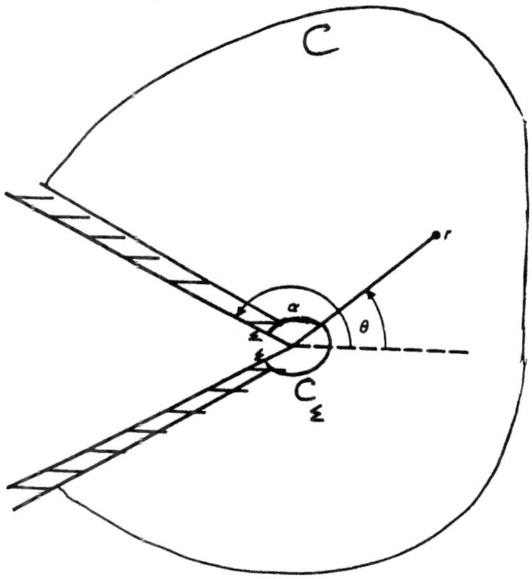

Fig. 12

The leading term in the expansion of $\bar{u}_3$ near the apex of the wedge will have the same form as (4.26) *i.e.*

$$\bar{u}_3 \sim K\, r^{\pi/2\alpha} \sin(\pi\theta/2\alpha) \tag{4.30}$$

for r small with higher order terms of the form

$$A_\lambda\, r^\lambda \sin(\lambda\theta) \quad \text{where } \lambda = \frac{(2N+1)\pi}{2\alpha}, \quad N = 1,2,\ldots \tag{4.31}$$

A formula for determining the unknown coefficient K in (4.30) in terms of far field data can be deduced by taking the integrals in (4.29) around a path such as that shown in Figure 12 and using the function $\hat{\bar{u}}_3$ defined in (4.27). The conditions satisfied by $\hat{\bar{u}}_3$ and the full solution on the wedge sides give zero contribution to the integrals and we are left with

$$I = \int_{C_\varepsilon} \left[\bar{u}_3 \frac{\partial \hat{\bar{u}}_3}{\partial r} - \hat{\bar{u}}_3 \frac{\partial \bar{u}_3}{\partial r} \right] \varepsilon \, d\theta = \int_{C_{AB}} \left[\bar{u}_3 \frac{\partial \hat{\bar{u}}_3}{\partial r} - \hat{\bar{u}}_3 \frac{\partial \bar{u}_3}{\partial r} \right] R \, d\theta \qquad (4.32)$$

where we assume the arcs C_ε and C_{AB} corresponds to arcs of circles radius ε and R transcribed anticlockwise. We further assume that the right hand side of (4.32) can be calculated with numerical data for $\bar{u}_3$. It is easy to see that combining (4.27) with expressions like (4.31) give zero contribution to the left hand side of (4.32) or ε tends to zero. The only nonzero term in this limit comes from combining (4.27) with (4.30) to give

$$\lim_{\varepsilon \to 0} I = -K \int_0^\alpha \frac{\pi}{2\alpha} \, d\theta = - \frac{K\pi}{2} \quad . \qquad (4.33)$$

Hence K can be determined in terms of far field numerical data. A method of this kind applies also to plane strain problems such as considered earlier in this lecture. See also the review of Atkinson and Craster (1994) for more complicated applications of this method.

5. INVERSE PROBLEMS OF CRACK DETECTION IN VISCOELASTIC MATERIALS

5.1. INTRODUCTION

In a recent paper Atkinson and Aparicio (1994) the problem of anti-plane crack detection in a two-dimensional infinite elastic medium was studied. The situation considered was that of an infinite medium with an internal closed boundary along which the displacements and stresses are known and the crack begins. This problem is a simplified version of sleeve fracturing in a borehole where a region of the borehole is pressurised by means of an inflatable packer and the pressure increased until a fracture is initiated in the adjacent rock. This situation has been modelled by Atkinson and Thiercelin (1993) who assume a known straight crack profile and use the displacement at the neck of the crack to infer properties of far field earth stresses. They considered the more realistic situation of plane strain. The object of Atkinson and Aparicio, however, was to detect *any unknown* crack profile using measurements of the displacements around the interior of the borehole wall. Using analytic function methods they developed a simple method for determining the parametric equations of each side of the crack using the corresponding displacements as parameters. In this note we show how their methods can also be applied to the situation where the material is modelled as a viscoelastic medium provided all processes are slow enough that the inertia terms in the equations of motion can be neglected. The simplicity of application of the above method should be compared with other recent methods that involve heavy numerical work. Examples are Santosa and Vogelius (1992) who use finite element methods for the Laplace equation problem (mode 3 elastic) of a body which may contain cracks in its interior which do not touch the surface. Sagami *et al* who used a boundary integral equation method to solve three dimensional direct problems (assuming a specified shape of crack) and then choose the crack shape to match the conditions around the hole with experimental data. Also Nishimura and Kobayashi (1991) use boundary integral and minimisation techniques for two and three dimensional problems.

5.2. ANTI-PLANE STRAIN DEFORMATION

For this mode of deformation there is only one non-zero component of displacement u_3 which acts in the x_3 direction but is a function of x_1 and x_2. The only non-zero components of stress are σ_{13} and σ_{23} and the equilibrium equations reduce to

$$\frac{\partial \sigma_{13}}{\partial x_1} + \frac{\partial \sigma_{23}}{\partial x_2} = 0 \quad . \tag{5.1}$$

We also have the constitutive equations in the relaxation, or creep forms,

$$\sigma_{i3} = \int_{-\infty}^{t} G_1(t-\tau) \frac{de_{i3}(\tau)}{d\tau} d\tau \tag{5.2}$$

$$e_{i3} = \int_{-\infty}^{t} J_1(t-\tau) \frac{d\sigma_{i3}(\tau)}{d\tau} d\tau \tag{5.3}$$

i = 1 or 2

with $$s^2\bar{J}_1(s)\bar{G}_1(s) = 1 \tag{5.4}$$

and $$e_{i3} = \frac{1}{2} \frac{\partial u_3}{\partial x_i} \tag{5.5}$$

The situation considered is that shown in Figure 13.

Fig. 13 A crack emanating from a hole.

An infinite two dimensional medium has a hole in it *e.g.* cross-section of a borehole in the earth. Loads are applied to the inside of the hole and as a consequence a fracture is initiated on the periphery of the hole and propagates into the medium. A measuring device is available which measures the displacements of the hole wall as the fracture progresses. Our object is to devise a method of deducing from these measurements the location and extent of the crack. This has been done for the elastic case by Atkinson and Aparicio (1994). We show here, how their method can be applied to the viscoelastic problem and hence the time dependent position of the crack monitored.

5.2.1 Description of the Problem

Since the hole displacement is measured and the traction prescribed on the hole (H), then we know σ_{n3} and u_3 on the hole boundary. Thus

$$u_3 = 2g_1(s,t), \quad \sigma_{n3} = f_2\,(s,t) \tag{5.6}$$

along H where $g_1(s,t)$ and $f_2(s,t)$ are known functions of s and time (t); s being the length measured along the contour H (anticlockwise) and n the normal to the contour H measured inward. From (5.6) we can compute $\partial u_3/\partial s = \dfrac{2\partial g_1}{\partial s}$ and thus on the hole boundary

$$\sigma_{s3}(s,t) = \int_{-\infty}^{t} G_1(t-\tau) \frac{\partial}{\partial s} \left(\frac{dg_1(s,\tau)}{d\tau} \right) d\tau \tag{5.7}$$

We now define a stress function $\psi(x_1,x_2,t)$ where (x_1,x_2) are orthogonal cartesian coordinates so that

$$\sigma_{13} = \frac{\partial\psi}{\partial x_1} \quad , \quad \sigma_{23} = \frac{\partial\psi}{\partial x_2} \tag{5.8}$$

so the equilibrium equation (5.1) becomes

$$\nabla^2\psi = 0 \quad . \tag{5.9}$$

Integrating (5.7) with respect to s thus gives

$$\psi(s,t) = f_1(s) = \int_{-\infty}^{t} G_1(t-\tau) \frac{dg_1(s,\tau)}{d\tau} d\tau \tag{5.10}$$

on the hole boundary H. We also note that because the displacements are known all along H, the beginning of the crack is also known. We define the values of the parameter s which correspond to both sides of the crack as s_0 and s_f, s being measured anticlockwise from s_0 to s_f around the hole H. Then because the displacement u_3 is discontinuous across the crack we expect ψ from (5.10) to be discontinuous there also so defining ψ_0 and ψ_f the values of ψ at these points also we have

$$f_1(s_0) = \psi_0, \quad f_1(s_f) = \psi_f \tag{5.11}$$

where

$$f_1(s) = \int_{-\infty}^{t} G_1(t-\tau) \frac{dg_1(s,\tau)}{d\tau} d\tau \tag{5.12}$$

and ψ_0,ψ_f will in general depend upon time. Following Atkinson and Aparicio (1994) the "crack" is defined as two lines one starting at $s = s_0$ and the other at $s = s_f$, their shapes are to be determined.

We restrict attention here to situations where the crack is finite in length. It is however, evolving with time.

Because the stress function ψ satisfies Laplace's equations (5.9) we define a complex function

$$\Phi(z) = \psi(x,y) + i\phi(x,y) \tag{5.13}$$

with (x,y) equivalent to (x_1,x_2) and z = z + iy.

Using (5.8), (5.6) and the Cauchy-Riemann equations gives along H, the relation

$$\frac{\partial\phi}{\partial s} = -\frac{\partial\psi}{\partial n} = -f_2(s) \quad . \tag{5.14}$$

where s measured anticlockwise and n the inward normal to the hole. Furthermore since the crack is stress free σ_{N3} is zero on it and thus $\frac{\partial\phi}{\partial s} = 0$ so ϕ is constant on the crack. Taking this constant to be zero enables (5.14) to be integrated as

$$\phi = -\int_{s_0}^{s} f_2(s)ds \quad . \tag{5.15}$$

Also from the displacement boundary condition (5.6) and (5.12) we have

$$\psi = f_1(s) \quad \text{on} \quad H \quad . \tag{5.16}$$

Thus the function $\Phi(z)$ (3.13) takes known values on the periphery of the hole. These values being assumed known for all times up to the present. We assume further that the displacement $\psi(x,y)$ tends to zero as $|z| \to \infty$. Then $\Phi(z)$ may tend to a complex constant which may depend on time as $|z| \to \infty$ thus we write

$$\Phi(z) \to iC \quad \text{as} \quad |z| \to \infty \tag{5.17}$$

where C is real and is to be determined by the solution method. See Atkinson and Aparicio (1994) for a fuller discussion of this issue. Also with $\phi = 0$ on both sides of the crack (see the argument following equation (5.14)) then he complex function $\Phi(z)$ takes real values (the displacement) on both sides of the crack.

We now consider the function

$$F(z) = [\Phi(z) - iC]^2 \cos\left(\frac{2n\pi\Phi(z)}{\psi_0 - \psi_f}\right) \tag{5.18}$$

where n is an integer. The function $\Phi(z)$ is analytic throughout the space occupied by the elastic medium, it is known along H and $\Phi(z)$ - iC is O(1/z) as $(z) \to \infty$.

Using Cauchy's theorem

$$\int_R F(z)dz = 0 \tag{5.19}$$

where R is the elastic medium exterior to the hole and crack and closed by a large contour at infinity. F(z) is analytic in R since it is the result of the composition of an entire function and a function analytic in R. The integral around a big circle of radius ϵ is of order $O(1/\epsilon)$ as $\epsilon \to \infty$ and since $\Phi(z)$ is bounded at the tip of the crack (finite displacements at the crack tip) we get

$$\int_H F(z)dz = \int_{\psi_0}^{\psi_1} (\psi - iC)^2 \frac{dz_-}{d\psi} \cos\left(\frac{2N\pi\psi}{\psi_0 - \psi_f}\right) d\psi + \int_{\psi_1}^{\psi_f} (\psi - iC)^2 \frac{dz_+}{d\psi} \cos\left(\frac{2N\pi\psi}{\psi_0 - \psi_f}\right) d\psi \tag{5.20}$$

where ψ_1 is the value of ψ at the tip of the crack. z_- and z_+ are used to distinguish the parametric equations of each side of the crack, z_- being the 'upper' and z_+ the 'lower' surface. The integral around H is taken anticlockwise. Following Atkinson and Aparicio (1994) we define the function

$$g(\psi) = \begin{cases} (\psi - iC)^2 (dz_+/d\psi) & \text{if } (\psi_f - \psi_1)(\psi - \psi_1) > 0 \\ (\psi - iC)^2 (dz_-/d\psi) & \text{if } (\psi_0 - \psi_1)(\psi - \psi_1) > 0 \end{cases} \tag{5.21}$$

then

$$\int_H F(z)dz = \int_{\psi_0}^{\psi_f} g(\psi) \cos\left(\frac{2N\pi\psi}{\psi_0 - \psi_f}\right) d\psi \tag{5.22}$$

Since from the known stress and displacement fields on the hole (equations (5.15) and (5.16)) F(z) is determined there and the left hand side of (5.22) can thus be calculated. Using the same idea with the function

$$G(z) = [\Phi(z)-iC]^2 \sin\left[\frac{2N\pi\Phi(z)}{\psi_0-\psi_f}\right] \tag{5.23}$$

we can generate a Fourier series for $g(\psi)$ as

$$g(\psi) = \frac{a_0}{2} + \sum_{N=1}^{\infty}\left[a_N \cos\left(\frac{2N\pi\psi}{\psi_0-\psi_f}\right) + b_N \sin\left(\frac{2N\pi\psi}{\psi_0-\psi_f}\right)\right] \tag{5.24}$$

with

$$a_N = \frac{2}{(\psi_f-\psi_0)}\int_H (\Phi(z)-iC)^2 \cos\left(\frac{2N\pi\Phi(z)}{\psi_0-\psi_f}\right)dz \tag{5.25}$$

$$b_N = \frac{2}{(\psi_f-\psi_0)}\int_H (\Phi(z)-iC)^2 \sin\left(\frac{2N\pi\Phi(z)}{\psi_0-\psi_f}\right)dz$$

Noting that $z_+(\psi_f) = z_-(\psi_0)$ (both sides of the crack begin at the same point on the borehole wall) and that $z_+(\psi_1) = z_-(\psi_1)$ (at the crack tip) the function $f(\psi)$ is defined as

$$f(\psi) = z_0 + \int_{\psi_f}^{\psi} [g(\psi)/(\psi-iC)^2]d\psi$$

$$= z_0 + \int_{\psi_0}^{\psi} [g(\psi)/(\psi-iC)^2]d\psi$$

where z_0 denotes the (known) beginning of the crack.

This function is also equal to

$$f(\psi) = \begin{cases} z_+(\psi) \text{ if } (\psi_f-\psi_1)(\psi-\psi_1) > 0 \\ z_-(\psi) \text{ if } (\psi_0-\psi_1)(\psi-\psi_1) > 0 \end{cases}$$

The value of ψ_1 and the position of the crack can easily be determined from these equations. It remains to estimate C which will in general depend on time as a parameter since f_1 and f_2 will also do so in general both from the possibility of time dependent conditions on the hole and of time dependence introduced through the relaxation of the medium.

Atkinson and Aparicio (1994) suggest the use of Parseval's theorem. Thus a necessary condition for the series (5.24) to converge is

$$\lim_{N\to\infty} (\alpha_N^2+\beta_N^2) = 0 \tag{5.28}$$

with α_N and β_N the real parts of a_N and b_N given in (5.25). From these expressions the equation $\alpha_N^2 + \beta_N^2 = 0$ gives a polynomial of fourth degree in C for each N. An alternative polynomial equation can be obtained by taking the imaginary parts of the integrals in equation (5.25).

C can be estimated at each time if there is a common point of convergence, as N increases, of at least two roots one for each polynomial.

Once C is estimated and the position of the crack determined the value of $\Phi(z)$ at infinity can be checked by integrating the function $\Phi(z)/z$ along the internal boundary (where we are assuming the origin is not in the region R). Thus

$$\int_{\psi_0}^{\psi_1} \frac{\psi}{z_-}\frac{dz_-}{d\psi} + \int_{\psi_1}^{\psi_f} \frac{\psi}{z_+}\frac{dz_+}{d\psi} \cdot d\psi - \int_H \frac{\Phi(z)}{z}\, dz = 2\pi i\, \Phi(\infty) \tag{5.29}$$

or

$$\int_{\psi_0}^{\psi_f} \ln\left(\frac{f(\psi)}{z_0}\right) d\psi + \int_H \frac{\Phi(z)}{z}\, dz = 2\pi C$$

where the cut of the logarithm is chosen such that it does not intersect the "normalised" crack $([x(\psi)+iy(\psi)]/z_0)$. This analysis also holds for C complex so the behaviour of ψ at infinity can be checked.

References

1. Akinson, C. (1975), Int. J. Engng. Sci., 13, 31-44.
2. Atkinson, C. and Appleby, S. (1994), Int. J. Engng. Sci., 32, 955-977.
3. Atkinson, C. (1977), Int. J. Fracture, 13, 807-820.
4. Atkinson, C. and Bourne, J.P. (1989), Quart. J. Mechs. Appl. Math. 42, 385-412.
5. England, A.H. (1971), Int. J. Engng. Sci., 9, 571-585.
6. Williams, M.L. (1959), Bulletin Seismological Society of America, 49, 199-204.
7. Achenbach and Chao (1962), J. Mech. Phys. Solids, 10, 245-252.
8. Christensen, R.M. (1971), Theory of viscoelasticity. An Introduction, Academic Press (1971).
9. Bourne, J.P. and Atkinson, C. (1990), IMA Jnl. Appl. Maths, 4, 163-180.
10. Atkinson, C. and Bourne, J.P. (1990), Int. J. Engng. Sci., 28, 615-630.
11. Eftaxiopulos, D.A. and Atkinson, C. (1991), Int. J. Engng. Sci., 29, 1569-1584.
12. Atkinson, C. and Craster, R.V. (1994), To appear in Jnl. Aerospace Science.
13. Atkinson, C. and Thiercelin, M. (1993), Int. J. Frac., 59, 23-40.
14. Atkinson, C. and Aparicio, N.D. (1994), Proc. Roy. Soc. A445, 637-652.
15. Sagami, T. et al. (1990), Trnas. J.S.M.E., 56, 27-32.
16. Santosa, F. and Vogelius, M. (1992), Technical Report No. 90-3, Centre for the Mathematics of Waves, University of Delaware, USA.
17. Nishimura, N. and Kobayashi, S. (1991), Boundary Elements XII, Vol. 2, 425-434, Southampton, Computational Mechanics.

EXISTENCE AND UNIQUENESS RESULTS FOR VISCOELASTIC MATERIALS

M. Fabrizio
University of Bologna, Bologna, Italy

Introduction

In this course some analytical problems present in the mathematical theory of the viscoelasticity will be studied. In particular we consider:

-the definition of materials with fading memory, without the use of a-priori topologies on the history space, but only impose directly this condition of fading memory on the constitutive functional,

-the derivation of free energies for the linear problem and their connection with the norms of history spaces

-a theorem on the domain of dependence proved by means of the free energies properties; this theorem asserts the finite velocity of the signal and assures the hyperbolicity of the integrodifferential system

-theorems of existence, uniqueness and asymptotic stability for the quasi-static and dynamical problem of linear viscoelasticity.

The aim of this course is to emphasize the strict connection between these results and the restrictions which the principles of the thermodynamics impose on the constitutive equations.

Finally, the research reported here has been in part extracted from the works [1],[2].

1. Materials with fading memory

Consider a body which occupies the region $\mathcal{R} \subset \mathcal{E}^3$, in an unstressed reference placement, and denote by $\mathbf{x} \in \mathcal{R}$ the position vector of any point of the body. The motion of the body is described by the displacement $\mathbf{u}(\mathbf{x}, t)$ as a function of the position $\mathbf{x}$ and the time t. The deformation is described by the strain tensor $\mathbf{F} = \partial \mathbf{u}/\partial \mathbf{x}$. The history of $\mathbf{F}$ related at the point $\mathbf{x}$, up to time t is defined by the function $\mathbf{F}^t(x, s) := \mathbf{F}(x, t-s)$ where $s \in [0, \infty)$. As is well known, an elastic material is described through a function of the type:

$$\mathbf{T}(\mathbf{x}, t) = \tilde{\mathbf{T}}(\mathbf{F}(\mathbf{x}, t)) \tag{1.1}$$

In the theory of materials with fading memory the constitutive equation (1.1) is replaced by a functional (see[3-6]):

$$\mathbf{T}(\mathbf{x}, t) = \hat{\mathbf{T}}(\mathbf{F}^t(\mathbf{x}, s)) \tag{1.2}$$

Moreover, because the present value (or instantaneous value) $\mathbf{F}(\mathbf{x},t)$, provides a significant contribution to the constitutive relation (1.2), (1.2) is often also written in the form:

$$\mathbf{T}(\mathbf{x},t) = \hat{\mathbf{T}}(\mathbf{F}(\mathbf{x},t), \mathbf{F}^t(\mathbf{x},s)) \tag{1.3}$$

In viscoelasticity, besides to the hereditarity effect, namely that the stress tensor $\hat{\mathbf{T}}$ at time t depends on the history of the deformation gradient $\mathbf{F}$, we observe a fading memory effect, that Volterra in [7] states in the following form:"the modulus of the variation of the quantity $z = F\Big[\overset{t}{\underset{-\infty}{f(\tau)}}\Big]$ when $f(\tau)$ varies in any way (its modulus being always $< M$) in the interval $(-\infty, t_1)$ (with $t_1 < t$), can be made as small as we please by taking the interval (t_1, t) sufficiently large".

One general and mathematical formulation of fading memory, was made by Coleman and Noll[C.N],who place on the space $\mathcal{H}$ of histories $\mathbf{F}^t$, the norm:

$$\|\mathbf{F}^t\| = |\mathbf{F}^t(0)| + \Big(\int_0^\infty h(s)[\mathbf{F}^t(s)]^2 ds\Big)^{1/2}. \tag{1.4}$$

where the function $h : [0,\infty) \to R^+$, called influence function is in $L^1(0,\infty)$. Then the constitutive functional (1.2) is supposed to be continuous respect to this norm. For a history $\mathbf{F}^t$, we denote by $(\mathbf{F}^t_a)$ it static continuation, namely the new history $\mathbf{F}^t_a = \begin{cases} \mathbf{F}^t(s)\ , & s > a \\ \mathbf{F}(t)\ , & s \le a \end{cases}$. Now , from the properties of the function h we have: $\lim_{a\to\infty} (\mathbf{F}^t_a) = \mathbf{F}^\dagger(t)$, The fading memory condition is then given by means of the following limit:

$$\lim_{a\to\infty} \hat{\mathbf{T}}(\mathbf{F}^t_a) = \hat{\mathbf{T}}(\mathbf{F}^\dagger(t)). \tag{1.5}$$

This approach requires the introduction of a norm on the history space. As a consequence, the theory itself depends on the norm (1,4), which does seem any connected to the feature of the viscoelastic material.

Moreover it has been proved in [1] that several natural norms, i.e. norms deducible from energies, exist. In fact, for material with fading memory, there exist several energies and each of them induces a topology on the space of histories, that makes the stress constitutive functional continuous. These topologies are not equivalent and none of them is to be preferred. Therefore it is interesting to state a fading memory property without making use of any norm. Let $\mathcal{D}$ denotes the set of the histories that make the functional (1.2) well defined. We suppose $\mathcal{D}$ to be endowed with no topological structure. The properties of an elastic material with fading memory are described by means of the following conditions:

i. The set $\mathcal{D}$ contains $L^\infty(0,\infty)$

ii. If $\mathbf{F}^t \in \mathcal{D}$, then its static continuation $\mathbf{F}^t_a$ also belongs to $\mathcal{D}$,

iii. There exists a function $\tilde{\mathbf{T}}(\mathbf{F}(t))$, describing the behaviour of an elastic body, such that:

$$\lim_{a\to\infty} \hat{\mathbf{T}}(\mathbf{F}^t_a) = \tilde{\mathbf{T}}(\mathbf{F}(t)), \tag{1.6}$$

These three conditions enable one to characterize the properties of a material with fading memory, at least in the linear case, when the functional (1.3) is of the form:

$$\mathbf{T}(x,t) = \mathbf{G}_0(x)\mathbf{E}(x,t) + \int_0^\infty \dot{\mathbf{G}}(x,s)\mathbf{E}^t(x,s)ds, \tag{1.7}$$

where $\mathbf{E}$ represents the infinitesimal strain tensor, that takes the place of $\mathbf{F}$ in linear approximations. $\mathbf{G}_0$ and $\dot{\mathbf{G}}$ are fourth-order tensors that are assumed to be symmetric and depend on the material. It is easy to prove that condition i. implies that $\dot{\mathbf{G}} \in L^1(\mathbb{R}^+)$ in order that the integral in (1.7) is finite for every $\mathbf{E}^t \in L^\infty(\mathbb{R}^+)$. Furthermore, conditions ii. and iii. ensure that

$$\lim_{a\to\infty} \hat{\mathbf{T}}(\mathbf{E}_a^t) = \mathbf{G}_\infty \mathbf{E}(t)$$

Hence, on the basis of i.i., $\mathbf{G}_\infty$ must be a positive definite tensor.

2. Linear viscoelastic solids

A linear viscoelastic solid is characterized by a stress-strain relation of the form

$$\mathbf{T}(t) = \mathbf{G}_0\mathbf{E}(t) + \int_0^\infty \mathbf{G}'(s)\mathbf{E}^t(s)ds, \tag{2.1}$$

where we shall suppose $\mathbf{G}_0 \in \mathrm{Lin}(\mathrm{Sym}, \mathrm{Sym})$ and

$$\mathbf{G}' \in L^1(\mathbb{R}^+) \cap L^2(\mathbb{R}^+) \tag{2.2}$$

with values in Lin(Sym, Sym); this means that $\mathbf{A} \cdot \mathbf{G}'\mathbf{A} \in L^1(\mathbb{R}^+) \cap L^2(\mathbb{R}^+)$, $\forall \mathbf{A} \in \mathrm{Sym}$. The relaxation function

$$\mathbf{G}(s) = \mathbf{G}_0 + \int_0^s \mathbf{G}'(\xi)d\xi$$

is then well defined along with $\mathbf{G}_\infty = \lim_{s\to\infty} \mathbf{G}(s)$; the solid character of the body is specified by letting

$$\mathbf{G}_\infty > 0, \tag{2.3}$$

namely, $\mathbf{G}_\infty$ is positive definite in Lin(Sym, Sym). This is a minimal set of properties for modelling a viscoelastic solid. Here some additional properties are considered.

The relaxation function is taken to be symmetric, namely $\mathbf{A} \cdot \mathbf{G}(s)\mathbf{B} = \mathbf{B} \cdot \mathbf{G}(s)\mathbf{A}$, $\forall \mathbf{A}, \mathbf{B} \in$ Sym, for any $s \in \mathbb{R}^+$. In this regard we observe that, as is known, thermodynamics implies the symmetry of the instantaneous elastic modulus $\mathbf{G}_0$ and the equilibrium elastic modulus $\mathbf{G}_\infty$, but not the symmetry of $\mathbf{G}(s)$ for other values of s.

For any function $f \in L^2(\mathbb{R})$ we denote by f_F the Fourier transform

$$f_F(\omega) = \int_{-\infty}^{\infty} f(\xi)\exp(-i\omega\xi)d\xi.$$

Functions defined on $\mathbb{R}^+$ are identified with functions on $\mathbb{R}$ which vanish identically on $(-\infty, 0)$. For such functions, $f_F = f_c - i f_s$ where f_s and f_c are the Fourier sine and cosine transforms

$$f_s(\omega) = \int_0^\infty f(\xi) \sin \omega\xi \, d\xi,$$

$$f_c(\omega) = \int_0^\infty f(\xi) \cos \omega\xi \, d\xi.$$

As shown in [8], a consequence of the second law of thermodynamics for cyclic processes is that

$$\mathbf{G}'_s(\omega) < 0, \quad \forall \omega \in \mathbb{R}^{++}, \tag{2.4}$$

whereby the sine transform of the Boltzmann function $\mathbf{G}'$ is negative definite for any strictly-positive value of the argument. Here the inequality (2.4) is taken as an assumption; but in §6 we show that (2.4) is a consequence of an appropriate statement of the second law (in the new version).

Remarkable consequences follow from (2.4). By the Fourier inversion formula,

$$\mathbf{G}'(\xi) = \frac{2}{\pi} \int_0^\infty \sin \omega\xi \, \mathbf{G}'_s(\omega) \, d\omega,$$

and an integration with respect to ξ gives

$$\mathbf{G}(\xi) - \mathbf{G}_0 = \frac{2}{\pi} \int_0^\infty \frac{1 - \cos \omega\xi}{\omega} \mathbf{G}'_s(\omega) \, d\omega. \tag{2.5}$$

Assume that the limit of ξ^{-1} times (2.5), as $\xi \to 0_+$, exists and is bounded; this limit is the initial derivative $\mathbf{G}'_0 := \mathbf{G}'(0)$. The boundedness of $\mathbf{G}'_0$ and the inequality (2.4) imply that $\mathbf{G}'_s$ is continuous on $\mathbb{R}^+$ and is in $L^1(\mathbb{R}^+)$ (cf. [9]). Then by (2.4), the limit of (2.5) yields the negative semi-definiteness of $\mathbf{G}'_0$,

$$\mathbf{G}'_0 \le 0.$$

Now evaluate the limit of (2.5) as $\xi \to \infty$. Application of the Riemann-Lebesgue lemma yields

$$\mathbf{G}_\infty - \mathbf{G}_0 = \frac{2}{\pi} \int_0^\infty \frac{1}{\omega} \mathbf{G}'_s(\omega) d\omega. \tag{2.6}$$

The boundedness of the left-hand side and the definiteness of $\mathbf{G}'_s$, in the sense of (2.4), yield $\mathbf{G}'_s/\omega \in L^1(\mathbb{R}^+)$. Incidentally, by (2.4) and (2.6) we obtain also that

$$\mathbf{G}_0 > \mathbf{G}_\infty;$$

by (2.3) this inequality implies the positive definiteness of $\mathbf{G}_0$.

For later convenience observe that, for any two real-valued functions $g, h \in L^2(\mathbb{R})$, Plancherel's theorem for the Fourier transform gives

$$\int_{-\infty}^{\infty} g(\xi)h(\xi)d\xi = \frac{1}{2\pi}\int_{-\infty}^{\infty} g_F(\omega)h_F^*(\omega)d\omega$$

where $*$ signifies complex conjugate. As a consequence, since $\mathbf{G}', \mathbf{G}'_s \in L^1(\mathbb{R}^+) \cap L^2(\mathbb{R}^+)$, we have

$$\int_0^{\infty} \mathbf{G}'(\xi)\mathbf{E}^t(\xi)\, d\xi = \frac{2}{\pi}\int_0^{\infty} \mathbf{G}'_s(\omega)\mathbf{E}^t_s(\omega)\, d\omega \tag{2.7}$$

for all histories $\mathbf{E}^t \in L^2(\mathbb{R}^-)$. The factor $2/\pi$ instead of $1/2\pi$ originates from writing the first integral on $\mathbb{R}$, then using Plancherel's theorem and finally writing the integral in ω on $\mathbb{R}^+$.

For any $\mathbf{E}^t \in L^2(\mathbb{R}^+)$, the relation (2.7) allows the constitutive equation (2.1) to be written as

$$\mathbf{T}(t) = \mathbf{G}_0\mathbf{E}(t) + \frac{2}{\pi}\int_0^{\infty} \mathbf{G}'_s(\omega)\mathbf{E}^t_s(\omega)d\omega. \tag{2.8}$$

By this representation, a new interpretation is gained of the properties of materials with memory. According to (2.8), materials with memory are intrinsically dispersive in that the stress tensor $\mathbf{T}$ depends on the spectral distribution $\mathbf{E}^t_s(\omega)$ of monochromatic waves which constitute the history $\mathbf{E}^t$. In this sense, we may view the function $\mathbf{G}'_s$, along with $\mathbf{G}_0$, as the basic constitutive properties characterizing the material.

A generalization of (2.8) is possible for all histories $\mathbf{E}^t \notin L^2(\mathbb{R}^+)$ with

$$\left|\int_0^{\infty} \mathbf{G}'(s)\mathbf{E}^t(s)ds\right| < \infty \tag{2.9}$$

where $|\cdot|$ denotes a norm in Sym. It is natural to define the set Φ of admissible histories relative to the constitutive equation (2.1) as the class of all functions $\mathbf{E}^t$ such that (2.9) holds with $\mathbf{G}' \in L^1(\mathbb{R}^+) \cap L^2(\mathbb{R}^+)$. Each element of Φ may be considered as a linear continuous functional on $\mathcal{F}$, the normed vector space of functions defined by

$$\mathcal{F} = \{\mathbf{V} : \mathbb{R}^+ \to \mathrm{Lin}(\mathrm{Sym}, \mathrm{Sym});\ \mathbf{V} = \alpha\mathbf{G}' + \mathbf{W}, \alpha \in \mathbb{R}, \mathbf{W} \in C_0^{\infty}(\mathbb{R}^+)\};$$

of course $\mathbf{W}$ has values in Lin(Sym, Sym). In this way we can take Φ as large as possible by letting $\Phi = \mathcal{F}'$, the space of all linear continuous functionals on $\mathcal{F}$. Upon a straightforward calculation, $\mathcal{F}'$ turns out to be the set of histories $\mathbf{E}^t \in \mathcal{D}'$ (the dual space of $\mathcal{D} = C_0^{\infty}(\mathbb{R}^+)$) such that (2.9) holds. Thus, $\mathcal{F}' \subset \mathcal{D}'$ and its elements are Fourier-transformable in the sense of distributions [9]. Let $\mathcal{F}'_s$ be the set of distributions $\mathbf{E}^t_s$ that are half-range Fourier sine transforms of elements $\mathbf{E}^t$ of $\mathcal{F}'$. Then the relation (2.7) can be carried over into the set of histories $\mathcal{F}'$. Accordingly, the constitutive equation (2.8) is valid for all elements $\mathbf{E}^t_s \in \mathcal{F}'_s$.

3. Free energy

Definition 3.1. A free energy, relative to the given continuous stress functional (2.1), is a functional $\Psi : \Phi \to \mathbb{R}$ with the following properties.

1) The set $\Phi \subset \mathcal{F}'$ is such that if $\mathbf{E}_0^t \in \Phi$ then each history $\mathbf{E}^{t+\tau}$ for which $\mathbf{E}^{t+\tau}(s) = \mathbf{E}_0^t(s-\tau), s \geq \tau$, is an element of Φ.

2) The functional Ψ is continuous on $\Phi = \mathrm{Sym} \times \Phi_r$, is differentiable with respect to the first argument and

$$\mathbf{T} = \partial_{\mathbf{E}} \Psi(\mathbf{E}, {}_r\mathbf{E}^t) \tag{3.1}$$

where $\mathbf{T}$ stands for the given stress functional.

3) For each value of τ such that $\dot{\mathbf{E}}(t+\tau)$ is continuous Ψ satisfies the inequality

$$\dot{\Psi}(\mathbf{E}^{t+\tau}) \leq \mathbf{T}(\mathbf{E}^{t+\tau}) \cdot \dot{\mathbf{E}}(t+\tau) \tag{3.2}$$

4) The functional Ψ is minimal in correspondence with the constant histories in that

$$\Psi(\mathbf{E}^t) \geq \Psi(\mathbf{E}^\dagger), \quad \forall \mathbf{E}^t \in \Phi_{\mathbf{E}}, \tag{3.3}$$

and equality holds if and only if $\mathbf{E}^t$ is a constant history, $\mathbf{E}^t = \mathbf{E}^\dagger$.

Consider the functional (cf. [1])

$$\begin{aligned} \Psi_M(\mathbf{E}(t), {}_r\mathbf{E}^t) &= \tfrac{1}{2}\mathbf{E}(t) \cdot \mathbf{G}_\infty \mathbf{E}(t) \\ &+ \tfrac{1}{\pi} \int_0^\infty \{(\mathbf{E}_s^t(\omega) - \frac{\mathbf{E}(t)}{\omega}) \cdot \mathbf{H}(\omega)(\mathbf{E}_s^t(\omega) - \frac{\mathbf{E}(t)}{\omega}) + \mathbf{E}_c^t(\omega) \cdot \mathbf{H}(\omega)\mathbf{E}_c^t(\omega)\} d\omega \end{aligned} \tag{3.4}$$

on $\mathcal{H}$. It is a simple matter to ascertain that the properties 1 to 4 of Definition 1 hold, so that Ψ_M is a free energy.

For, since $\mathbf{H}(\omega) = -\omega \mathbf{G}_s'(\omega)$, the partial derivative of Ψ_M with respect to $\mathbf{E}(t)$ is

$$\mathbf{G}_\infty \mathbf{E}(t) + \tfrac{2}{\pi} \int_0^\infty \mathbf{G}_s'(\omega)(\mathbf{E}_s^t(\omega) - \frac{\mathbf{E}(t)}{\omega}) d\omega;$$

hence (3.1) holds.

To prove the validity of (3.2) we observe that, letting $\Psi_M(t)$ stand for $\Psi_M(\mathbf{E}(t), {}_r\mathbf{E}^t)$, upon time differentiation we have

$$\dot{\psi}_M(t) = \dot{\mathbf{E}}(t) \cdot \mathbf{G}_\infty \mathbf{E}(t) + \tfrac{2}{\pi} \int_0^\infty \{(\dot{\mathbf{E}}_s^t(\omega) - \frac{\dot{\mathbf{E}}(t)}{\omega}) \cdot \mathbf{H}(\omega)(\mathbf{E}_s^t(\omega) - \frac{\mathbf{E}(t)}{\omega}) + \dot{\mathbf{E}}_c^t(\omega) \cdot \mathbf{H}(\omega)\mathbf{E}_c^t(\omega)\} d\omega.$$

Integrations by parts yield

$$\dot{\mathbf{E}}_s^t(\omega) = \omega \mathbf{E}_c^t(\omega), \qquad \dot{\mathbf{E}}_c^t(\omega) = \mathbf{E}(t) - \omega \mathbf{E}_s^t(\omega);$$

hence, by substitution,

$$\dot{\psi}_M(t) = \dot{\mathbf{E}}(t) \cdot \{\mathbf{G}_\infty \mathbf{E}(t) + \frac{2}{\pi} \int_0^\infty \mathbf{G}'_s(\omega)(\mathbf{E}^t_s(\omega) - \frac{\mathbf{E}(t)}{\omega})d\omega\}.$$

Accordingly, by use of (3.5) we have

$$\dot{\psi}_M(t) = \mathbf{T}(t) \cdot \dot{\mathbf{E}}(t). \tag{3.5}$$

For constant histories the integral in (3.4) vanishes. Since the integral in (3.4) is a positive-definite, quadratic form, by (2.4) it follows that, among the histories ending at a given tensor $\mathbf{E}(t)$, only the constant history $\mathbf{E}^\dagger(u) = \mathbf{E}(t)$, $u \in \mathbb{R}^+$, makes Ψ_M minimal, namely

$$\Psi_M(\mathbf{E}(t), {}_r\mathbf{E}^t) \geq \Psi_1(\mathbf{E}(t), \mathbf{E}^\dagger) = \tfrac{1}{2}\mathbf{E}(t) \cdot \mathbf{G}_\infty \mathbf{E}(t),$$

and this proves (3.3).

In the theory of materials with memory, the fading memory property of the norm is a consequence of the fact that any static continuation, equal to the final value $\mathbf{E}$ from t_1 to t_2, and constant history $\mathbf{E}^\dagger$ are closer and closer as $t_2 - t_1 \to \infty$. Here this is not true. Letting

$$\mathbf{E}^{t_2}(s) - \mathbf{E}^\dagger(s) = \begin{cases} 0, & s \in [0, t_2 - t_1], \\ \hat{\mathbf{E}}(s - t_2 + t_1), & s \in (t_2 - t_1, \infty), \end{cases} \tag{3.6}$$

with $\hat{\mathbf{E}}$ a given function on $\mathbb{R}^+$ and $\hat{\mathbf{E}}(0) = 0$, we have

$$\|\mathbf{E}^{t_2} - \mathbf{E}^\dagger\| = \frac{1}{\pi} \int_0^\infty [\hat{\mathbf{E}}_s(\omega) \cdot \mathbf{H}(\omega)\hat{\mathbf{E}}_s(\omega) + \hat{\mathbf{E}}_c(\omega) \cdot \mathbf{H}(\omega)\hat{\mathbf{E}}_c(\omega)]d\omega.$$

Hence $\|\mathbf{E}^{t_2} - \mathbf{E}^\dagger\|$ remains constant as $t_2 - t_1 \to \infty$.

Remark 3.1 The free energy functional (3.4) is not continuous relative to any norm endowed with the fading memory property. Indeed, fading memory means implies that the history $\mathbf{E}^{t_2}$, defined by (3.6), approaches the constant history $\mathbf{E}^\dagger(s) = \mathbf{E}$ as $t_2 - t_1 \to \infty$, while $\Psi_M(\mathbf{E}^{t_2}) \neq \Psi_M(\mathbf{E}^\dagger)$ is independent of $t_2 - t_1$. Thus the continuity property

$$\lim_{t_2 - t_1 \to \infty} \Psi_M(\mathbf{E}^{t_2}) = \Psi_M(\mathbf{E}^\dagger)$$

does not hold.

A drawback of the previous scheme is that oscillating histories do not belong to the history space $\mathcal{H}$, so that and then is not defined on oscillating histories. To improve the scheme we introduce, in the next section, a history space that involves a weight function that alloows for oscillating histories. On the other hand, $\mathcal{H}$ and the corresponding free energy functional Ψ_M for the stress functional (2.1) can be defined such that the properties (3.1), (3.2) and (3.3) hold for every $\mathbf{G}'$ obeying (2.2) to (2.4).

4. An enlarged history space

Let $\mathbf{M} \in \mathcal{H}(\mathbb{R}^+)$ with values in Lin(Sym, Sym) and, moreover, let $\mathbf{M}(u)$ be symmetric and positive definite $\forall u \in \mathbb{R}^+$, and $\mathbf{M}_s(\omega) > 0$, $\forall \omega \in \mathbb{R}^{++}$. We strengthen (2.2) by assuming that

$$\mathbf{G}'\mathbf{M}^{-1} \in L^2(\mathbb{R}^+), \tag{4.1}$$

$\mathbf{M}^{-1}$ being the inverse of $\mathbf{M}$. Letting $\mathbf{K} = -(\mathbf{G}'\mathbf{M}^{-1})_s$, we assume that

$$\mathbf{K}(\omega) > 0, \qquad \forall \omega \in \mathbb{R}^{++}. \tag{4.2}$$

Such functions $\mathbf{M}(u), u \in \mathbb{R}^+$, do exist. A simple example is

$$\mathbf{M} = \mathbf{1},$$

$\mathbf{1}$ being the identity of Lin(Sym, Sym), whence $\mathbf{K} = -\mathbf{G}'_s$, and (4.1)-(4.2) are satisfied. A second example involves $\mathbf{G}$ subject to the conditions $\mathbf{G}' < 0$, $\mathbf{G}'' \geq 0$. We let

$$\mathbf{M} = (-\mathbf{G}')^{1/2};$$

whence

$$\mathbf{K} = [(-\mathbf{G}')^{1/2}]_s = \mathbf{M}_s.$$

Since $(-\mathbf{G}')^{1/2} \in L^2(\mathbb{R}^+)$, (4.1)-(4.2) again hold true.

To prove (4.2) we must show that the half-range Fourier sine transform of a positive, monotone-decreasing (scalar, vector or tensor-valued) function is positive definite a. e. on $\mathbb{R}^+$. This is the content of the following theorem, whose proof is given in ([9], p. 108).

Theorem.4.1. Let $F > 0, F' \leq 0$ a. e. on $\mathbb{R}^+$ and $F \in L^2(\mathbb{R}^+)$. Then $F_s(\omega) > 0$, $\forall \omega \in \mathbb{R}^{++}$.

Because $\mathbf{M}_s, \mathbf{K} > 0$, $\forall \omega \in \mathbb{R}^{++}$, it follows that

$$\mathbf{J} := \mathbf{K}^{1/2}\mathbf{M}_s^{-1}\mathbf{K}^{1/2} > 0, \ \forall \omega \in \mathbb{R}^{++}.$$

Consider the Hilbert space $\mathcal{H}_\mathbf{M}$ obtained as the completion of

$$\{\mathbf{E}^t : \ \mathbf{M}\mathbf{E}^t \in L^2(\mathbb{R}^+), \int_0^\infty [(\mathbf{M}\mathbf{E}^t)_s \cdot \mathbf{J}(\mathbf{M}\mathbf{E}^t)_s + (\mathbf{M}\mathbf{E}^t)_c \cdot \mathbf{J}(\mathbf{M}\mathbf{E}^t)_c] d\omega < \infty\}$$

relative to the inner product $(\ ,\ )_\mathbf{M}$ defined on any pair of elements $\mathbf{E}_1^t, \mathbf{E}_2^t$ as

$$\begin{aligned}(\mathbf{E}_1^t, \mathbf{E}_2^t)_\mathbf{M} =& \tfrac{1}{2}\mathbf{E}_1(t) \cdot \mathbf{G}_\infty \mathbf{E}_2(t) \\ &+ \tfrac{1}{\pi}\int_0^\infty \{[(\mathbf{M}\mathbf{E}_1^t)_s - \mathbf{M}_s\mathbf{E}_1(t)] \cdot \mathbf{J}[(\mathbf{M}\mathbf{E}_2^t)_s - \mathbf{M}_s\mathbf{E}_2(t)] + (\mathbf{M}\mathbf{E}_1^t)_c \cdot \mathbf{J}(\mathbf{M}\mathbf{E}_2^t)_c\}(\omega) d\omega.\end{aligned}$$

We denote by $\|\cdot\|_\mathbf{M}$ the corresponding norm. We see that $\|\cdot\|_\mathbf{M} = \|\cdot\|$ if $\mathbf{M} = \mathbf{1}$.

Observe that

$$\begin{aligned}\mathbf{T}(t) &= \mathbf{G}_\infty\mathbf{E}(t) + \int_0^\infty \mathbf{G}'(u)[\mathbf{E}^t(u) - \mathbf{E}(t)]du\\ &= \mathbf{G}_\infty\mathbf{E}(t) + \int_0^\infty (\mathbf{G}'\mathbf{M}^{-1})(u)[(\mathbf{M}\mathbf{E}^t)(u) - \mathbf{M}(u)\mathbf{E}(t)]du.\end{aligned}$$

Since $\mathbf{G}'\mathbf{M}^{-1} \in L^2(\mathbb{R}^+)$ and $\mathbf{M}\mathbf{E}^t \in L^2(\mathbb{R}^+)$, we can apply Plancherel's theorem to obtain

$$\int_0^\infty \mathbf{G}'(u)[\mathbf{E}^t(u) - \mathbf{E}(t)]du = -\frac{2}{\pi}\int_0^\infty \mathbf{K}(\omega)[(\mathbf{M}\mathbf{E}^t)_s(\omega) - \mathbf{M}_s(\omega)\mathbf{E}(t)]d\omega. \tag{4.3}$$

Hence

$$|\mathbf{T}(t)| \le |\mathbf{G}_\infty^{1/2}|(\mathbf{E}(t)\cdot\mathbf{G}_\infty\mathbf{E}(t))^{1/2} + \tfrac{2}{\pi}\Big|\int_0^\infty \mathbf{K}(\omega)[(\mathbf{M}\mathbf{E}^t)_s(\omega) - \mathbf{M}_s(\omega)\mathbf{E}(t)]d\omega\Big|,$$

and application of the inequality (3.2) yields

$$\begin{aligned}|\mathbf{T}(t)| \le&|\mathbf{G}_\infty^{1/2}|(\mathbf{E}(t)\cdot\mathbf{G}_\infty\mathbf{E}(t))^{1/2}\\ &+\frac{2}{\pi}\Big(\int_0^\infty |\mathbf{K}^{1/2}\mathbf{M}_s^{1/2}|_2^2(\omega)d\omega\Big)^{1/2}\Big(\int_0^\infty \{[(\mathbf{M}\mathbf{E}^t)_s - \mathbf{M}_s\mathbf{E}(t)]\cdot\mathbf{J}[(\mathbf{M}\mathbf{E}^t)_s - \mathbf{M}_s\mathbf{E}(t)]\\ &\qquad + (\mathbf{M}\mathbf{E}^t)_c\cdot\mathbf{J}(\mathbf{M}\mathbf{E}^t)_c\}(\omega)d\omega\Big)^{1/2}.\end{aligned}$$

This inequality shows that the functional (3.4) is continuous relative to the norm $\|\cdot\|_{\mathbf{M}}$.

Consider the functional

$$\begin{aligned}\Psi_{\mathbf{M}}(\mathbf{E}(t), {}_r\mathbf{E}^t) =& \tfrac{1}{2}\mathbf{E}(t)\cdot\mathbf{G}_\infty\mathbf{E}(t)\\ &+\tfrac{1}{\pi}\int_0^\infty \{[(\mathbf{M}\mathbf{E}^t)_s - \mathbf{M}_s\mathbf{E}(t)]\cdot\mathbf{J}[(\mathbf{M}\mathbf{E}^t)_s - \mathbf{M}_s\mathbf{E}(t)] + (\mathbf{M}\mathbf{E}^t)_c\cdot\mathbf{J}(\mathbf{M}\mathbf{E}^t)_c\}(\omega)d\omega\end{aligned} \tag{4.4}$$

on $\mathcal{H}_{\mathbf{M}}$ and denote by $\psi_{\mathbf{M}}(t)$ the corresponding function on $\mathbb{R}^+$. Relative to the known literature on free energy in linear viscoelasticity, prominent particular cases are

$$\mathbf{M} = 1 \quad\Longrightarrow\quad \mathbf{K} = -\mathbf{G}'_s, \quad \mathbf{J} = -\omega\mathbf{G}'_s$$

and, if $\mathbf{G}' < 0, \mathbf{G}'' \ge 0$,

$$\mathbf{M} = (-\mathbf{G}')^{1/2} \quad\Longrightarrow\quad \mathbf{K} = \mathbf{M}_s, \quad \mathbf{J} = 1.$$

In the first case we have the functional

$$\begin{aligned}(\mathbf{E}(t), {}_r\mathbf{E}^t) =& \tfrac{1}{2}\mathbf{E}(t)\cdot\mathbf{G}_\infty\mathbf{E}(t)\\ &+\tfrac{1}{\pi}\int_0^\infty [(\mathbf{E}_s^t(\omega) - \frac{\mathbf{E}(t)}{\omega})\cdot\mathbf{H}(\omega)(\mathbf{E}_s^t(\omega) - \frac{\mathbf{E}(t)}{\omega}) + \mathbf{E}_c^t(\omega)\cdot\mathbf{H}(\omega)\mathbf{E}_c^t(\omega)]d\omega,\end{aligned}$$

namely the free energy (3.4). Because of (3.5), $\dot{\psi}_M = \mathbf{T}\cdot\dot{\mathbf{E}}$. Then, by (7.2), the free energy is maximal among all other free energies (4.4), viz.

$$\Psi_M(\mathbf{E}(t), {}_r\mathbf{E}^t) \geq \Psi_{\mathbf{M}}(\mathbf{E}(t), {}_r\mathbf{E}^t), \quad \mathbf{E}^t \in \mathcal{H}. \tag{4.5}$$

In the second case we have the functional

$$\begin{aligned}\Psi_G(\mathbf{E}(t), {}_r\mathbf{E}^t) &= \tfrac{1}{2}\mathbf{E}(t)\cdot\mathbf{G}_\infty\mathbf{E}(t)\\ &+\tfrac{1}{\pi}\int_0^\infty \{[(-\mathbf{G}')^{1/2}(\mathbf{E}^t-\mathbf{E}(t))]_s\cdot[(-\mathbf{G}')^{1/2}(\mathbf{E}^t-\mathbf{E}(t))]_s\\ &\qquad + ((-\mathbf{G}')^{1/2}\mathbf{E}^t)_c\cdot((-\mathbf{G}')^{1/2}\mathbf{E}^t)_c\}(\omega)d\omega.\end{aligned}$$

Hence, by Plancherel's theorem,

$$\Psi_G(\mathbf{E}(t), {}_r\mathbf{E}^t) = \tfrac{1}{2}\mathbf{E}(t)\cdot\mathbf{G}_\infty\mathbf{E}(t) - \tfrac{1}{2}\int_0^\infty [\mathbf{E}^t(s)-\mathbf{E}(t)]\cdot\mathbf{G}'(s)[\mathbf{E}^t(s)-\mathbf{E}(t)]ds,$$

which is the "Graffi-Volterra free energy" (cf. [10-11]). Incidentally, Ψ_G is a free energy in the sense of Definition 3.1 if $\mathbf{G}'<0, \mathbf{G}''\geq 0$.

Not every choice of $\mathbf{M}$ obeying (4.1) and (4.2) leads to a free energy $\Psi_{\mathbf{J}}$ in the sense of Definition 1. Indeed, we can easily verify that the condition (4.3) holds because $\mathbf{J}>0$ implies that the integral in (4.4) is positive definite. Condition (3.1) holds too as a consequence of (4.3), viz.

$$\partial_{\mathbf{E}}(\mathbf{E}(t), {}_r\mathbf{E}^t) = \mathbf{G}_\infty\mathbf{E}(t) - \frac{2}{\pi}\int_0^\infty \mathbf{K}(\omega)[(\mathbf{M}\mathbf{E}^t)_s - \mathbf{M}_s\mathbf{E}(t)](\omega)d\omega.$$

But, in order that (3.2) hold for $\Psi_{\mathbf{J}}$ it is necessary that $\mathbf{M}$ obey the inequality

$$\int_0^\infty \{(\dot{\mathbf{M}}\mathbf{E}^t)_s\cdot\mathbf{J}[(\mathbf{M}\mathbf{E}^t)_s - \mathbf{M}_s\mathbf{E}(t)] + [(\dot{\mathbf{M}}\mathbf{E}^t)_c - \dot{\mathbf{M}}_c\mathbf{E}(t)]\cdot\mathbf{J}(\mathbf{M}\mathbf{E}^t)_c\}(\omega)d\omega \leq 0 \tag{4.6}$$

for any history $\mathbf{E}^t \in \mathcal{H}_{\mathbf{M}}$. Integrating by parts,

$$[(\mathbf{M}\mathbf{E}^t)^{\cdot}]_s = \omega(\mathbf{M}\mathbf{E}^t)_c + (\dot{\mathbf{M}}\mathbf{E}^t)_s, \qquad [(\mathbf{M}\mathbf{E}^t)^{\cdot}]_s = \mathbf{M}(0)\mathbf{E}(t) - \omega(\mathbf{M}\mathbf{E}^t)_s + (\dot{\mathbf{M}}\mathbf{E}^t)_c;$$

hence, by substitution,

$$\begin{aligned}\dot{\psi}_{\mathbf{M}}(t) &= \mathbf{T}(t)\cdot\dot{\mathbf{E}}(t) + \frac{2}{\pi}\int_0^\infty [(\dot{\mathbf{M}}\mathbf{E}^t)_s\cdot\mathbf{J}(\mathbf{M}\mathbf{E}^t)_s + (\dot{\mathbf{M}}\mathbf{E}^t)_c\cdot\mathbf{J}(\mathbf{M}\mathbf{E}^t)_c](\omega)d\omega\\ &+\frac{2}{\pi}\mathbf{E}(t)\cdot\int_0^\infty\{[\mathbf{M}(0)-\omega\mathbf{M}_s]\mathbf{J}(\mathbf{M}\mathbf{E}^t)_c - \mathbf{M}_s\mathbf{J}(\dot{\mathbf{M}}\mathbf{E}^t)_s\}(\omega)d\omega.\end{aligned}$$

Accordingly, because of the relation

$$\omega\mathbf{M}_s = \mathbf{M}(0) + \dot{\mathbf{M}}_c,$$

we have

$$\dot{\psi}_{\mathbf{J}}(t) = \mathbf{T}(t) \cdot \dot{\mathbf{E}}(t) + \frac{2}{\pi}\int_0^\infty \{(\dot{\mathbf{M}\mathbf{E}^t})_s \cdot \mathbf{J}[(\mathbf{M}\mathbf{E}^t)_s - \mathbf{M}_s\mathbf{E}(t)] + [(\dot{\mathbf{M}\mathbf{E}^t})_c - \dot{\mathbf{M}}_c\mathbf{E}(t)] \cdot \mathbf{J}(\mathbf{M}\mathbf{E}^t)_c\}(\omega)d\omega.$$

Thus (3.2) is a consequence of (4.6).

We characterize the set of admissible tensors $\mathbf{M}$ by $\mathcal{M}$, the space of all symmetric, positive definite elements of Lin(Sym, Sym) such that $\mathbf{M}_s > 0$ on $\mathbb{R}^{++}$ and (4.1), (4.2) and (4.6) hold.

Consistent with the representation (4.4) we have

$$\|\mathbf{E}^t\|^2_{\mathbf{M}} = \Psi_{\mathbf{J}}(\mathbf{E}(t), {}_r\mathbf{E}^t). \tag{4.7}$$

It is therefore natural to view $\mathbf{M}$ and $\mathbf{J}$ in (4.4) as two related intrinsic influence functions; $\mathbf{M}$ is relative to the time domain, $\mathbf{J}$ is relative to the frequency domain (and depends on $\mathbf{M}$ and $\mathbf{G}'$). Although both of them model the memory properties of the material, previous examples suggest that in fact $\mathbf{M}$ is regarded as the fading memory kernel. If $\mathbf{M} = \mathbf{1}$, as in the maximal free energy , there is no fading memory effect in the corresponding norm. Instead, if $\mathbf{J} = \mathbf{1}$ as in Graffi-Volterra's free energy Ψ_G, the memory of the norm fades.

It does not seem obvious how to obtain from (4.4), as a particular case, the free energy functional determined by Day [12] upon evaluation of the maximum recoverable work. Day's investigation on free energy in linear viscoelasticity is very general (the relaxation function is only required to have values in the set of symmetric tensors) but the explicit expression is given for scalar relaxation functions of exponential type. It is well-known that exponential-type, isotropic relaxation functions $\mathbf{G}$ are sufficient for the functional

$$\Psi_D(\mathbf{E}(t), {}_r\mathbf{E}^t) = \tfrac{1}{2}\mathbf{E}(t)\mathbf{G}_\infty\mathbf{E}(t) + \tfrac{1}{2}\{(\mathbf{G}_0 - \mathbf{G}_\infty)^{-1/2}\int_0^\infty \mathbf{G}'(s)[\mathbf{E}(t) - \mathbf{E}(t-s)]ds\}^2$$

to be a free energy, viz. to satisfy (3.1), (3.2), and (3.3) of Definition 3.1

It is of interest to prove that the exponential dependence of the relaxation function is also necessary for (3.1), (3.2) and (3.3) to hold.

The properties (3.1) and (3.3) hold identically, since

$$\frac{\partial \Psi_D}{\partial \mathbf{E}}(t) = \mathbf{G}_\infty\mathbf{E}(t) + (\mathbf{G}_0 - \mathbf{G}_\infty)^{-1/2}\int_0^\infty \mathbf{G}'(s)[\mathbf{E}(t) - \mathbf{E}(t-s)]ds$$
$$= \mathbf{G}_0\mathbf{E}(t) + \int_0^\infty \mathbf{G}'(s)\mathbf{E}(t-s)ds = \mathbf{T}(\mathbf{E}(t), {}_r\mathbf{E}^t).$$

It remains to prove that the validity of (5.2) implies exponential dependence. Observe that

$$d\Psi_D(\mathbf{E}(t), {}_r\mathbf{E}^t|{}_r\dot{\mathbf{E}}^t) = -\int_0^\infty \mathbf{G}'(s)[\mathbf{E}(t) - \mathbf{E}(t-s)]ds \cdot (\mathbf{G}_0 - \mathbf{G}_\infty)^{-1}\int_0^\infty \mathbf{G}'(s)\dot{\mathbf{E}}(t-s)ds$$
$$= \int_0^\infty \mathcal{K}'(s)\mathbf{e}(s)ds \cdot \int_0^\infty \mathcal{K}''(s)\mathbf{e}(s)ds$$

where $\mathcal{K}'(s) = (\mathbf{G}_0 - \mathbf{G}_\infty)^{-1/2}\mathbf{G}'(s)$ and, for any value of t, $\mathbf{e}(s) = \mathbf{E}(t) - \mathbf{E}(t-s)$. If $\mathcal{K}'(s)$ is isotropic, then

$$\mathcal{K}'(s)\mathbf{e}(s) = k(s)\mathring{\mathbf{e}}(s) + \kappa(s)\mathrm{tr}\,\mathbf{e}(s)\,\mathbf{1},$$

where a superposed ring denotes the trace-free part and tr the trace. Hence

$$d\Psi_D = \int_0^\infty k(s)\mathring{\mathbf{e}}(s)ds \cdot \int_0^\infty k'(s)\mathring{\mathbf{e}}(s)ds + 3\int_0^\infty \kappa(s)\mathrm{tr}\,\mathbf{e}(s)ds \int_0^\infty \kappa'(s)\mathrm{tr}\,\mathbf{e}(s)ds.$$

Thus $d\Psi_D \leq 0$ if and only if both products of integrals are non-positive. To save writing we examine the first product in a way that applies formally also to the second. Define the function f on $\mathbb{R}^+$ such that

$$k'(s) = f(s)k(s).$$

Then

$$d\Psi_0 := \int_0^\infty k(s)\mathring{\mathbf{e}}(s)ds \cdot \int_0^\infty f(s)k(s)\mathring{\mathbf{e}}(s)ds \leq 0 \tag{3.8}$$

holds for any piecewise continuous function $\mathbf{e}$ if and only if f is a (non-positive) constant whence $k(s) = k(0)\exp(fs)$.

$\gamma\mathbf{E}, s \in (\beta-\epsilon, \beta), 0,$ otherwis*where*tr E$=0$*and*E$\cdot\mathbf{E} = 1$. Upon substitution and letting $g(s) = f(s)/f(\alpha) - 1$, we can write

$$d\Psi_0 = f(\alpha)\Big[\int_\alpha^{\alpha+\epsilon} k(s)ds + \gamma\int_{\beta-\epsilon}^{\beta} k(s)ds\Big]^2\Big[1 + \frac{\int_\alpha^{\alpha+\epsilon} g(s)k(s)ds + \gamma\int_{\beta-\epsilon}^{\beta} g(s)k(s)ds}{\int_\alpha^{\alpha+\epsilon} k(s)ds + \gamma\int_{\beta-\epsilon}^{\beta} k(s)ds}\Big].$$

By allowing ϵ to be arbitrarily small we can keep only the leading terms to obtain

$$\frac{\int_\alpha^{\alpha+\epsilon} g(s)k(s)ds + \gamma\int_{\beta-\epsilon}^{\beta} g(s)k(s)ds}{\int_\alpha^{\alpha+\epsilon} k(s)ds + \gamma\int_{\beta-\epsilon}^{\beta} k(s)ds} \to \frac{\gamma g(\beta)k(\beta)}{k(\alpha)+\gamma k(\beta)}.$$

The choice $\gamma = 0$ shows that $d\Psi_0 \leq 0$ only if $f(\alpha) \leq 0$. Now let $k(\alpha) + \gamma k(\beta) = \lambda$, whence $\gamma k(\beta) = \lambda - k(\alpha)$. Substitution yields

$$\frac{\gamma g(\beta)k(\beta)}{k(\alpha)+\gamma k(\beta)} = g(\beta)\Big[1 - \frac{k(\alpha)}{\lambda}\Big].$$

Since λ is arbitrary, (3.8) is contradicted unless $g(\beta) = 0$, and since β is arbitrary, $f(s)$ is a non-positive constant.

5. History space

Given a history $\mathbf{E}^t : \mathbb{R}^+ \to \mathrm{Sym}$ we denote by $_r\mathbf{E}^t$ the past history which is obtained by restriction of $\mathbf{E}^t$ to $\mathbb{R}^{++}$. Let Φ be a set of admissible histories for the system under consideration and let Φ_r be the set of past histories obtained by restriction of the histories of Φ to $\mathbb{R}^{++}$. Henceforth we denote by $\Phi_{\mathbf{E}}$ the set of admissible histories with present value $\mathbf{E}$, viz. $\mathbf{E}^t(0) = \mathbf{E}$, and by $\mathbf{E}^\dagger$ the constant history taking the value $\mathbf{E}$, viz. $\mathbf{E}^\dagger(s) = \mathbf{E}, s \in \mathbb{R}^+$. A norm on Φ is a function $\|\cdot\| : \Phi \to \mathbb{R}^+$ such that

$$\|\mathbf{E}^t\| = |\mathbf{E}(t)| + \|\mathbf{E}^t\|_r$$

where $\|\mathbf{E}^t\|_r$ is a norm defined on Φ_r. We then have $\Phi = \mathrm{Sym} \oplus \Phi_r$.

For convenience, let $\mathbf{H}(\omega) = -\omega\mathbf{G}'_s(\omega)$; $\mathbf{H}$ maps $\mathbb{R}^{++}$ into the set of symmetric, positive definite elements of Lin(Sym, Sym). Since $\mathbf{G}'_s$ is continuous, summable and vanishing at infinity, $\mathbf{H}$ is bounded on $\mathbb{R}^+$. Consider the space of functions (histories) on $\mathbb{R}^+$, with values in Sym, defined as

$$= \{(\mathbf{E}(t), {}_r\mathbf{E}^t) \in \mathrm{Sym} \times L^2(\mathbb{R}^{++}),$$
$$\int_0^\infty [(\mathbf{E}^t_s(\omega) - \frac{\mathbf{E}(t)}{\omega}) \cdot \mathbf{H}(\omega)(\mathbf{E}^t_s(\omega) - \frac{\mathbf{E}(t)}{\omega}) + \mathbf{E}^t_c(\omega) \cdot \mathbf{H}(\omega)\mathbf{E}^t_c(\omega)]d\omega < \infty\}.$$

The space becomes a pre-Hilbert space by considering the inner product of two elements $\mathbf{E}^t_1, \mathbf{E}^t_2$ as

$$(5.3)\qquad \begin{aligned}(\mathbf{E}^t_1, \mathbf{E}^t_2) =& \tfrac{1}{2}\mathbf{E}_1(t) \cdot \mathbf{G}_\infty \mathbf{E}_2(t)\\ &+\tfrac{1}{\pi}\int_0^\infty [(\mathbf{E}^t_{1s}(\omega) - \frac{\mathbf{E}_1(t)}{\omega}) \cdot \mathbf{H}(\omega)(\mathbf{E}^t_{2s}(\omega) - \frac{\mathbf{E}_2(t)}{\omega}) + \mathbf{E}^t_{1c}(\omega) \cdot \mathbf{H}(\omega)\mathbf{E}^t_{2c}(\omega)]d\omega.\end{aligned}$$

We denote by $\|\cdot\|$ the corresponding norm.

Theorem 5.1. The functional (2.1), namely

$$(5.4)\qquad \mathbf{T}(t) = \mathbf{G}_\infty \mathbf{E}(t) + \int_0^\infty \mathbf{G}'(s)[\mathbf{E}^t(s) - \mathbf{E}(t)]ds,$$

is continuous on with respect to $\|\cdot|$.

Proof. By (2.1), (2.6), and (2.8), letting ${}_r\mathbf{E}^t \in L^2(\mathbb{R}^{++})$ we have

$$(5.5)\qquad \mathbf{T}(t) = \mathbf{G}_\infty \mathbf{E}(t) + \tfrac{2}{\pi}\int_0^\infty \mathbf{G}'_s(\omega)(\mathbf{E}^t_s(\omega) - \frac{\mathbf{E}(t)}{\omega})d\omega.$$

Then, by (5.1),

$$|\mathbf{T}(t)| \leq |\mathbf{G}^{1/2}_\infty|(\mathbf{E}(t) \cdot \mathbf{G}_\infty \mathbf{E}(t))^{1/2} + \tfrac{2}{\pi}\Big|\int_0^\infty \frac{1}{\omega}\mathbf{H}(\omega)(\mathbf{E}^t_s(\omega) - \frac{\mathbf{E}(t)}{\omega})d\omega\Big|.$$

On applying (5.2), with $\mathbf{N}(\omega) = \mathbf{1}/\omega$ and $\mathbf{A}(\omega) = \mathbf{H}(\omega)/\omega$, we obtain

$$(5.6)\qquad |\mathbf{T}(t)|_2 \leq \kappa\|\mathbf{E}^t\|,$$

where

$$\kappa = \max\{|\mathbf{G}_\infty^{1/2}|, \tfrac{2}{\pi}\Big(\int_0^\infty \frac{c}{\omega^2}|\mathbf{H}(\omega)|d\omega\Big)^{1/2}\}.$$

Since $\mathbf{G}'_s(\omega)/\omega \in L^1(\mathbb{R}^+)$, it follows that $|\mathbf{H}(\omega)|/\omega^2 \in L^1(\mathbb{R}^+)$ and κ is bounded. □

The space is not complete. Consider the space $\mathcal{H}$ obtained by completion of relative to the norm $\|\cdot\|$. Obviously $\mathcal{H}$ is a Hilbert space relative to the inner product (5.3).

Remark 5.1. The space $\mathcal{H}$ contains constant histories and histories which differ from constant histories by -functions. For constant histories $\mathbf{E}^\dagger(s) = \mathbf{E}$, though, we need to consider the Fourier transform in the set of tempered distributions for which

$$\omega \mathbf{E}^\dagger_s(\omega) = \mathbf{E}, \qquad \omega \mathbf{E}^\dagger_c(\omega) = 0.$$

In such a case (5.3) reduces to

$$(\mathbf{E}^\dagger_1, \mathbf{E}^\dagger_2) = \tfrac{1}{2}\mathbf{E}_1 \cdot \mathbf{G}_\infty \mathbf{E}_2.$$

Also, each constant history can be obtained as a limit in $\mathcal{H}$ of the -histories

$$\mathbf{E}^t_\alpha(s) = \begin{cases} \mathbf{E}, & s \in [0,t], \\ \mathbf{E}\exp[\alpha(t-s)], & s \in (t,\infty) \end{cases}$$

as the positive parameter α approaches zero; indeed, as $\mathbf{G}'_s/\omega \in L^1(\mathbb{R}^+)$ and $\alpha^2/(\alpha^2+\omega^2)$ converges to the zero function in $L^\infty(\mathbb{R}^+)$ as $\alpha \to 0$, we have

$$\lim_{\alpha\to 0} \|\mathbf{E}^t_\alpha - \mathbf{E}^\dagger\|^2 = -\lim_{\alpha\to 0}\frac{1}{2\pi}\int_{-\infty}^{\infty} \frac{\alpha^2 \mathbf{E}\cdot \mathbf{G}'_s(\omega)\mathbf{E}}{\omega(\alpha^2+\omega^2)}d\omega = 0.$$

For later convenience, we observe that the inner product (5.3) can be written as

$$(\mathbf{E}^t_1, \mathbf{E}^t_2) = \tfrac{1}{2}\mathbf{E}_1(t)\cdot \mathbf{G}_\infty \mathbf{E}_2(t) + \frac{1}{2\pi}\int_{-\infty}^{\infty} [i\omega \mathbf{E}^t_{1F}(\omega) - \mathbf{E}_1(t)] \cdot (\omega)[i\omega \mathbf{E}^t_{2F}(\omega) - \mathbf{E}_2(t)]^* d\omega,$$

where

$$(\omega) = -\frac{\mathbf{G}'_s(\omega)}{\omega} = \mathrm{Re}\frac{\mathbf{G}'_F(\omega)}{i\omega} > 0, \quad \forall \omega \in \mathbb{R}^{++}.$$

Because $\in L^1(\mathbb{R}^+)$, the space $\mathcal{H}$ contains all histories $\mathbf{E}^t$ such that $|i\omega \mathbf{E}^t_F(\omega)|$ is bounded for almost all $\omega \in \mathbb{R}^+$.

Remark 5.2. Not all bounded functions belong to $\mathcal{H}$. Indeed, sinusoidal histories do not belong to $\mathcal{H}$.

Remark 5.3. Unlike the standard theories of materials with memory, here the norm $\|\cdot\|$ does not exhibit the customary fading memory property. In this regard consider a history $\mathbf{E}^t_a$ in the form

$$\mathbf{E}^t_a(u) = \begin{cases} \mathbf{E}, & u \in [0,a], \\ 0, & u \in (a,\infty), \end{cases}$$

and evaluate the integral

$$I = \int_0^\infty \{(\mathbf{E}_s^t(\omega) - \frac{\mathbf{E}(t)}{\omega}) \cdot \mathbf{H}(\omega)(\mathbf{E}_s^t(\omega) - \frac{\mathbf{E}(t)}{\omega}) + \mathbf{E}_c^t(\omega) \cdot \mathbf{H}(\omega)\mathbf{E}_c^t(\omega)\} d\omega.$$

Now,

$$\mathbf{E}_s^t(\omega) = \int_0^\infty \mathbf{E}_a^t(u) \sin \omega u \, du = \frac{1}{\omega}(1 - \cos \omega a)\mathbf{E},$$

$$\mathbf{E}_c^t(\omega) = \int_0^\infty \mathbf{E}_a^t(u) \cos \omega u \, du = \frac{1}{\omega} \sin \omega a \, \mathbf{E}.$$

Substitution yields the result

$$I = \mathbf{E} \cdot [\int_0^\infty \frac{1}{\omega^2} \mathbf{H}(\omega) d\omega] \mathbf{E} = \frac{\pi}{2} \mathbf{E} \cdot (\mathbf{G}_0 - \mathbf{G}_\infty) \mathbf{E}$$

and then

$$\|\mathbf{E}_a^t\| = \tfrac{1}{2} \mathbf{E} \cdot \mathbf{G}_0 \mathbf{E}.$$

The independence of a shows a lack of memory in the sense of standard theories. In §6 we consider a weaker form of the principle of fading memory such that the norm $\|\cdot\|$ satisfies the fading memory property.

Consider the linear functionals

$$\mathbf{L}_1(\mathbf{E}^t) = \int_0^\infty \mathbf{G}'(u)\mathbf{E}^t(u) \, du,$$

$$\mathbf{L}_2(\mathbf{E}^t) = \int_0^\infty \mathbf{G}'_s(\omega)\mathbf{E}_s^t(\omega) \, d\omega.$$

The domain of $\mathbf{L}_1$, $\mathcal{D}(\mathbf{L}_1)$, consists of all histories $\mathbf{E}^t$ such that the definition of $\mathbf{L}_1$ makes sense, that is $\mathcal{D}(\mathbf{L}_1) = \mathcal{F}'$. The domain of $\mathbf{L}_2$, $\mathcal{D}(\mathbf{L}_2)$, coincides with , and $\mathcal{D}(\mathbf{L}_1) \supset \mathcal{D}(\mathbf{L}_2)$.

Using (5.6) we can prove that $\mathbf{L}_1$ and $\mathbf{L}_2$ are continuous on . Moreover, by (2.7) we have $\mathbf{L}_1(\mathbf{E}^t) = \mathbf{L}_2(\mathbf{E}^t)$, for all histories $\mathbf{E}^t$ in . It is less obvious that $\mathbf{L}_1$ and $\mathbf{L}_2$ are continuous and equal to each other on $\mathcal{H}$.

Theorem 5.2. The extensions of the linear functionals $\mathbf{L}_1, \mathbf{L}_2$ to $\mathcal{H}$ are unique, continuous, and such that

$$\mathbf{L}_1(\mathbf{E}^t) = \mathbf{L}_2(\mathbf{E}^t), \qquad \mathbf{E}^t \in \mathcal{H}. \tag{5.7}$$

Proof. As is dense in $\mathcal{H}$, we may use the continuity (5.6) to show that: for every sequence $\mathbf{E}_n^t$ of histories in that converges to $\mathbf{E}^t \in \mathcal{H}$, there exist the limits

$$\lim_{n\to\infty} \mathbf{L}_1(\mathbf{E}_n^t), \ \lim_{n\to\infty} \mathbf{L}_2(\mathbf{E}_n^t).$$

Accordingly we can define $\mathbf{L}_1(\mathbf{E}^t), \mathbf{L}_2(\mathbf{E}^t)$ in $\mathcal{H}$ as

$$\mathbf{L}_1(\mathbf{E}^t) = \lim_{n\to\infty} \mathbf{L}_1(\mathbf{E}^t_n), \qquad \mathbf{L}_2(\mathbf{E}^t) = \lim_{n\to\infty} \mathbf{L}_2(\mathbf{E}^t_n).$$

Consider the linear functional $\mathbf{L}(\mathbf{E}^t) = \mathbf{L}_1(\mathbf{E}^t) - \mathbf{L}_2(\mathbf{E}^t)$. As $\mathbf{L}(\mathbf{E}^t) = 0, \forall \mathbf{E}^t \in$, by the unique extension of $\mathbf{L}$ to $\mathcal{H}$ we have $\mathbf{L}(\mathbf{E}^t) = 0$, $\forall \mathbf{E}^t \in \mathcal{H}$, whence (5.7) follows. □

6. Dynamics of the viscoelastic solid

The thermodynamic restrictions derived for the (linear) viscoelastic solid are essential to the investigation of existence, uniqueness, and stability properties of solutions to the dynamics of the viscoelastic solid. Indeed, this number is devoted to the connection between thermodynamic restrictions and the dynamic behaviour of the solid.

Let $\Omega \subset \mathcal{E}^3$ be a smooth domain occupied by the viscoelastic body in the reference placement and, as usual, let $\mathbf{u} : \Omega \to V$ be the displacement field relative to the reference placement. Cauchy's problem for the viscoelastic solid is expressed as:

$$\rho\ddot{\mathbf{u}}(\mathbf{x},t) = \nabla\cdot\mathbf{T}(\mathbf{x},t) + \rho\mathbf{b}(\mathbf{x},t) \quad \text{in} \quad \Omega \times \mathbb{R}^{++}, \tag{6.1}$$

$$\mathbf{T}(\mathbf{x},t) = \mathbf{G}_0(\mathbf{x})\,\nabla\mathbf{u}(\mathbf{x},t) + \int_0^\infty \mathbf{G}'(\mathbf{x},s)\,\nabla\mathbf{u}(\mathbf{x},t-s)\,ds \quad \text{in} \quad \Omega \times \mathbb{R}^{+}, \tag{6.2}$$

$$\mathbf{u}(\mathbf{x},t) = 0 \quad \text{on} \quad \partial\Omega \times \mathbb{R}^{+}, \tag{6.3}$$

$$\mathbf{u}(\mathbf{x},t) = \mathbf{u}_0(\mathbf{x},t), \quad \text{in} \quad \Omega \times \mathbb{R}^{-}. \tag{6.4}$$

The relaxation function $\mathbf{G}$ is assumed to be continuous on $\bar{\Omega} \times \mathbb{R}^+$ and to meet the following conditions.

(a) $\mathbf{G} \in C^1(\Omega \times \mathbb{R}^{++}, \text{Lin(Sym)})$,

(b) $\mathbf{G}'(\mathbf{x},\cdot) \in L^1(\mathbb{R}^+)$.

(c) $\mathbf{G}_0(\mathbf{x}) > 0$, $\mathbf{x} \in \Omega$.

On the basis of the conditions (a) to (c), theorems of existence, uniqueness, and continuous dependence on data were proved by Duvaut and Lions [13] and Dafermos [14-15] for Cauchy's problem (6.1)-(6.4). The main result is stated as follows.

Theorem 6.1. *If the conditions (a), (b), (c) hold then Cauchy's problem (6.1)-(6.4) with* $\mathbf{b} \in L^2(0,T;L^2(\Omega))$ *and* $\mathbf{u}_0(\mathbf{x},t)$ *such that*

$$\nabla\cdot\int_t^\infty \mathbf{G}'(\mathbf{x},s)\nabla\mathbf{u}_0(\mathbf{x},t-s)\,ds \in L^2(0,T;L^2(\Omega)), \tag{6.5}$$

has one and only one weak solution $\mathbf{u} \in H^1(0,T;L^2(\Omega)) \cap L^2(0,T;H_0^1(\Omega))$.

Observe that ρ is a given function of $\mathbf{x} \in \mathcal{R}$. For formal simplicity, in this we let ρ be constant and then we can set $\rho = 1$.

7. Ill-posed problems

In the last decade Fichera emphasized, through illuminating papers, that the conditions (a), (b), (c) are not enough to guarantee the well-posedness of the problem (6.1)-(6.4). Indeed, he exhibited various counterexamples which have addressed the attention to some aspects of the theory neglected up to that time. It is then worth giving an outline of Fichera's counterexamples.

First Fichera [16] considered the one-dimensional relaxation function

$$G(s) = 1 - \lambda + \lambda \exp(-s), \tag{7.1}$$

λ being a real-valued parameter. Such relaxation function satisfies the conditions (a), (b), (c) and then Theorem 6.1 applies. Nevertheless, Fichera showed that the spectrum of (6.2) is the half-space

$$\Re\lambda \geq 1. \tag{7.2}$$

This means that for any λ satisfying (7.2) we can have non-uniqueness for the solution to the quasi-static problem (See §4.6.) in the class $\mathbf{u} \in L^1(\mathbb{R}; H_0^1(\Omega))$.

For a while it was believed (cf. [17]), where account is given of a letter to Fichera by Capriz and Gurtin. that the counterexample was scarcely significant in that $\lambda \geq 1$ corresponds to physically unreasonable forms of the relaxation function because $\lambda > 1$ implies $G(s) < 0$ for sufficiently large values of s while $\lambda = 1$ and then $G_\infty = 0$, corresponds to fluids. In particular Capriz and Gurtin observed that to them "using the strange behavior associated with $\lambda > 1$ as an argument against fading memory is equivalent to using the lack of uniqueness in elasticity outside of the positive-definite range as an argument against elasticity". Yet, through a second counterexample, Fichera showed that the requirement $G(s) > 0$ is not enough to guarantee existence and uniqueness. Precisely he considered the relaxation function (cf. [18])

$$G(s) = \frac{1}{2} - s\ \exp(-s) \tag{7.3}$$

for which nonuniqueness is ascertained very easily Capriz and Gurtin sent the letter to Fichera on May 29, 1980. Fichera answered them immediately by providing the further counterexample (7.3) on June 12, 1980..

With a view similar to Capriz & Gurtin's, it was believed [19] that the counterexample (7.3) was scarcely worrisome in that it does not comply with the second law of thermodynamics, notably the inequality (2.4).

Setting aside remarks of physical character, the elaboration of these counterexamples by Fichera constituted an important step. It was then clear that, despite the validity of Theorem

6.1, even the linear problem (6.1)-(6.4) might be ill-posed. In this regard we mention that the quasi-static problem (cf. §4.3) has been shown to be ill-posed [16-17] and the solution to the dynamic problem (6.1)-(6.4) to be asymptotically unstable [20].

Soon after the appearance of [19] concerning the compatibility with thermodynamics, Fichera elaborated a third counterexample based on the relaxation function

$$G(s) = G_\infty + (G_0 - G_\infty)\exp(-\lambda s), \tag{7.4}$$

where G_∞, $G_0 - G_\infty$, $\lambda \in \mathbb{R}^{++}$, which apparently meets all mechanical and thermodynamic requirements (Examined in ch. 6.). Yet the strain function $E(t) = \exp[-(\lambda G_\infty/G_0)t]$ is an eigensolution of

$$G_0 E(t) + \int_0^\infty G'(s)E(t-s)ds = 0$$

thus rendering the solution to (the one-dimensional version of) the quasi-static problem nonunique. Accordingly, the physically well-grounded relaxation function (7.4) allows for nonuniqueness if strain functions are considered which are unbounded at the past infinity.

The topic of existence and uniqueness for the solution to various problems involving viscoelastic bodies has been further investigated, especially among the Italian researchers on continuum mechanics and related topics (Cf.,e. g., [21]). Some developments are delineated shortly. Here we observe that, concerning the questions raised by Fichera's counterexamples, we agree on what Fichera wrote to one of us Letter to A. Morro, on August 1, 1985., namely that in essence the difficulties about existence and uniqueness occur because of the infinite extent of the memory of the material. These difficulties are such that we can speak of the function space where the problem might be well-posed only if the relaxation function is given. Alternatively, given at the outset the function space we can ask for the class of relaxation functions which allow for the well-posedness of the problem. In this chapter we let **G** be given and look for the function space where the problem is well-posed.

The difficulties cannot be removed merely via thermodynamic restrictions. Yet, and this may be seen as the viewpoint at the bottom of this book, we regard the thermodynamic analysis as a preliminary task toward any modelling within continuum mechanics or, more generally, physics. Compatibility of the constitutive properties with thermodynamics does *not* imply per se the well-posedness of the possible (dynamic) problems but addresses the attention to mathematical topics which concern physically sound situations.

8. The quasi-static problem

The relaxation functions (7.1), (7.3), and (7.4) have been elaborated as counterexamples to the existence and/or uniqueness of the solution of the so-called quasi-static problem. In such a formulation the inertial term is neglected and then the problem is expressed through the differential equation

$$\nabla\cdot\left[\mathbf{G}_0\,\nabla\mathbf{u}(t)+\int_0^\infty \mathbf{G}'(s)\,\nabla\mathbf{u}(t-s)\,ds\right]-\mathbf{b}(t)=0, \tag{8.1}$$

in the unknown function $\mathbf{u}:\Omega\times\mathbb{R}\to V$, along with the boundary condition

$$\mathbf{u}=0 \quad \text{on} \quad \partial\Omega. \tag{8.2}$$

The relaxation function $\mathbf{G}:\bar{\Omega}\times\mathbb{R}^+\to \mathrm{Lin(Sym)}$ is taken to be symmetric The symmetry condition is not essential to later developments on existence and uniqueness (cf. [22]); it is assumed merely for the sake of formal simplicity., i. e. $\mathbf{G}=\mathbf{G}^T$, and to meet the requirements (a), (b), (c), and
(d) $\mathbf{G}_\infty>0$,
(e) $\mathbf{G}'_s(\omega)<0,\ \omega\in\mathbb{R}^{++}$.
As shown in a moment, the physical requirements (d), (e) are not only necessary conditions but, along with (a), (b), and (c), also sufficient for the existence and uniqueness of the solution to (8.1)-(8.2). Hence the viscoelastic solid is here required to satisfy the conditions (a) to (e). According to the literature, stronger monotonicity conditions on the relaxation function are often assumed primarily as necessary conditions to gurantee that (e) holds. Indeed, it seems that no experimental evidence would suggest that stronger monotonicity conditions are physically undesirable. However, (e) is a necessary condition placed by thermodynamics. The next developments show how (e) is also sufficient to gurantee the well-posedness of the quasi-static problem and the asymptotic stability of the dynamic problem.

To begin with we give the meaning of the term solution.

Definition 8.1. A function $\mathbf{u}\in L^2(\mathbb{R};H_0^1(\Omega))$ is said to be a weak *solution to the quasi-static problem* (8.1)-(8.2), with source function $\mathbf{b}\in L^2(\mathbb{R};L^2(\Omega))$, if $\mathbf{u}$ satisfies (8.1)-(8.2), in the distributional sense, in $\Omega\times\mathbb{R}$.

Observe that, the formal application of the Fourier transform, with respect to t, to (8.1) and (8.2) provides

$$\nabla\cdot[(\mathbf{G}_0(\mathbf{x})+\mathbf{G}'_F(\mathbf{x},\omega))\nabla\mathbf{u}_F(\mathbf{x},\omega)]=\mathbf{b}_F(\mathbf{x},\omega), \qquad \mathbf{x}\in\Omega, \quad \omega\in\mathbb{R}, \tag{8.3}$$

$$\mathbf{u}_F(\mathbf{x},\omega)=0 \quad \text{on} \quad \partial\Omega, \qquad \omega\in\mathbb{R}. \tag{8.4}$$

The subscript F denotes the Fourier transform, namely

$$\mathbf{u}_F(\mathbf{x},\omega)=\int_{-\infty}^{\infty}\mathbf{u}(\mathbf{x},t)\exp(-i\omega t)\,dt.$$

Since $\mathbf{G}(\mathbf{x},\cdot)$ is defined on $\mathbb{R}^+$ only, the Fourier transform $\mathbf{G}'_F$ involves the integral on $\mathbb{R}^+$ instead of $\mathbb{R}$. To save writing, henceforth we omit specifying that the conditions (a) to (e) are assumed to hold.

Lemma 8.1. *For any value of* ω *it is*

$$\mathbf{G}_0(\mathbf{x}) + \mathbf{G}'_F(\mathbf{x},\omega) \neq 0, \qquad \mathbf{x} \in \Omega, \quad \omega \in \mathbb{R}. \tag{8.5}$$

Proof. Because of (e), when $\omega \neq 0$

$$\Im[\mathbf{G}_0(\mathbf{x}) + \mathbf{G}'_F(\mathbf{x},\omega)] = -\mathbf{G}'_s(\mathbf{x},\omega) \neq 0, \tag{8.6}$$

and, because of (d), when $\omega = 0$

$$\Re[\mathbf{G}_0(\mathbf{x},0) + \mathbf{G}'_F(\mathbf{x},0)] = \mathbf{G}_0(\mathbf{x}) + \int_0^\infty \mathbf{G}'(\mathbf{x},s)\,ds = \mathbf{G}_\infty > 0, \tag{8.7}$$

which completes the proof. □

In view of (8.3) we have to investigate the properties of the operator $L_0(\omega)$, in $H_0^1(\Omega)$, defined as

$$L_0(\omega)\mathbf{u}_F = \nabla\cdot[(\mathbf{G}_0(\mathbf{x}) + \mathbf{G}'_F(\mathbf{x},\omega))\nabla\mathbf{u}_F].$$

In this regard the following result is of decisive importance.

Lemma 8.2. *There exist two constants* $\gamma_1, \gamma_2 \in \mathbb{R}^{++}$, *independent of* ω, *such that, for every* $\mathbf{A} \in Sym$, *at least one of the inequalities*

$$\mathbf{A}^* \cdot [\mathbf{G}_0(\mathbf{x}) + \int_0^\infty \mathbf{G}'(\mathbf{x},s)\cos\omega s\,ds]\mathbf{A} \geq \gamma_1\,\mathbf{A}^*\cdot\mathbf{A}, \tag{8.8}$$

$$-\mathbf{A}^* \cdot [\omega \int \mathbf{G}'(\mathbf{x},s)\sin\omega s\,ds]\mathbf{A} \geq \gamma_2\,\mathbf{A}^*\cdot\mathbf{A}, \tag{8.9}$$

holds with $\mathbf{A}^*$ *complex conjugate of* $\mathbf{A}$.

Proof. The inequality (e) and the continuity property (a) imply that, at any $\mathbf{x} \in \Omega$, (8.9) holds for any ω in the intervals $\omega_0 < |\omega| < \omega_\infty$, with ω_0, ω_∞ strictly positive and finite. Meanwhile, since

$$\lim_{\omega\to\pm\infty} \int_0^\infty \mathbf{G}'(\mathbf{x},s)\cos\omega s\,ds = 0, \tag{8.10}$$

$$\lim_{\omega\to 0} \int_0^\infty \mathbf{G}'(\mathbf{x},s)\cos\omega s\,ds = \mathbf{G}_\infty(\mathbf{x}) - \mathbf{G}_0(\mathbf{x}), \tag{8.11}$$

upon a suitable choice of ω_0 and ω_∞ the inequality (8.8) holds as $|\omega| \in [0,\omega_0]$ and $|\omega| \in [\omega_\infty, \infty)$ too. □

Theorem 8.1. *The solution to the quasi-static problem (8.1)-(8.2) exists and is unique.*

Proof. (Uniqueness) By Lemma 8.1 the (vector) equation (8.3) turns out to be elliptic and then, for any $\omega \in \mathbb{R}$, the homogeneous problem

$$\nabla\cdot\{[\mathbf{G}_0(\mathbf{x}) + \mathbf{G}'_F(\mathbf{x},\omega)]\nabla\mathbf{u}_F(\mathbf{x},\omega)\} = 0, \qquad \mathbf{x} \in \Omega, \quad \omega \in \mathbb{R}, \tag{8.12}$$

$$\mathbf{u}_F(\mathbf{x},\omega) = 0 \quad \text{on} \quad \partial\Omega, \qquad \omega \in \mathbb{R}, \tag{8.13}$$

has the unique solution

$$\mathbf{u}_F(\mathbf{x},\omega) = 0, \qquad \mathbf{x} \in \Omega, \quad \omega \in \mathbb{R}.$$

This implies that the solution to the problem (8.1)-(8.2) is unique in that the inverse Fourier transform $\mathbf{u}$ of $\mathbf{u}_F$ necessarily vanishes.

(Existence) To prove the existence of the solution to (8.1)-(8.2) we need a uniform ellipticity condition for the (real and imaginary) coefficients of (8.12). By Lemma 8.2, the validity of (8.8) or (8.9) and the theorems on elliptic equations allow us to say that there exists a unique solution $\mathbf{u}_F(\mathbf{x},\omega) \in H_0^1(\Omega)$, to the problem (8.12)-(8.13), for any $\omega \in \mathbb{R}$. To complete the proof we have to show the existence of the solution to the given problem (8.1)-(8.2).

Consider the function $\mathbf{H}(\mathbf{x},\mathbf{x}',\omega)$, whose values are second-order tensors, such that

$$\nabla'\cdot\{[\mathbf{G}_0(\mathbf{x}) + \mathbf{G}'_F(\mathbf{x},\omega)]\nabla'\mathbf{H}(\mathbf{x},\mathbf{x}',\omega)\} = \delta(\mathbf{x}-\mathbf{x}')\mathbf{1}, \tag{8.14}$$

where ∇' stands for the gradient with respect to $\mathbf{x}'$, and

$$\mathbf{H}(\mathbf{x},\mathbf{x}',\omega) = 0 \quad \text{as} \quad \mathbf{x}' \in \partial\Omega. \tag{8.15}$$

Again, as a consequence of [23-24], there exists a unique solution $\mathbf{H}(\mathbf{x},\cdot,\omega) \in H_0^1(\Omega)$ and then we have

$$\mathbf{u}_F(\mathbf{x},\omega) = \int_\Omega \mathbf{H}(\mathbf{x},\mathbf{x}',\omega)\,(\rho\mathbf{b})_F(\mathbf{x}',\omega)\,d\mathbf{x}'. \tag{8.16}$$

Moreover, since $\mathbf{G}_0(\mathbf{x}) + \mathbf{G}'_F(\mathbf{x},\omega)$ is continuous in ω then the function $\mathbf{H}(\mathbf{x},\mathbf{x}',\omega)$ as well is continuous in ω. Then, because $\mathbf{G}'_F(\mathbf{x},\omega) \to 0$ as $\omega \to \pm\infty$, it follows that the Green's function $\mathbf{H}(\mathbf{x},\mathbf{x}',\pm\infty)$ exists and is the solution to the problem

$$\nabla'\cdot[\mathbf{G}_0(\mathbf{x})\nabla'\mathbf{H}(\mathbf{x},\mathbf{x}',\pm\infty)] = \delta(\mathbf{x}-\mathbf{x}'), \tag{8.17}$$

$$\mathbf{H}(\mathbf{x},\mathbf{x}',\pm\infty) = 0 \quad \text{as} \quad \mathbf{x}' \in \partial\Omega. \tag{8.18}$$

Hence $\mathbf{H}(\mathbf{x},\cdot,\omega) \in H_0^1(\Omega)$ and $\|\mathbf{H}(\mathbf{x},\cdot,\omega)\|_{H_0^1(\Omega)}$ is bounded as $\omega \in \mathbb{R}$. Thus, because $\mathbf{b}_F(\mathbf{x}',\cdot) \in L^2(\mathbb{R})$, by (8.16) and the properties of $\mathbf{H}$ we have $\mathbf{u}_F \in L^2(\mathbb{R}; H_0^1(\Omega))$. Then there exists the inverse Fourier transform

$$\mathbf{u}(\mathbf{x},t) = \frac{1}{2\pi}\int_{-\infty}^{\infty} \mathbf{u}_F(\mathbf{x},\omega)\exp(i\omega t)\,d\omega$$

such that $\mathbf{u} \in L^1(\mathbb{R}; H_0^1(\Omega))$ and is the solution to the problem (8.1)-(8.2). □

Still about the existence and uniqueness of the solution to the quasi-static problem, it is worth emphasizing how the hypothesis (e) is essential. The weaker restriction $\mathbf{G}'_s(\omega) \leq 0, \omega \in \mathbb{R}^+$ does not guarantee existence and/or uniqueness of the solution. In fact, if $\mathbf{G}'_s(\omega) \leq 0, \omega \in \mathbb{R}^+$ then Lemma 8.1 does not apply and $\mathbf{G}_0(\mathbf{x}) + \mathbf{G}'_F(\mathbf{x}, \omega)$ can vanish for some $\omega \in \mathbb{R}^{++}$. Hence the ellipticity condition may cease to hold. This is shown through a family of counterexamples [25] as follows.

Let $\mathbf{C} \in \mathrm{Lin(Sym)}$ be positive definite and possibly dependent on the position $\mathbf{x}$. Then we represent the relaxation function in the form

$$\mathbf{G}(s) = \mathbf{C}\, G(s)$$

and confine the attention to the scalar function $G(s)$ on $\mathbb{R}^+$. Consider the two-parameter family of functions

$$G(s) = \frac{\kappa(\kappa-2)}{16\,\nu^3} - \int_0^s \left(r^2 - \frac{\kappa-1}{\nu} r + \frac{\kappa^2}{8\nu^2}\right) \exp(-\nu r)\, dr, \tag{8.19}$$

where $\kappa \in (6,8), \nu > 0$. Of course

$$G(0) = \frac{\kappa(\kappa-2)}{16\,\nu^3}, \qquad G'(s) = -\left(s^2 - \frac{\kappa-1}{\nu}s + \frac{\kappa^2}{8\,\nu^2}\right)\exp(-\nu s).$$

Upon direct evaluation we have

$$G(\infty) = G(0) + \int_0^\infty G'(r)\, dr = -\frac{(\kappa-6)(\kappa-8)}{16\,\nu^3};$$

because $\kappa \in (6,8)$ it is $G(\infty) > 0$. Now we evaluate G'_s to get

$$G'_s(\omega) = -\frac{\omega\kappa^2}{8\nu^2(\nu^2+\omega^2)^3}\left(\omega^2 - \frac{8-\kappa}{\kappa}\nu^2\right)^2$$

whereby $G'_s(\omega) \leq 0, \omega \in \mathbb{R}^+$. Indeed, letting

$$\bar{\omega} = \nu\sqrt{\frac{8-\kappa}{\kappa}}$$

we have

$$G'_s(\omega) = 0 \quad \text{as} \quad \omega = 0, \bar{\omega}; \qquad G'_s(\omega) < 0 \quad \text{as} \quad \omega \neq 0, \bar{\omega}.$$

A direct integration yields

$$\int_0^\infty G'(s) \cos\bar{\omega}s\, ds = -\frac{\kappa(\kappa-2)}{16\,\nu^3}$$

and then

$$G(0) + G'_c(\bar{\omega}) = 0.$$

As an aside we notice that the functions (8.19) are not monotone in that $G'(s) > 0$ as s runs between the two (positive) roots of $G'(s)$. However this does not prevent (4.8.19) from being compatible with thermodynamics.

Now consider the equation of motion, in the quasi-static approximation, with the further assumption that the body force vanishes, namely

$$\nabla\cdot\{\mathbf{C}[G(0)\mathbf{E}(\mathbf{x},t) + \int_0^\infty G'(s)\mathbf{E}(\mathbf{x},t-s)\,ds]\} = 0. \tag{8.20}$$

Examine the possibility of non-unique solutions to (8.20). Evidently, $\mathbf{E}(\mathbf{x},t) = 0$ is a solution. Let

$$\mathbf{E}(\mathbf{x},t) = \mathbf{E}_0(\mathbf{x})\exp(i\bar{\omega}t)$$

with $\mathbf{E}_0$ differentiable but otherwise arbitrary. A non-zero $\mathbf{E}_0$ satisfies (8.20) if

$$G(0)\exp(i\bar{\omega}t) + \int_0^\infty G'(s)\exp[i\bar{\omega}(t-s)]\,ds = 0,$$

namely

$$G(0) + \int_0^\infty G'(s)\cos\bar{\omega}s\,ds = 0, \qquad \int_0^\infty G'(s)\sin\bar{\omega}s\,ds = 0.$$

The functions (8.19) meet these conditions and hence they allow nonuniqueness of the solution to the quasi-static problem.

9. The quasi-static problem with time-harmonic body force

Theorem 8.1 deals with body forces $\mathbf{b} \in L^2(\mathbb{R}; L^2(\Omega))$. In many problems though the body force $\mathbf{b}$ is oscillatory in time and this dependence sets $\mathbf{b}$ outside $L^2(\mathbb{R}; L^2(\Omega))$. The same occurs, more generally, for forces which do not vanish at infinity. For formal convenience, we begin by investigating the time-harmonic dependence.

Let $\mathbf{b}(\mathbf{x},t) = \mathbf{b}_0(\mathbf{x})\exp(i\omega t)$ and look for solutions $\mathbf{u}(\mathbf{x},t)$ to the quasi-static problem (8.1)-(8.2) of the form $\mathbf{u}(\mathbf{x},t) = \mathbf{u}_0(\mathbf{x})\exp(i\omega t)$. Substitution in (8.1) yields

$$\nabla\cdot[\mathbf{G}_0(\mathbf{x}) + \int_0^\infty \mathbf{G}'(\mathbf{x},s)\exp(-i\omega s)\,ds]\nabla\mathbf{u}_0(\mathbf{x})\exp(i\omega t) = \mathbf{b}_0(\mathbf{x})\exp(i\omega t).$$

Then the quasi-static problem is equivalent to

$$\nabla\cdot[\mathbf{G}_0(\mathbf{x}) + \int_0^\infty \mathbf{G}'(\mathbf{x},s)\exp(-i\omega s)\,ds]\nabla\mathbf{u}_0(\mathbf{x}) = \mathbf{b}_0(\mathbf{x}), \tag{9.1}$$

$$\mathbf{u}_0 = 0 \quad \text{on} \quad \partial\Omega, \tag{9.2}$$

in the unknown function $\mathbf{u}_0 \in H^2(\Omega)$. We know already that $\mathbf{G}_0 + \mathbf{G}'_F \neq 0$ and then that (9.1) is elliptic. This proves the following result.

Theorem 9.1. *For any body force* $\mathbf{b}(\mathbf{x},t) = \mathbf{b}_0(\mathbf{x})\exp(i\omega t)$, *with* $\mathbf{b}_0 \in L^2(\Omega)$, *there exists a unique weak solution* $\mathbf{u}(\mathbf{x},t) = \mathbf{u}_0(\mathbf{x})\exp(i\omega t)$, *to the problem (9.1)-(9.2), with* $\mathbf{u}_0 \in H_0^1(\Omega)$.

Now consider the case when, for any position $\mathbf{x}$, the body force $\mathbf{b}(\mathbf{x},t)$ has a finite limit in the past and future infinity. Accordingly, let

$$\lim_{t\to\pm\infty} \mathbf{b}(\mathbf{x},t) = \mathbf{b}_\infty(\mathbf{x}),$$

with $\mathbf{b}_\infty(\mathbf{x}) \in L^2(\Omega)$. Moreover let $\mathbf{u}_\infty(\mathbf{x})$ be the solution to Dirichlet's problem

$$\nabla\cdot[\mathbf{G}_\infty(\mathbf{x})\nabla\mathbf{u}_\infty(\mathbf{x})] = \mathbf{b}_\infty(\mathbf{x}), \tag{9.3}$$

$$\mathbf{u}_\infty(\mathbf{x}) = 0 \quad \text{on} \quad \partial\Omega. \tag{9.4}$$

Then the functions $\mathbf{u}(\mathbf{x},t)$ are considered which make the difference $\hat{\mathbf{u}}(\mathbf{x},t) = \mathbf{u}(\mathbf{x},t) - \mathbf{u}_\infty(\mathbf{x})$ belong to $L^2(\mathbb{R}; H^2(\Omega))$. Letting $\hat{\mathbf{b}}(\mathbf{x},t) = \mathbf{b}(\mathbf{x},t) - \mathbf{b}_\infty(\mathbf{x})$ look at the problem

$$\nabla\cdot\big[\mathbf{G}_0(\mathbf{x})\nabla\hat{\mathbf{u}}(\mathbf{x},t) + \int_0^\infty \mathbf{G}'(\mathbf{x},s)\nabla\hat{\mathbf{u}}(\mathbf{x},t-s)\,ds\big] = \hat{\mathbf{b}}(\mathbf{x},t), \tag{9.5}$$

$$\hat{\mathbf{u}} = 0 \quad \text{on} \quad \partial\Omega. \tag{9.6}$$

It is a simple matter to prove the following

Theorem 9.2. *Let* $\mathbf{u}_\infty \in H_0^1(\Omega)$ *be the weak solution to the problem (9.3)-(9.4), with* $\mathbf{b}_\infty \in L^2(\Omega)$, *and* $\hat{\mathbf{u}} \in L^2(\mathbb{R}; H_0^1(\Omega))$ *the weak solution to the problem (9.5)-(9.6) with* $\hat{\mathbf{b}} \in L^2(\mathbb{R}; L^2(\Omega))$. *Then* $\mathbf{u}(\mathbf{x},t) = \hat{\mathbf{u}}(\mathbf{x},t) + \mathbf{u}_\infty(\mathbf{x})$ *is the unique weak solution to the problem*

$$\nabla\cdot\big[\mathbf{G}_0(\mathbf{x})\nabla\mathbf{u}(\mathbf{x},t) + \int_0^\infty \mathbf{G}'(\mathbf{x},s)\nabla\mathbf{u}(\mathbf{x},t-s)\,ds\big] = \mathbf{b}(\mathbf{x},t), \tag{9.7}$$

$$\mathbf{u} = \mathbf{u}_\infty(\mathbf{x}) \quad \text{on} \quad \partial\Omega, \tag{9.8}$$

where $\mathbf{b}(\mathbf{x},t) = \hat{\mathbf{b}}(\mathbf{x},t) + \mathbf{b}_\infty(\mathbf{x})$.

Proof. The positive definiteness of $\mathbf{G}_\infty(\mathbf{x})$ makes the equation (9.3) to be elliptic and Dirichlet's problem (9.3)-(9.4) to have a unique solution $\mathbf{u}_\infty \in H_0^1(\Omega)$. Now, the problem (9.5)-(9.6), for $\hat{\mathbf{u}}$, is formally the same as the problem (9.1)-(9.2) for which we have proved, through Theorem 8.1, existence and uniqueness of the solution in $L^2(\mathbb{R}; H_0^1(\Omega)$. Then there exists a unique solution $\mathbf{u} = \mathbf{u}_\infty + \hat{\mathbf{u}}$, $\hat{\mathbf{u}} \in L^2(\mathbb{R}; H_0^1(\Omega))$, to the problem (9.7)-(9.8). □

10. The dynamic problem

Really the dynamics of a viscoelastic body is governed by (6.1) and (6.2) (in the case $\rho = 1$); relative to the quasi-static problem examined so far we have to account also for the inertial term. Accordingly, the remaining part of this chapter is devoted to the investigation of the dynamic problem

$$\ddot{\mathbf{u}}(\mathbf{x},t) = \nabla\cdot\big[\mathbf{G}_0(\mathbf{x})\nabla\mathbf{u}(\mathbf{x},t) + \int_0^\infty \mathbf{G}'(\mathbf{x},s)\nabla\mathbf{u}(t-s)\,ds\big] + \mathbf{b}(\mathbf{x},t), \quad \text{on} \quad \Omega\times\mathbb{R}, \tag{10.1}$$

$$\mathbf{u}(\mathbf{x},t) = 0 \qquad \text{on} \qquad \partial\Omega\times\mathbb{R}, \tag{10.2}$$

To such equations one should adjoin the initial conditions as in (6.4). Alternatively, one can make suitable hypotheses on $\mathbf{b}$ and look for the properties of the solution $\mathbf{u}(\mathbf{x},t)$ with $t\in\mathbb{R}$. It is just this type of problem that is investigated in this section.

Incidentally, besides being of interest on its own, the dynamic problem may be connected to the questions raised by Fichera's counterexamples in that, as conjectured by Capriz (Cf. [26]), nonexistence and nonuniqueness might originate from the quasi-static approximation. By accounting for the inertial term as well (dynamic problem), some, if not all, drawbacks might disappear. With this view, a research has recently been undertaken by Virga & Capriz [27] that is outlined as follows.

The displacement decays exponentially in the past. Consider the one-dimensional version of (10.1) with $\Omega = [0,l], l>0$. Let $G\in C^2(\mathbb{R}^+,\mathbb{R}^+)$. Moreover let $u(x,t)\in C^2(\Sigma)$, with

$$\Sigma = \{(x,t); x\in[0,l], t\in\mathbb{R}^-\},$$

and

$$u(0,t) = u(l,t) = 0,\ t\in\mathbb{R}^-.$$

The counterpart of the initial data is expected to be a suitable restriction on the behaviour of the solution at the past infinity. Owing to linearity, we can say that the solution to (10.1) is unique if and only if there exists only the zero solution to (10.1) when $\mathbf{b}$ vanishes.

Given $\varepsilon\in\mathbb{R}^{++}$ consider the Banach spaces

$$U_\varepsilon = \{u\in C(\Sigma) : \sup\exp(-\varepsilon t)|u(x,t)| < \infty\},$$

$$V_\varepsilon = \{v\in C(\mathbb{R}^-) : \sup\exp(-\varepsilon t)|v(t)| < \infty\},$$

$$V'_\varepsilon = \{v\in C(\mathbb{R}^+) : \int_0^\infty |v(t)|\exp(-\varepsilon t)\,dt < \infty\}$$

with the norms $\|u\|_\varepsilon$, $\|v\|_\varepsilon$, $\|v\|'_\varepsilon$ as the supremums and the integral appearing in the definition of U_ε, V_ε, V'_ε, respectively. For later purposes it is worth proving the following

Lemma 10.1. *If f is absolutely integrable on $\mathbb{R}^-$ then the differential equation*

$$\alpha^2\ddot{v}(t) + \beta^2 v(t) = f(t), \qquad t \in \mathbb{R}^-, \quad \alpha, \beta \neq 0, \tag{10.3}$$

has only one solution of class C^2 such that $v, \dot{v} \in V_\varepsilon$, that is

$$v(t) = \frac{1}{\alpha\beta}\int_{-\infty}^{t} \sin\frac{\beta}{\alpha}(t-\tau)\, f(\tau)\, d\tau. \tag{10.4}$$

Proof. For any $s \in (-\infty, t)$ we can write the solution to (10.3) as

$$v(t) = v(s)\cos\frac{\beta}{\alpha}(t-s) + \frac{\alpha}{\beta}\sin\frac{\beta}{\alpha}(t-s) + \frac{1}{\alpha\beta}\int_{s}^{t}\sin\frac{\beta}{\alpha}(t-\tau)\, f(\tau)\, d\tau.$$

The fact that $v, \dot{v} \in V_\varepsilon$ implies that $v(s) \to 0$ as $s \to -\infty$. Then the limit $s \to -\infty$ provides the solution in the form (10.4). □

We are now in a position to prove the desired uniqueness result.

Theorem 10.1. *Let $G(x, \cdot)$ be such that $G', G'' \in V_\varepsilon'$ and*

$$G_0 > \int_0^\infty |G'(s)| \exp(-\varepsilon s)\, ds + \frac{1}{\varepsilon}(|G_0'| + \int_0^\infty |G''(s)| \exp(-\varepsilon s)\, ds). \tag{10.5}$$

Then the solution to the problem (10.1)-(10.2) is unique.

Proof. Let $u \in C^2(\Sigma)$. For any $t \in \mathbb{R}^-$ extend $u(\cdot, t)$ to $[-l, l]$ by letting

$$\bar{u}(x,t) = \begin{cases} u(x,t), & x \in (0, l], \\ -u(x,t), & x \in [-l, 0); \end{cases}$$

$\bar{u}$ is of class C^2 in $\bar{\Sigma} = \{(x,t) : |x| \in (0, l], t \in \mathbb{R}^-\}$. We require $\bar{u}$ to be solution to the homogeneous problem

$$-\bar{u}_{tt}(x,t) + G_0\bar{u}_{xx}(x,t) + \int_{-\infty}^{t} G'(t-\tau)\bar{u}_{xx}(x,\tau)\, d\tau = 0, \tag{10.6}$$

$$\bar{u}(-l,t) = \bar{u}(0,t) = \bar{u}(l,t) = 0, \quad t \in \mathbb{R}^-,$$

the subscripts t, x denoting partial derivatives. Let

$$v_n(t) = \frac{1}{l}\int_0^l u(x,t)\sin\frac{n\pi x}{l}\, dx, \quad t \in \mathbb{R}^-,$$

with $n = 1, 2, ...$; of course $v_n \in C^2(\mathbb{R}^{--})$ for any n. Moreover $u, u_t \in U_\varepsilon$ implies $v, \dot{v} \in V_\varepsilon$. Multiplication of (10.6) by $\sin(n\pi x/l)$ and integration gives

$$\left(\frac{l}{n\pi}\right)^2 \ddot{v}_n(t) + G_0 v_n(t) = -\int_{-\infty}^{t} G'(t-\tau)v_n(\tau)\, d\tau, \quad t \in \mathbb{R}^-.$$

Then, letting

$$\alpha = \frac{l}{n\pi}, \qquad \beta = \sqrt{G_0}, \qquad f(t) = -\int_{-\infty}^{t} G'(t-\tau)\, v_n(\tau)\, d\tau$$

and applying Lemma 10.1 we have

$$v_n(t) = -\frac{\lambda_n}{G_0}\int_{-\infty}^{t} d\sigma \, \sin\lambda_n(t-\sigma) \int_{-\infty}^{\sigma} G'(\sigma-\tau)\, v_n(\tau)\, d\tau,$$

where

$$\lambda_n = \frac{n\pi}{l}\sqrt{\frac{G_0}{\rho}}.$$

Since $G'' \in V_\varepsilon'$ an integration by parts yields

$$v_n = \mathcal{F}_n(v_n)$$

where $\mathcal{F}_n$ is the linear operator on V_ϵ defined by

$$\mathcal{F}_n(v) = \frac{1}{G_0}\Big[-\int_{-\infty}^{t} G'(t-\tau)\, v(\tau)\, d\tau + G'(0)\int_{-\infty}^{t} \cos\lambda_n(t-\tau)\, v(\tau)\, d\tau$$
$$+ \int_{-\infty}^{t} d\sigma \cos\lambda_n(t-\sigma)\int_{-\infty}^{\sigma} G''(\sigma-\tau)\, v(\tau)\, d\tau\Big].$$

Hence it follows that, for any $t \in \mathbb{R}^{--}$,

$$\exp(-\varepsilon t)|\mathcal{F}_n(v)| \le \frac{1}{G_0}\Big[\int_0^\infty |G'(s)|\exp(-\varepsilon s)\, ds$$
$$+ \big(|G'(0)| + \int_0^\infty |G''(s)|\exp(-\varepsilon s)\, ds\big)\int_0^\infty \exp(-\varepsilon s)|\cos\lambda_n(t-s)|\, ds\Big]\|v\|_\varepsilon.$$

By virtue of (10.5) $\mathcal{F}_n$ turns out to be a contraction mapping in V_ε and then v_n vanishes identically, for each n, in $\mathbb{R}^{--}$. □

By way of example look at the relaxation function (7.4.), namely

$$G(s) = G_\infty + (G_0 - G_\infty)\exp(-\lambda s), \qquad G_\infty,\ G_0 - G_\infty,\ \lambda > 0,$$

which satisfies the constitutive requirements (a) to (e). A trivial calculation shows that the condition (10.5) holds for

$$\frac{\lambda}{\varepsilon} \in \Big(0, \frac{G_0}{2(G_0 - G_\infty)}\Big).$$

As shown in particular by this example, according to the condition (10.5) counterexamples to uniqueness can be elaborated once G is given, by choosing ε sufficiently small; of course we let

$$|G_0'| + \int_0^\infty |G''(s)|\exp(-\varepsilon s)\, ds > 0.$$

As a by-product, elaborating a counterexample in the dynamic case may shed light on the connection with the quasi-static approximation.

Consider again the relaxation function (7.4) and look for solutions of the form

$$u(x,t) = u_0 \sin \frac{\pi x}{l} \exp[(\mu + i\nu)t]$$

where μ, $\nu \in \mathbb{R}$. Substitution into (10.6) gives

$$-\frac{l^2}{\pi^2}(\mu^2 - \nu^2) + G_0 - (G_0 - G_\infty)\frac{\lambda(\lambda+\mu)}{(\lambda+\mu)^2+\nu^2}$$

and

$$-2\frac{l^2}{\pi^2}\mu\nu + (G_0 - G_\infty)\frac{\lambda\nu}{(\lambda+\mu)^2+\nu^2} = 0.$$

In the quasi-static case (formally, $\rho = 0$) we have

$$\nu = 0, \quad \mu = -\frac{G_\infty}{G_0}\lambda, \ -\lambda,$$

one being just the third counterexample provided by Fichera. Since $\lambda > 0$, this means that eigensolutions are unbounded at past infinity. In the dynamic case, instead, we have $\mu > 0, \nu \neq 0$.

A suggestive way of elaborating a counterexample was adopted in [28] by letting ε be a small parameter and considering

$$\nu = \nu_0 + \varepsilon\nu_1 + o(\varepsilon), \qquad \mu = \varepsilon\mu_1 + o(\varepsilon).$$

Furthermore, the density was taken as $\rho = \rho_0\varepsilon$, thus regarding the density as the perturbative parameter. Upon substitution, at the zero-th and first order in ε we have

$$G_0 + \int_0^\infty G'(s) \cos \nu_0 s \, ds = 0,$$

$$\int_0^\infty G'(s) \sin \nu_0 s \, ds = 0$$

and

$$\rho_0 \frac{l^2}{\pi^2}\nu_o^2 + (\mu_1 + i\nu_1)\int_0^\infty sG'(s) \exp(-i\nu_0 s)\, ds = 0.$$

Then μ_1 is allowed to be negative provided that

$$\int_0^\infty s\, G'(s) \cos \nu_0 s \, ds < 0.$$

This is viewed as the condition ensuring the existence of counterexamples Unfortunately the procedure does not deliver the possible explicit solution for μ and ν.. Capriz and Virga considered some examples of relaxation functions compatible with this condition, such as

$$G(s) = (G_\infty + 2G_0)\exp(-2\lambda s) - (2G_\infty + G_0)\exp(-\lambda s), \quad \lambda > 0, \ 4G_\infty > G_0 > 0,$$

with $\nu_0 = \pm\lambda\sqrt{2G_\infty/G_0}$. However they were unable to determine relaxation functions compatible with the thermodynamic requirement (e), namely $G_s'(\omega) < 0,\ \omega \in \mathbb{R}^{++}$.

The Fourier transform of the displacement is in H_0^1. Back to the three-dimensional case, now we follow [91] and prove an existence and uniqueness result. By analogy with the quasi-static problem, consider the operator

$$L(\omega)\mathbf{u}_F = \omega^2\mathbf{u}_F + \nabla\cdot[(\mathbf{G}_0(\mathbf{x}) + \mathbf{G}_F'(\mathbf{x},\omega))\nabla\mathbf{u}_F]$$

in H_0^1. Upon the application of the Fourier transform, the problem (10.1)-(10.2) leads to

$$L(\omega)\mathbf{u}_F(\mathbf{x},\omega) = \mathbf{b}_F(\mathbf{x},\omega), \tag{10.7}$$

$$\mathbf{u}_F(\mathbf{x},\omega) = 0 \quad \text{on} \quad \partial\Omega. \tag{10.8}$$

As with $L_0(\omega)$, $L(\omega)$ is uniformly coercive in $H_0^1(\Omega)$ in the following sense.

Lemma 10.2. *There exist two constants $\nu_1, \nu_2 \in \mathbb{R}^{++}$, independent of ω, such that at least one of the inequalities*

$$\int_\Omega \{\nabla\mathbf{u}_F^* \cdot [\mathbf{G}_0(\mathbf{x}) + \mathbf{G}_c'(\mathbf{x},\omega)]\nabla\mathbf{u}_F + \omega^2\mathbf{u}_F^*\cdot\mathbf{u}_F\}dx \geq \nu_1\int_\Omega(|\nabla\mathbf{u}_F|^2 + |\mathbf{u}_F|^2)dx, \tag{10.9}$$

$$-\omega\int_\Omega \nabla\mathbf{u}_F^*\cdot\mathbf{G}_s'(\mathbf{x},\omega)\nabla\mathbf{u}_F\,dx \geq \nu_2\int_\Omega(|\nabla\mathbf{u}_F|^2 + |\mathbf{u}_F|^2)dx, \tag{10.10}$$

holds.

Proof. By virtue of Lemma 8.2, the application of Poincaré's inequality and the observation that $\mathbf{u}_F \in H_0^1(\Omega)$, implies the validity of (10.10) for $|\omega| \in (\omega_0, \omega_\infty)$ with ω_0, ω_∞ strictly positive and finite. Now we prove that (10.9) holds as ω is close or equal to zero. When $\omega = 0$ the left-hand side of (10.9) reduces to

$$\int_\Omega \nabla\mathbf{u}_F^*\cdot\mathbf{G}_\infty(\mathbf{x})\,\nabla\mathbf{u}_F\,dx.$$

Then the positive definiteness of $\mathbf{G}_\infty$ allows us to write

$$\int_\Omega \nabla\mathbf{u}_F^*\cdot\mathbf{G}_\infty(\mathbf{x})\,\nabla\mathbf{u}_F\,dx \geq \nu_0\int_\Omega|\nabla\mathbf{u}_F|^2dx,$$

for a suitable ν_0. Again we have recourse to Poincaré's inequality and, owing to the continuity of $\mathbf{G}_c'(\mathbf{x},\cdot)$, we obtain (10.9) for $|\omega| \leq \omega_0$. Finally, by means of (8.10) and the positive definiteness of $\mathbf{G}_0$ we obtain (10.9) for $|\omega| \geq \omega_\infty$. □

To prove the existence of a weak solution to the problem (10.7)-(10.8) it is convenient to examine preliminarily the problem

$$\omega^2\mathbf{H}(\mathbf{x},\mathbf{x}',\omega) - \nabla'\cdot\{[\mathbf{G}_0(\mathbf{x}') + \mathbf{G}_F'(\mathbf{x}',\omega)]\nabla'\mathbf{H}(\mathbf{x},\mathbf{x}',\omega)\} = \delta(\mathbf{x}-\mathbf{x}')\mathbf{1}, \tag{10.11}$$

$$\mathbf{H}(\mathbf{x},\mathbf{x}',\omega) = 0 \qquad \text{as} \qquad \mathbf{x}' \in \partial\Omega, \tag{10.12}$$

in the unknown Green's tensor function $\mathbf{H}(\mathbf{x},\mathbf{x}',\omega)$; in indicial notation $[\nabla' \cdot (\mathbf{G}_0\nabla'\mathbf{H})]_{ij} = \partial[(G_0)_{ihlm}\partial H_{mj}/\partial x'_l]/\partial x'_h$.

Lemma 10.6. *There exists a unique solution* $\mathbf{H}(\mathbf{x},\mathbf{x}',\omega)$ *to the problem (10.11)-(10.12) such that* $\mathbf{H}(\mathbf{x},\cdot,\omega) \in H_0^1(\Omega)$, $\mathbf{x} \in \Omega$ *; moreover* $\mathbf{H}(\mathbf{x},\mathbf{x}',\cdot)$ *is continuous and bounded with* $\mathbf{H}(\mathbf{x},\mathbf{x}',\omega) = O(\omega^{-2+\varepsilon})$ *as* $\omega \to \infty$.

Proof. We know that $L(\omega)$ is a coercive operator in $H_0^1(\Omega)$. Then by Lax-Milgram's theorem $L(\omega)$ is an isomorphism from $H_0^1(\Omega)$ onto $H^{-1}(\Omega)$ (Cf. [22], Lemma 26.1.). Since $\delta \in H^{-1}(\Omega)$ then the tensor-valued function $\mathbf{H}$ exists, is unique, and belongs to $H_0^1(\Omega)$.

By the continuous dependence of $L(\omega)$ on ω, application of Lemma 44.1 of [22] provides the continuous dependence of $\mathbf{H}$ on ω.

Now examine the behaviour of $\mathbf{H}$ as $\omega \to \infty$. Let $\phi \in C_0^\infty(\Omega)$ be real-valued. By (10.11) we have

$$\int_\Omega \mathbf{H}(\mathbf{x},\mathbf{x}',\omega)\phi(\mathbf{x}')\,dx' = \frac{1}{\omega^2}\{\int_\Omega \nabla' \cdot [(\mathbf{G}_0(\mathbf{x}') + \mathbf{G}'_F(\mathbf{x}',\omega))\nabla'\mathbf{H}(\mathbf{x},\mathbf{x};,\omega)]\phi(\mathbf{x}')\,dx' + \phi(\mathbf{x})\}.$$

Application of the divergence theorem yields

$$\int_\Omega \mathbf{H}(\mathbf{x},\mathbf{x}',\omega)\{\phi(\mathbf{x}') - \frac{1}{\omega^2}\nabla' \cdot [(\mathbf{G}_0(\mathbf{x}') + \mathbf{G}'_F(\mathbf{x}',\omega))\nabla'\phi(\mathbf{x}')]\}dx' = \frac{1}{\omega^2}\phi(\mathbf{x}).$$

Hence we have

$$\lim_{\omega\to\infty} \omega^{2-\varepsilon} \int_\Omega \mathbf{H}(\mathbf{x},\mathbf{x}',\omega)\phi(\mathbf{x}')\,dx' = \omega^{-\varepsilon}\int_\Omega \nabla' \cdot [\mathbf{G}_0(\mathbf{x}')\nabla'\phi(\mathbf{x}')]\,dx' + \phi(\mathbf{x}).$$

The observation that the function ϕ is arbitrary completes the proof. □

We are now in a position to prove the main result.

Theorem 10.2. *For every body force* $\mathbf{b} \in L^1(\mathbb{R}; L^2(\Omega))$ *there exists a unique solution*

$$\mathbf{u} \in L^1(\mathbb{R}; H_0^1(\Omega)) \cap W^{1,1}(\mathbb{R}; L^2(\Omega))$$

to the problem (10.1)-(10.2).

Proof. We know that, upon Fourier transformation, the problem (10.1)-(10.2) leads to (10.7)-(10.8) in the unknown function $\mathbf{u}_F$. By Lemma 10.3, $\mathbf{H}(\mathbf{x},\mathbf{x}',\omega)$ is the unique solution to the problem (10.11)-(10.12) with $\mathbf{H}(\mathbf{x},\cdot,\omega) \in H_0^1(\Omega)$ while $\mathbf{H}(\mathbf{x},\mathbf{x}',\cdot)$ is bounded and continuous and $\mathbf{H}(\mathbf{x},\mathbf{x}',\omega) = O(\omega^{-2+\varepsilon})$ as $\omega \to \infty$. Then we can write the solution $\mathbf{u}_F$ to (10.7)-(10.8) as

$$\mathbf{u}_F(\mathbf{x},\omega) = \int_\Omega \mathbf{H}(\mathbf{x},\mathbf{x}',\omega)\,\mathbf{b}_F(\mathbf{x}',\omega)\,dx'. \tag{10.13}$$

Since $\rho\mathbf{b} \in L^1(\mathbb{R}; L^2(\Omega))$ then we have $(\rho\mathbf{b})_F \in L^1(\mathbb{R}; L^2(\Omega))$. Accordingly there exists the inverse Fourier transform of the function (10.13), that is

$$\begin{aligned}\mathbf{u}(\mathbf{x},t) &= \frac{1}{2\pi}\int_{-\infty}^{\infty} \mathbf{u}_F(\mathbf{x},\omega)\,\exp(-i\omega t)\,d\omega \\ &= \frac{1}{2\pi}\int_{\Omega}\int_{-\infty}^{\infty} \mathbf{H}(\mathbf{x},\mathbf{x}',\omega)\,\mathbf{b}_F(\mathbf{x}',\omega)\exp(-i\omega t)\,d\omega\,dx',\end{aligned}$$

which proves the existence of the solution to the problem (10.1)-(10.2).

As regards uniqueness, let $\mathbf{u}_1, \mathbf{u}_2$ be two solutions to (10.1)-(10.2) corresponding to the same body force $\mathbf{b} \in L^1(\mathbb{R}; L^2(\Omega))$. The Fourier transform $\mathbf{u}_F$ of the difference $\mathbf{u} = \mathbf{u}_1 - \mathbf{u}_2$ is a solution to the homogeneous problem

(10.14) $$L(\omega)\,\mathbf{u}_F(\mathbf{x},\omega) = 0, \qquad \mathbf{u}_F(\mathbf{x},\omega) = 0 \quad \text{on} \quad \partial\Omega.$$

The analogue of (10.13) shows that $\mathbf{u}_F(\mathbf{x},\omega) = 0$ is the unique solution to (10.14). Upon inverse Fourier transforming we conclude that $\mathbf{u}(\mathbf{x},t) = 0$ is the unique solution to the homogeneous problem for $\mathbf{u}$. □

11. Cauchy's problem: existence, uniqueness, and stability

In this section we go back to Cauchy's problem with Dirichlet's conditions, as expressed by (6.1)-(6.4), and investigate the properties of the solution $\mathbf{u}$ in the space-time domain $\Omega \times \mathbb{R}^+$.

The relaxation function $\mathbf{G}$ is still required to satisfy the conditions (a) to (e) of the previous section. This allows us to prove the following, preliminary result [29].

Lemma 11.1. *If* $\mathbf{G}$ *satisfies (6.1.1) and (2.4) then, for every pair of real parameters* α, β, *with* $\alpha \geq 0$ *and* $\beta \neq 0$, *there exist two positive-valued functions* $\gamma_1(\alpha)$ *and* $\gamma_2(\alpha,\beta)$ *such that, for each* $\mathbf{x} \in \Omega$ *and* $\mathbf{A} \in \mathrm{Sym}$

(11.1) $$\mathbf{A}\cdot\Big[\mathbf{G}_0(\mathbf{x}) + \int_0^{\infty} \exp(-\alpha s)\,\mathbf{G}'(\mathbf{x},s)\,ds\Big]\cdot\mathbf{A} \geq \gamma_1(\alpha)\mathbf{A}\cdot\mathbf{A},$$

(11.2) $$-\beta\int_0^{\infty} \exp(-\alpha s)\sin\beta s\;\mathbf{G}'(\mathbf{x},s)\,ds\}\mathbf{A} \geq \gamma_2(\alpha,\beta)\mathbf{A}\cdot\mathbf{A}.$$

Proof. By Theorem 6.4.2 and the continuity of $\mathbf{G}'$ (property (a) of §4.1) we obtain at once (11.1). By the same token, Theorem 6.4.1 yields (11.2). □

Now we improve previous results in that, besides proving existence and uniqueness, we show that stability properties hold. This is accomplished by letting

(11.3) $$\mathbf{b} \in L^2(\mathbb{R}^+; H^{-1}(\Omega)),$$

(11.4) $$\mathbf{u}_0(\cdot,\tau) \in H_0^1(\Omega), \qquad \dot{\mathbf{u}}_0(\cdot,\tau) \in L^2(\Omega), \qquad \tau \in \mathbb{R}^-,$$

and, moreover, the initial history $\mathbf{u}_0$ be such that the function

$$\mathbf{U}(\mathbf{x},t) = \nabla\cdot\int_t^\infty \mathbf{G}'(\mathbf{x},s)\nabla\mathbf{u}_0(\mathbf{x},t-s)\,ds \in L^2(\mathbb{R}^+, H^{-1}(\Omega)), \tag{11.5}$$

namely

$$\int_0^\infty \Big|\int_\Omega \Big[\int_t^\infty \mathbf{G}'(\mathbf{x},s)\nabla\mathbf{u}_0(\mathbf{x},t-s)\,ds\Big]\cdot\nabla\mathbf{v}(\mathbf{x})\Big|^2 dt < \infty, \qquad \forall\mathbf{v}\in H_0^1(\Omega).$$

To study the stability properties of the solution $\mathbf{u}$ it is convenient to investigate the behaviour of the Laplace transform $\mathbf{u}_L(p) = \int_0^\infty \exp(-pt)\mathbf{u}(t)\,dt$ of $\mathbf{u}$, p being a complex-valued parameter. Application of the Laplace transform operator to (6.1)-(6.4) leads to

$$\begin{aligned} p^2\mathbf{u}_L(\mathbf{x},p) - \nabla\cdot\{[\mathbf{G}_0(\mathbf{x}) + \mathbf{G}_L'(\mathbf{x},p)]\nabla\mathbf{u}_L(\mathbf{x},p)\} \\ = \mathbf{b}_L(\mathbf{x},p) + p\,\mathbf{u}(\mathbf{x},0) + \dot{\mathbf{u}}(\mathbf{x},0) + \mathbf{U}_L(\mathbf{x},p) =: \mathbf{B}_L(\mathbf{x},p), \end{aligned} \tag{11.6}$$

$$\mathbf{u}_L(\mathbf{x},p) = 0 \qquad \text{on} \qquad \partial\Omega \tag{11.7}$$

where $\mathbf{U}_L(\mathbf{x},p)$ is the Laplace transform of $\mathbf{U}(\mathbf{x},t)$. We have the following result.

Lemma 11.2. *If* $\mathbf{b}$ *and* $\mathbf{u}_0$ *satisfy (11.3)-(11.5) then there exists a unique solution to the problem (11.6)-(11.7) in* $H_0^1(\Omega)$ *for every* $p \in \mathbb{C}^+$.

Proof. Since $\mathbf{b} \in L^2(\mathbb{R}^+; H^{-1}(\Omega))$ and $\mathbf{u}_0$ satisfies (11.4), (11.5), the Laplace transforms $\mathbf{b}_L(\mathbf{x},p)$ and $\mathbf{U}_L(\mathbf{x},p)$ are well defined for every $p \in \mathbb{C}^{++}$. If, though, $p = i\beta$, $\beta \in \mathbb{R}$, then $\mathbf{b}_L(\mathbf{x},i\beta)$ and $\mathbf{U}_L(\mathbf{x},i\beta)$ can be considered as L^2-Fourier transforms of $\mathbf{b}$ and $\mathbf{u}_0$. Accordingly we conclude that the Laplace transforms $\mathbf{b}_L(\mathbf{x},p)$ and $\mathbf{U}_L(\mathbf{x},p)$ are well defined for every $p \in \mathbb{C}^+$.

Following known theorems on elliptic equations (Cf. [23].), to prove the desired result we need only showing that the operator

$$L(p)\mathbf{u}_L = -p^2\mathbf{u}_L - \nabla\cdot\{[\mathbf{G}_0(\mathbf{x}) + \mathbf{G}_L(\mathbf{x},p)]\nabla\mathbf{u}_L\}$$

is coercive in that the bilinear form

$$a(\mathbf{u},\mathbf{v};p) = \int_\Omega \{p^2\mathbf{u}(\mathbf{x})\cdot\mathbf{v}^*(\mathbf{x}) + \nabla\mathbf{v}^*(\mathbf{x})\cdot[\mathbf{G}_0(\mathbf{x}) + \mathbf{G}_L'(\mathbf{x},p)]\nabla\mathbf{u}(\mathbf{x})\}dx$$

satisfies the inequality

$$|a(\mathbf{u},\mathbf{u};p)| > k(p)\|\mathbf{u}\|^2_{H_0^1(\Omega)}$$

where $k(p)$ is a strictly positive constant, for every $\mathbf{u} \in H_0^1(\Omega)$ and $p \in \mathbb{C}^+$. Let, as usual, $p = \alpha + i\beta$ and look first at the case $\beta = 0$. We have

$$\mathbf{G}_0(\mathbf{x}) + \mathbf{G}_L'(\mathbf{x},\alpha) = \mathbf{G}_0(\mathbf{x}) + \int_0^\infty \exp(-\alpha s)\,\mathbf{G}'(\mathbf{x},s)\,ds,$$

and hence the bilinear form becomes

$$a(\mathbf{u},\mathbf{u};\alpha) = \int_\Omega \{\alpha^2 \mathbf{u}\cdot\mathbf{u}^* + \nabla\mathbf{u}\cdot[\mathbf{G}_0(\mathbf{x}) + \mathbf{G}'_L(\mathbf{x},\alpha)]\,\nabla\mathbf{u}^*\,\}dx.$$

Then by the positive definiteness expressed by (11.1) and Poincaré's inequality we have

$$|a(\mathbf{u},\mathbf{u},\alpha)| > \nu(\alpha)\|\mathbf{u}\|^2_{H^1_0(\Omega)},$$

where $\nu(\alpha) > 0$ as $\alpha \in \mathbb{R}^+$.

Now let $\beta \neq 0$. It follows at once that

$$\beta\Im a(\mathbf{u},\mathbf{u},p) = \int_\Omega (2\,\alpha\beta^2\mathbf{u}\cdot\mathbf{u}^* + \beta\nabla\mathbf{u}^*\cdot\Im\mathbf{G}_L\nabla\mathbf{u})dx.$$

By (11.2) we can write

$$|\Im a(\mathbf{u},\mathbf{u};p)| \geq \gamma_2(p)\Big(\int_\Omega |\nabla\mathbf{u}(\mathbf{x})|^2\,dx + \int_\Omega |\mathbf{u}(\mathbf{x})|^2\,dx\Big)$$

where $\gamma_2(p) > 0$ as $\beta \neq 0$. Hence there exists a positive-valued function $\gamma(p)$ such that

$$|a(\mathbf{u},\mathbf{u};p)| \geq \gamma(p)\Big(\int_\Omega \nabla\mathbf{u}(\mathbf{x})|^2\,dx + \int_\Omega |\mathbf{u}(\mathbf{x})|^2\,dx\Big) \geq \gamma(p)\|\mathbf{u}\|^2_{H^1_0(\Omega)}.$$

This proves the coerciveness of $L(p)$. □

With the purpose of investigating stability properties of the solution $\mathbf{u}_L(\mathbf{x},p)$ to (11.6)-(11.7) it is convenient to consider the representation of $\mathbf{u}_L$ in terms of the Green's function$\mathbf{H}(\mathbf{x},\mathbf{x}';p)$ that is the solution to the problem

$$p^2\mathbf{H}(\mathbf{x},\mathbf{x}';p) - \nabla'\cdot\{[\mathbf{G}_0(\mathbf{x}') + \mathbf{G}'_L(\mathbf{x}',p)]\nabla'\mathbf{H}(\mathbf{x},\mathbf{x}';p)\} = \delta(\mathbf{x}-\mathbf{x}')\mathbf{1}, \tag{11.8}$$

$$\mathbf{H}(\mathbf{x},\mathbf{x}';p) = 0, \qquad \mathbf{x}' \in \partial\Omega \tag{11.9}$$

where $\nabla' = \partial/\partial\mathbf{x}'$. The representation is

$$\mathbf{u}_L(\mathbf{x},p) = \int_\Omega \mathbf{H}(\mathbf{x},\mathbf{x}';p)\,\mathbf{B}_L(\mathbf{x}',p)\,dx. \tag{11.10}$$

Concerning $\mathbf{H}(\mathbf{x},\mathbf{x}';p)$ we have the following Throughout, when p is complex-valued the writing $p \to \infty$ stands for $|p| \to \infty$.

Lemma 11.3. *If* $\mathbf{G}$ *satisfies the conditions (11.1)-(11.2) then there exists a unique solution* $\mathbf{H}(\mathbf{x},\mathbf{x}';p)$*, to the problem (11.8)-(11.9), such that* $\mathbf{H}(\mathbf{x},\cdot;p) \in H^1_0(\Omega)$; $\mathbf{H}(\mathbf{x},\mathbf{x}';\cdot)$ *is continuous;* $\lim_{p\to\infty} p^{2-\varepsilon}\mathbf{H}(\mathbf{x},\mathbf{x}';p) = 0$ *a. e. in* Ω, $\varepsilon > 0$.

Proof. Since $L(p)$ is a coercive operator, by Lax-Milgram's theorem it follows that $L(p)$ is an isomorphism of $H_0^1(\Omega)$ onto $H^{-1}(\Omega)$. Then, since $\delta \in H^{-1}(\Omega)$, the existence and uniqueness of $\mathbf{H}(\mathbf{x},\cdot;p) \in H_0^1(\Omega)$ are proved.

Because $L(p)$ is a continuous function of p, by [22], Lemma 44.1, we can say that $\mathbf{H}(\mathbf{x},\mathbf{x}';\cdot)$ is a continuous function and hence is bounded in every bounded set.

To determine the behaviour of $\mathbf{H}(\mathbf{x},\mathbf{x}';\cdot)$ at infinity observe that by (11.8)

$$\int_\Omega \mathbf{H}(\mathbf{x},\mathbf{x}';p)\boldsymbol{\phi}(\mathbf{x}')\,dx' = \frac{1}{p^2}\Big\{\int_\Omega \nabla'\cdot[(\mathbf{G}_0(\mathbf{x}')+\mathbf{G}'_L(\mathbf{x}',p))\nabla'\mathbf{H}(\mathbf{x},\mathbf{x}';p)]\boldsymbol{\phi}(\mathbf{x}')\,dx' + \boldsymbol{\phi}(\mathbf{x})\Big\}$$

for every real-valued $\boldsymbol{\phi} \in C_0^\infty(\Omega)$. Application of the divergence theorem gives

$$\int_\Omega \mathbf{H}(\mathbf{x},\mathbf{x}';p)\{\boldsymbol{\phi}(\mathbf{x}') - \frac{1}{p^2}\nabla\cdot[(\mathbf{G}_0(\mathbf{x}')+\mathbf{G}'_L(\mathbf{x}',p))\nabla'\boldsymbol{\phi}(\mathbf{x}')]\}dx' = \frac{1}{p^2}\boldsymbol{\phi}(\mathbf{x}).$$

Since

$$\lim_{p\to\infty} p^2\int_\Omega \mathbf{H}(\mathbf{x},\mathbf{x}';p)\boldsymbol{\phi}(\mathbf{x}')\,dx' = \int_\Omega \mathbf{H}(\mathbf{x},\mathbf{x}';p)\nabla\cdot[\mathbf{G}_0(\mathbf{x}')\nabla'\boldsymbol{\phi}(\mathbf{x}')]\,dx' + \boldsymbol{\phi}(\mathbf{x})$$

then

$$\lim_{p\to\infty} p^{2-\varepsilon}\int_\Omega \mathbf{H}(\mathbf{x},\mathbf{x}';p)\boldsymbol{\phi}(\mathbf{x}')\,dx' = 0.$$

The arbitrariness of $\boldsymbol{\phi} \in C_0^\infty(\Omega)$ provides the desired result. □

Incidentally, the behaviour of $\mathbf{H}$ as $p \to \infty$ allows us to write

$$\lim_{p\to\infty}\int_\Omega p^2\mathbf{H}(\mathbf{x},\mathbf{x}';p)\boldsymbol{\phi}(\mathbf{x}')\,dx' = \boldsymbol{\phi}(\mathbf{x}). \tag{11.11}$$

We are now in a position to prove the stability result we are looking for.

Theorem 11.1. *If* $\mathbf{b}$ *and* $\mathbf{u}_0$ *satisfy (11.3)-(11.5) then there exists a unique solution*

$$\mathbf{u}(\mathbf{x},t) \in H^1(\mathrm{I\!R}^+;L^2(\Omega)) \cap L^2(\mathrm{I\!R}^+;H_0^1(\Omega))$$

to the problem (6.1)-(6.4) such that

$$\int_0^\infty\int_\Omega [|\nabla\mathbf{u}(\mathbf{x},t)|^2 + |\dot{\mathbf{u}}(\mathbf{x},t)|^2]dx < \infty.$$

Proof. For $p \in \mathbb{C}^+$, by (11.6) we have

$$\begin{aligned}\lim_{p\to\infty}\mathbf{B}_L(\mathbf{x},p) &= \lim_{p\to\infty}\,[\mathbf{b}_L(\mathbf{x},p) + p\,\mathbf{u}(\mathbf{x},0) + \dot{\mathbf{u}}(\mathbf{x},0) + \mathbf{U}_L(\mathbf{x},p)]\\ &= \lim_{p\to\infty}\,[p\,\mathbf{u}(\mathbf{x},0) + \dot{\mathbf{u}}(\mathbf{x},0)]\end{aligned}$$

Then, for every $\varepsilon > 0$, the representation (11.10) of $\mathbf{u}_L$ in terms of $\mathbf{H}$ and the asymptotic behaviour of $\mathbf{H}$ yield

$$\lim_{p\to\infty} p^{1-\varepsilon}\mathbf{u}_L(\mathbf{x},p) = \lim_{p\to\infty} p^{1-\varepsilon}\int_\Omega \mathbf{H}(\mathbf{x},\mathbf{x}';p)[p\mathbf{u}(\mathbf{x}',0) + \dot{\mathbf{u}}(\mathbf{x}',0)]dx' = 0.$$

Moreover, (11.11) gives

$$\lim_{p\to\infty} p\,\mathbf{u}_L(\mathbf{x},p) = \lim_{p\to\infty}\Big[\int_\Omega p^2\mathbf{H}(\mathbf{x},\mathbf{x}';p)\mathbf{u}(\mathbf{x}',0)\,dx' + p\int_\Omega \mathbf{H}(\mathbf{x},\mathbf{x}';p)\dot{\mathbf{u}}(\mathbf{x}',0)\,dx'\Big] = \mathbf{u}(\mathbf{x},0).$$

Hence we have

$$\lim_{p\to\infty} p^{1-\varepsilon}[p\mathbf{u}_L(\mathbf{x},p) - \mathbf{u}(\mathbf{x},0)] = 0.$$

Letting $p = i\beta$, $\beta \in \mathbb{R}$, we can regard $\mathbf{u}_F(\mathbf{x},p)$ and $\beta\mathbf{u}_F(\mathbf{x},\beta) - \mathbf{u}(\mathbf{x},0)$ as L^2-functions with respect to β. Indeed, they are the L^2-Fourier transforms of the function

$$\tilde{\mathbf{u}}(\mathbf{x},t) = \begin{cases} \mathbf{u}(\mathbf{x},t), & \text{as} \quad t \geq 0,\\ 0, & \text{as} \quad t < 0,\end{cases}$$

and of its first-order time derivative. Then Plancherel's theorem gives

$$\infty > \frac{1}{2\pi}\int_{-\infty}^{\infty} |\mathbf{u}_F(\mathbf{x},\beta)|^2 d\beta = \int_{-\infty}^{\infty} |\tilde{\mathbf{u}}(\mathbf{x},t)|^2 dt = \int_0^\infty |\mathbf{u}(\mathbf{x},t)|^2 dt,$$

$$\infty > \frac{1}{2\pi}\int_{-\infty}^{\infty} |i\beta\mathbf{u}_F(\mathbf{x},\beta) + \mathbf{u}(\mathbf{x},0)|^2\, d\beta = \int_0^\infty |\dot{\mathbf{u}}(\mathbf{x},t)|^2\, dt$$

a. e. in Ω.

Now we need an estimate for $\nabla\mathbf{u}_L$. We have

$$\nabla\mathbf{u}_L(\mathbf{x},p) = \int_\Omega \nabla\mathbf{H}(\mathbf{x},\mathbf{x}';p)\mathbf{b}_L(\mathbf{x}',p)\,dx';$$

of course $\nabla\mathbf{H}(\mathbf{x},\mathbf{x}';p)$ is a solution to

$$p^2\,\nabla\mathbf{H}(\mathbf{x},\mathbf{x}';p) - \nabla'\cdot\{[\mathbf{G}_0(\mathbf{x}') + \mathbf{G}'_L(\mathbf{x}',p)]\nabla'\nabla\mathbf{H}(\mathbf{x},\mathbf{x}';p)\} = \nabla\delta(\mathbf{x}-\mathbf{x}')\mathbf{1}.$$

By paralleling Lemma 11.3 (Cf. also [22], Lemma 26.2.) we can prove that

$$\nabla\mathbf{H}(\mathbf{x},\cdot;p) \in L^2(\Omega),$$

$$\nabla\mathbf{H}(\mathbf{x},\mathbf{x}';\cdot) \text{ is a continuous function on } \mathbb{C}^+,$$

$$\lim_{p\to\infty} p^{2-\varepsilon}\int_\Omega \nabla\mathbf{H}(\mathbf{x},\mathbf{x}';p)\boldsymbol{\phi}(\mathbf{x}')\,dx' = 0, \qquad \boldsymbol{\phi}\in C_0^\infty(\Omega),\ \varepsilon > 0.$$

Then Plancherel's theorem yields

$$\infty > \frac{1}{2\pi}\int_{-\infty}^{\infty} |\nabla\mathbf{u}_F(\mathbf{x},\beta)|^2\, d\beta = \int_0^\infty |\nabla\mathbf{u}(\mathbf{x},t)|^2\, dt$$

a. e. in Ω. In conclusion we have

$$\int_0^\infty\int_\Omega [|\nabla\mathbf{u}(\mathbf{x},t)|^2 + |\dot{\mathbf{u}}(\mathbf{x},t)|^2]\,dx\,dt < \infty,$$

which completes the proof. □

The last step of this analysis consists in showing which properties hold for the solution $\mathbf{u}$ to (6.1)-(6.4). In this regard, letting $H(\Omega)$ be a Hilbert space, consider the spaces

$$L^2_{loc}(\mathbb{R}^+; H(\Omega)) = \{\mathbf{u} : \mathbb{R}^+ \to H(\Omega);\ \|\mathbf{u}(t)\|_H \in L^2_{loc}(\mathbb{R}^+)\},$$

$$L^2_l(\mathbb{R}^+; H(\Omega)) = \{\mathbf{u} \in L^2_{loc}(\mathbb{R}^+; H(\Omega);\ \|\mathbf{u}_L(p)\|_H < \infty \quad \forall p \in \mathbb{C}^{++}\},$$

$$H^1_l(\mathbb{R}^+; H(\Omega)) = \{\mathbf{u}, \dot{\mathbf{u}} \in L^2_l(\mathbb{R}^+; H(\Omega))\}.$$

Theorem 11.2. *If* $\mathbf{b} \in L^2_l(\mathbb{R}^+; H^{-1}(\Omega))$, $\mathbf{u}_0(\mathbf{x}, t)$ *satisfies (11.4), and* $\mathbf{U} \in L^2_l(\mathbb{R}^+; H^{-1}(\Omega))$, *then there exists a unique solution* $\mathbf{u}$ *to the problem (6.1)-(6.4) such that*

$$\mathbf{u} \in H^1_l(\mathbb{R}^+; L^2(\Omega)) \cap L^2_l(\mathbb{R}^+; H^1_0(\Omega)).$$

Proof. The Laplace trasform $\mathbf{B}_L(\mathbf{x}, p)$ is well-defined for every $p \in \mathbb{C}^{++}$. Then there exist the inverse Laplace transforms of $\mathbf{u}_L(\mathbf{x}, p)$ and $p\mathbf{u}_L(\mathbf{x}, p) - \mathbf{u}(\mathbf{x}, 0)$ for every $p \in \mathbb{C}^{++}$, that is $\mathbf{u}(\mathbf{x}, t)$, $\dot{\mathbf{u}}(\mathbf{x}, t) \in L^2_l(\mathbb{R}^+; L^2(\Omega))$. By (11.10),

$$\nabla \mathbf{u}_L(\mathbf{x}, p) = \int_\Omega \nabla \mathbf{H}(\mathbf{x}, \mathbf{x}'; p)\, \mathbf{B}_L(\mathbf{x}', p)\, dx'.$$

The properties of $\nabla \mathbf{H}$, shown in Theorem 11.1, imply that there exists the inverse Laplace transform of $\nabla \mathbf{u}(\mathbf{x}, p)$ for every $p \in \mathbb{C}^{++}$, which means that $\mathbf{u}(\mathbf{x}, t) \in L^2_l(\mathbb{R}^+; H^1_0(\Omega))$. □

12. Asymptotic behaviour: exponential decay

Concerning the properties of the solution to the problem (6.1)-(6.4) it is of interest to investigate the asymptotic behaviour in terms of Liapunov functionals [30]. In this regard we recall that, as shown in §6.5, two forms are possible for the free energy of a linear viscoelastic material. Here we have recourse to the Volterra-type free energy and show that it is a profitable Liapunov functional for the problem (6.1)-(6.4). Preliminarly, though, we make it clear that the investigation is performed by assuming that the relaxation function $\mathbf{G}$ satisfy suitable restrictions. These are as follows. For any $\mathbf{v} \in C_0^\infty(\Omega)$,

$$-\int_\Omega \nabla \mathbf{v}(\mathbf{x}) \cdot \mathbf{G}'(\mathbf{x}, s) \nabla \mathbf{v}(\mathbf{x})\, dx > 0, \tag{12.1}$$

$$\int_\Omega \nabla \mathbf{v}(\mathbf{x}) \cdot \mathbf{G}''(\mathbf{x}, s) \nabla \mathbf{v}(\mathbf{x})\, dx > 0. \tag{12.2}$$

Further, there exists a positive constant κ such that

$$\int_\Omega \nabla \mathbf{v}(\mathbf{x}) \cdot [\mathbf{G}''(\mathbf{x}, s) + \kappa \mathbf{G}'(\mathbf{x}, s)] \nabla \mathbf{v}(\mathbf{x})\, dx > 0. \tag{12.3}$$

Finally,

$$\int_0^\infty |\mathbf{G}_\infty(\mathbf{x}) - \mathbf{G}(\mathbf{x},t)|\, dt < \infty. \tag{12.4}$$

The condition (12.4) ensures that initial histories $\mathbf{u}_0$ with constant gradient $\nabla\mathbf{u}(\mathbf{x},\cdot)$ in time are admissible. As a whole, the conditions (12.1)-(12.4) are not severely restrictive in that they are satisfied by a large class of relaxation functions. In particular relaxation functions of exponential type belong to this class.

Let $\hat{\mathbf{u}}^t$ be the difference history defined by

$$\hat{\mathbf{u}}^t(\mathbf{x},s) = \mathbf{u}^t(\mathbf{x},s) - \mathbf{u}(\mathbf{x},t)$$

and consider the free energy functional

$$\Psi(t) = \frac{1}{2}\int_\Omega \{\nabla\mathbf{u}(\mathbf{x},t)\cdot \mathbf{G}_\infty(\mathbf{x})\nabla\mathbf{u}(\mathbf{x},t) - \int_0^\infty \nabla\hat{\mathbf{u}}^t(\mathbf{x},s)\cdot \mathbf{G}'(\mathbf{x},s)\nabla\hat{\mathbf{u}}^t(\mathbf{x},s)\, ds\}dx. \tag{12.5}$$

Then we define the total energy of the body E as

$$E(t) = \frac{1}{2}\int_\Omega |\dot{\mathbf{u}}(\mathbf{x},t)|^2 dx + \Psi(t). \tag{12.6}$$

In terms of E we prove shortly the main result of this section. Preliminarily, though, it is convenient to write the problem (6.1)-(6.4) in the form

$$\frac{\partial}{\partial t}\mathbf{u}(\mathbf{x},t) = \mathbf{v}(\mathbf{x},t), \qquad \mathbf{u}(\mathbf{x},0) = \mathbf{u}_0(\mathbf{x},0), \tag{12.7}$$

$$\frac{\partial}{\partial t}\mathbf{v}(\mathbf{x},t) = \nabla\cdot\big[\mathbf{G}_\infty(\mathbf{x})\nabla\mathbf{u}(\mathbf{x},t) + \int_0^\infty \mathbf{G}'(\mathbf{x},s)\nabla\hat{\mathbf{u}}^t(\mathbf{x},s)\, ds\big], \qquad \mathbf{v}(\mathbf{x},0) = \frac{\partial\mathbf{u}_0}{\partial t}(\mathbf{x},0) \tag{12.8}$$

$$\frac{\partial}{\partial t}\hat{\mathbf{u}}^t(\mathbf{x},s) = -\frac{\partial}{\partial s}\hat{\mathbf{u}}^t(\mathbf{x},s) - \mathbf{v}(\mathbf{x},t), \qquad \hat{\mathbf{u}}^0(\mathbf{x},s) = \mathbf{u}_0(\mathbf{x},-s) - \mathbf{u}_0(\mathbf{x},0),\ s\in\mathbb{R}^+. \tag{12.9}$$

Equations (12.7)-(12.9) are in a form which is especially suited for the application of the theory of contraction semigroups in Hilbert spaces. For brevity let χ be the triplet

$$\chi(\mathbf{x},t) = (\mathbf{u}(\mathbf{x},t),\ \mathbf{v}(\mathbf{x},t),\ \hat{\mathbf{u}}^t(\mathbf{x},\cdot))$$

and let $\mathcal{H}$ be the Hilbert space of all triplets χ such that $\mathbf{u}(\cdot,t)\in H_0^1(\Omega)$, $\mathbf{v}(\cdot,t)\in L^2(\Omega)$ a. e. on $\mathbb{R}^+$ and

$$-\int_0^\infty\int_\Omega \nabla\hat{\mathbf{u}}^t(\mathbf{x},s)\cdot\mathbf{G}'(\mathbf{x},s)\nabla\hat{\mathbf{u}}^t(\mathbf{x},s)\, dx\, ds < \infty.$$

The inner product of two elements $\chi_1, \chi_2 \in \mathcal{H}$ is taken as

$$(\chi_1, \chi_2) = \int_\Omega [\nabla \mathbf{u}_1(\mathbf{x},t) \cdot \mathbf{G}_\infty(\mathbf{x}) \nabla \mathbf{u}_2(\mathbf{x},t) + \mathbf{v}_1(\mathbf{x},t) \cdot \mathbf{v}_2(\mathbf{x},t) \\ - \int_0^\infty \nabla \hat{\mathbf{u}}_1^t(\mathbf{x},s) \cdot \mathbf{G}'(\mathbf{x},s) \nabla \hat{\mathbf{u}}_2^t(\mathbf{x},s)\, ds]\, dx$$

and, of course, the norm is taken as the energy norm

$$\|\chi\| = (\chi, \chi)^{1/2}.$$

Now we prove some preliminary results about the system (12.7)-(12.9) [31].

Lemma 12.1. *If the body force* $\mathbf{b}$ *is zero then the energy (12.6) is an integrable function on* $\mathbb{R}^+$.

Proof. By (12.5), (12.6), and the assumption (12.2) we have

$$\frac{d}{dt} E(t) = -\frac{1}{2} \int_\Omega \int_0^\infty \nabla \hat{\mathbf{u}}^t(s) \cdot \mathbf{G}''(\mathbf{x},s) \nabla \hat{\mathbf{u}}^t(s)\, ds\, dx \leq 0 \tag{12.10}$$

whence

$$0 \leq E(t) \leq E(0), \qquad t \in \mathbb{R}^+.$$

Integration of (12.10) on $\mathbb{R}^+$ yields

$$\int_0^\infty \int_\Omega \int_0^\infty \nabla \hat{\mathbf{u}}^t(\mathbf{x},s) \cdot \mathbf{G}''(\mathbf{x},s) \nabla \hat{\mathbf{u}}^t(\mathbf{x},s)\, ds\, dx\, dt = 2 \lim_{t\to\infty} [E(0) - E(t)] \leq 2E(0),$$

which in turn, along with (12.3), yields

$$-\int_0^\infty \int_\Omega \int_0^\infty \nabla \hat{\mathbf{u}}^t(\mathbf{x},s) \cdot \mathbf{G}'(\mathbf{x},s) \nabla \hat{\mathbf{u}}^t(\mathbf{x},s)\, ds\, dx\, dt < \infty.$$

Then, because of Theorem 11.1, we have

$$\int_0^\infty E(t) dt = \int_0^\infty \int_\Omega [\nabla \mathbf{u}(\mathbf{x},t) \cdot \mathbf{G}_\infty \nabla \mathbf{u}(\mathbf{x},t) + |\dot{\mathbf{u}}(\mathbf{x},t)|^2 \\ - \int_0^\infty \nabla \hat{\mathbf{u}}^t(\mathbf{x},s) \cdot \mathbf{G}'(\mathbf{x},s) \nabla \hat{\mathbf{u}}^t(\mathbf{x},s) ds]\, dx\, dt < \infty,$$

which completes the proof. □

Observe that the system (12.7)-(12.9) can be written as

$$\dot{\chi} = A\chi, \qquad \chi(0) = \chi_0$$

where A stands for the operator defined by the right-hand side of the three equations. The domain $\mathcal{D}(A)$ of A is defined as

$$\mathcal{D}(A) = \{\chi \in \mathcal{H};\ \frac{\partial}{\partial s} \hat{\mathbf{u}}^t(\cdot,s) \in H_0^1(\Omega);\ \int_0^\infty \int_\Omega \frac{\partial}{\partial s} \nabla \hat{\mathbf{u}}(\mathbf{x},t-s) \cdot \mathbf{G}'(\mathbf{x},s) \frac{\partial}{\partial s} \nabla \hat{\mathbf{u}}(\mathbf{x},t-s)\, dx\, ds < \infty; \\ \nabla \cdot [\mathbf{G}_\infty(\cdot) \nabla \hat{\mathbf{u}}(\cdot,t) + \int_0^\infty \mathbf{G}'(\cdot,s) \nabla \hat{\mathbf{u}}(\cdot,t-s)\, ds] \in H^1(\Omega)\}.$$

Lemma 12.2. *If* $\mathbf{G}$ *satisfies (12.4) then any history* $\mathbf{u}^t$ *such that* $\chi(\mathbf{x},t) = (\mathbf{u}(\mathbf{x},t), \dot{\mathbf{u}}(\mathbf{x},t), \mathbf{u}^t(\mathbf{x},s) - \mathbf{u}(\mathbf{x},t)) \in \mathcal{D}(A)$ *satisfies (11.4)-(11.5) and* $A\chi \in \mathcal{H}$.

Proof. By the definition of $\mathcal{H}$, every $\chi \in \mathcal{H}$ meets the condition (11.4). Letting $\phi \in H_0^1(\Omega)$, by the Cauchy-Schwarz inequality we have

$$\int_0^\infty \Big| \int_\Omega \int_t^\infty -\nabla \mathbf{u}^t(\mathbf{x},s) \cdot \mathbf{G}'(\mathbf{x},s) \nabla \phi(\mathbf{x})\, ds\, dx \Big|^2 dt$$
$$\leq \int_0^\infty \Big[\int_\Omega \int_0^\infty -\nabla \mathbf{u}^t(\mathbf{x},s) \cdot \mathbf{G}'(\mathbf{x},s) \nabla \mathbf{u}^t(\mathbf{x},s) ds\, dx \int_\Omega \int_t^\infty -\nabla \phi(\mathbf{x}) \cdot \mathbf{G}'(\mathbf{x},s) \nabla \phi(\mathbf{x}) ds\, dx \Big] dt$$
$$= \int_0^\infty \Big[\int_\Omega \int_0^\infty -\nabla \mathbf{u}^t(\mathbf{x},s) \cdot \mathbf{G}'(\mathbf{x},s) \nabla \mathbf{u}^t(\mathbf{x},s) ds\, dx \int_\Omega \nabla \phi(\mathbf{x}) \cdot [\mathbf{G}(\mathbf{x},t) - \mathbf{G}_\infty(\mathbf{x})] \nabla \phi(\mathbf{x}) dx \Big] dt.$$

By the definition of $\mathcal{H}$ the integral on $\Omega \times \mathbb{R}^+$ is finite for every $t \in \mathbb{R}^+$. Then by the condition (12.4) it follows that the right-hand side is finite, whence we have (11.5). □

Since

$$(A\chi, \chi) = -\frac{1}{2} \int_0^\infty \int_\Omega \nabla \hat{\mathbf{u}}^t(\mathbf{x},s) \cdot \mathbf{G}''(\mathbf{x},s) \nabla \hat{\mathbf{u}}^t(\mathbf{x},s)\, ds \leq 0,$$

following Dafermos [34] we state the following properties of A.

Lemma 12.6. *The operator* A *is dissipative, namely* $(A\chi, \chi) \leq 0$, $\chi \in \mathcal{D}(A)$, *and the range of* $A - I$ *is* $\mathcal{H}$.

The properties established by Lemma 12.3 allow the application of Lumer-Phillips theorem whereby A generates a strongly continuous semigroup of linear contractions $S(t)$ on $\mathcal{H}$ relative to the norm $\|\chi\|$. Then we can write the solution to (12.7)-(12.9) as

$$\chi(t) = S(t)\, \chi_0$$

Since

(12.11) $$E(t) = \frac{1}{2}(S(t)\, \chi_0,\ S(t)\, \chi_0)$$

then by Lemma 12.1 we have

$$\int_0^\infty (S(t)\, \chi_0,\ S(t)\, \chi_o)\, dt < \infty,$$

which holds for every $\chi_0 \in \mathcal{H}$ in that $\mathcal{D}(A)$ is dense in $\mathcal{H}$.

This allows the application of Datko-Pazy's theorem whereby for a strongly continuous semigroup of linear operators $S(t)$ on a Hilbert space $\mathcal{H}$ such that

$$\int_0^\infty (S(t)\, \chi_0,\ S(t)\, \chi_0) dt < \infty, \qquad \chi_0 \in \mathcal{H},$$

there exist two positive constants M, μ such that

(12.12) $$(S(t)\, \chi_0,\ S(t)\, \chi_0) \leq M \exp(-\mu t)(\chi_0, \chi_0).$$

Remark 12.1. Quite often what we call Datko-Pazy's theorem is quoted as Pazy's theorem [32]. Really, by extending a well-known theorem of Liapunov concerning Hurwitzian matrices, Datko [36] proved the following theorem.

A necessary and sufficient condition that a strongly continuous semigroup of operators $S(t)$ of class C^0 defined on a complex Hilbert space X satisfy the condition $\|S(t)\| \leq M\exp(-\mu t)$, with $M \in [1,\infty)$, $\mu \in \mathbb{R}^{++}$, is the existence of an Hermitian endomorphism B on X, with $B \geq 0$, such that for all x in the domain of the infinitesimal generator A of $S(t)$ the relation $2(BAx, x) = -\|x\|^2$ holds.

He also gave the following corollary.

A necessary and sufficient condition that a strongly continuous semigroup $S(t)$ of class C^0 defined on a complex Hilbert space satisfy the condition $\|S(t)\| \leq M\exp(-\mu t)$ is that for each $x \in X$ the integral $\int_0^\infty \|S(t)\|^2 dt$ be convergent.

Later, while investigating Hermitian forms in X such that $(Bx, x) \geq b\|x\|^2$, Pazy [32] proved the following theorem.

Let $p \geq 1$ be fixed, X a Banach space, $S(T)$ a strongly continuous semigroup on X. The norm $\|x\|_p = (\int_0^\infty \|S(t)x\|^p dt)^{1/p}$ is finite for every $x \in X$ if and only if there exist constants $M \in [1,\infty)$ and $\mu \in \mathbb{R}^{++}$ such that $\|S(t)\| \leq M\exp(-\mu t)$.

So we have applied Datko's corollary or Pazy's theorem in the case $p = 2$.

We are now in a position to prove the desired result on the asymptotic behaviour of the solution [50].

Theorem 12.1. *If the body force* $\mathbf{b}$ *is zero then, for any solution* $\mathbf{u}$ *to the problem (6.1)-(6.4), there exist two positive constants* M, μ *such that*

$$E(t) \leq M\ \exp(-\mu t)E(0). \tag{12.13}$$

Proof. Because $S(0) = I$, by (12.11) and (12.12) we find the sought result (12.13). □

13. Existence, uniqueness, and stability for fluids

Cauchy's problem is now investigated for the viscoelastic fluid as modelled in §6.6. Really, as usual we confine the attention to incompressible fluids (with $\rho = 1$) and to the linear approximation of the equation of motion.

Let $\Omega \subset \mathcal{E}^3$ be a smooth domain occupied by the linear viscoelastic fluid and let $\mathbf{v} : \Omega \to V$ be the velocity field. Cauchy's problem consists in finding the fields $\mathbf{v}$ and p on $\Omega \times \mathbb{R}^+$ such that

$$\nabla\cdot\mathbf{v}(\mathbf{x}, t) = 0 \qquad \text{in} \quad \Omega \times \mathbb{R}^{++}, \tag{13.1}$$

$$\dot{\mathbf{v}}(\mathbf{x}, t) = -\nabla p(\mathbf{x}, t) + \int_0^\infty \mu(s)\,\Delta\mathbf{v}(\mathbf{x}, t-s)\,ds + \mathbf{b}(\mathbf{x}, t) \qquad \text{in} \quad \Omega \times \mathbb{R}^+, \tag{13.2}$$

$$\mathbf{v}(\mathbf{x},t) = 0 \quad \text{on} \quad \partial\Omega \times \mathrm{I\!R}^+, \tag{13.3}$$

$$\mathbf{v}(\mathbf{x},t) = \mathbf{v}_0(\mathbf{x},t), \qquad \text{in} \quad \Omega \times \mathrm{I\!R}^-. \tag{13.4}$$

The constant mass density ρ is understood and not written.

Concerning linear viscoelastic fluids Slemrod [33] proved that if, in our notation,

$$\mu \in C^2(\mathrm{I\!R}^+), \qquad \mu(s) \to 0 \quad \text{as} \quad s \to \infty$$

and

$$\mu(s) > 0, \qquad \mu'(s) < 0, \qquad \mu''(s) \geq 0, \qquad s \in \mathrm{I\!R}^+,$$

then the rest state of the fluid is stable, in the sense of a norm of "fading memory" type, and the solution to (13.1)-(13.4) converges to the rest state in this norm as $t \to \infty$. Next he proved that the additional assumption

$$-\int_0^\infty \mu'(s)\, s^2\, ds < \infty$$

yields asymptotic stability. Later [34] he showed that the further assumption

$$\mu'(s) + \xi\,\mu(s) \leq 0, \qquad \mu''(s) + \xi\,\mu'(s) \geq 0, \qquad s \in \mathrm{I\!R}^+,$$

for some $\xi \in \mathrm{I\!R}^{++}$, implies exponential decay in an appropriate fading memory norm.

Here different assumptions on the constitutive properties are made. The relaxation function μ is required to satisfy the principle of fading memory in the form

$$\mu \in C(\mathrm{I\!R}^+, \mathrm{I\!R}), \qquad \int_0^\infty |\mu(s)| ds < \infty, \tag{13.5}$$

and the second law of thermodynamics, viz.

$$\mu_c(\omega) > 0, \quad \omega \in \mathrm{I\!R} \tag{13.6}$$

As we show shortly, the two conditions (13.5), (13.6) alone are sufficient to enforce existence, uniqueness, and stability of the solution to the problem (13.1)-(13.4). This allows us to view (13.6) as the right physical condition on the shear relaxation function μ instead of the convexity condition.

For later convenience we introduce the following notation. Let ${}^2(\Omega)$ be the Hilbert space obtained by the completion of solenoidal vector fields $\mathbf{v}(\mathbf{x}) \in C_0^\infty(\Omega)$ in the $L^2(\Omega)$ inner product. Similarly, let ${}_0^1(\Omega)$ be the Hilbert space obtained by the completion of solenoidal vector fields in the $H_0^1(\Omega)$ norm. Moreover, let ${}^2(\Omega)$ be the Hilbert space obtained by the completion of irrotational vector fields $\mathbf{v}(\mathbf{x}) \in C_0^\infty(\Omega)$ in the $L^2(\Omega)$ norm. Then we have $L^2(\Omega) ={}^2(\Omega) \oplus{}^2(\Omega)$. The symbol ${}^{-1}(\Omega)$ denotes the dual of ${}_0^1(\Omega)$.

Letting, as usual, a subscript L denote the Laplace transform (in time), we write the problem (13.1)-(13.4) in the form

$$\nabla\cdot\mathbf{v}_L(\mathbf{x},p) = 0 \quad \text{in} \quad \Omega, \tag{13.6}$$

$$p\,\mathbf{v}_L(\mathbf{x},p) = -\nabla p_L(\mathbf{x},p) + \mu(p)\,\Delta\mathbf{v}_L(\mathbf{x},p)) + \tilde{\mathbf{b}}(\mathbf{x},p) \quad \text{in} \quad \Omega, \tag{13.7}$$

$$\mathbf{v}_L(\mathbf{x},p) = 0 \quad \text{on} \quad \partial\Omega \tag{13.8}$$

where

$$\tilde{\mathbf{b}}(\mathbf{x},p) = \mathbf{b}_L(\mathbf{x},p) + \mathbf{v}_0(\mathbf{x},0) + \int_0^\infty \exp(-ps)\int_s^\infty \mu(\tau)\,\Delta\mathbf{v}_0(\mathbf{x},s-\tau)\,d\tau\,ds.$$

Observe that the hypotheses $\mathbf{b}\in L^2(\mathbb{R}^+;H^{-1}(\Omega))$, $\mathbf{v}_0 \in_0^1 (\Omega)$, and

$$\mathbf{V}_0(\mathbf{x},t) := \int_t^\infty \mu(s)\Delta\mathbf{v}_0(\mathbf{x},t-s)\,ds \in L^2(\mathbb{R}^+;^{-1}(\Omega))$$

make $\tilde{\mathbf{b}}(\mathbf{x},p)$ well defined for any $p\in\mathbb{C}^+$.

Following [35], Lemma 2.1, we can say that if $\mathbf{v}_L$ is a solution to

$$\int_\Omega [\mu_L(p)\nabla\mathbf{v}_L(\mathbf{x},p)\cdot\nabla\mathbf{w}(\mathbf{x}) + p\,\mathbf{v}_L(\mathbf{x},p)\mathbf{w}(\mathbf{x})]dx = \int_\Omega \tilde{\mathbf{b}}(\mathbf{x},p)\cdot\mathbf{w}(\mathbf{x})dx, \qquad \mathbf{w}\in_0^1(\Omega), \tag{13.9}$$

then there exists a scalar field in $L^2(\Omega)$, say p_L, such that (13.6) and (13.7) hold in Ω in the distributional sense and, meanwhile, (13.8) holds. This motivates the following

Definition 13.1. A function $\mathbf{v}_L \in_0^1 (\Omega)$ is called weak solution to (13.6)-(13.8) if (13.9) holds.

By general theorems on elliptic systems of equations [35-22-23], the coerciveness of the bilinear form

$$a(\mathbf{v},\mathbf{w};p) = \int_\Omega [\mu_L(p)\nabla\mathbf{v}(\mathbf{x})\cdot\nabla\mathbf{w}^*(\mathbf{x}) + p\mathbf{v}(\mathbf{x})\cdot\mathbf{w}^*(\mathbf{x})]dx$$

in ${}_0^1(\Omega)$ enforces the existence of a (weak) solution to (13.9) for every $\tilde{\mathbf{b}}\in^{-1}(\Omega)$. Then to prove the existence of a solution to (13.9) we have to show that there exists $\alpha\in\mathbb{R}^{++}$, possibly dependent on p, such that

$$|a(\mathbf{v},\mathbf{v};p)| \geq \alpha(p)\|\mathbf{v}\|^2_{{}_0^1(\Omega)}. \tag{13.10}$$

In fact the coerciveness of a is a consequence of (13.6). To show that this is so we begin by recalling a property of Fourier integrals.

Lemma 13.1. *Let $\phi, \psi \in L^1(\mathbb{R})$ with $\phi\psi \in L^1(\mathbb{R})$. If*

$$\Phi(\omega) = \int_{-\infty}^{\infty} \phi_F(\tau)\, \psi_F(\omega - \tau)\, d\tau$$

is continuous in $\mathbb{R}$ then

$$\int_{-\infty}^{\infty} \exp(-i\omega s)\, \phi(s)\, \psi(s) = \frac{1}{2\pi} \int_{-\infty}^{\infty} \phi_F(\tau)\, \psi_F(\omega - \tau)\, d\tau. \tag{13.11}$$

The proof is immediate and can be found, e.g., in [6], ch. III.

Lemma 13.2. *If (13.5) holds and $\mu \in L^1(\mathbb{R}^+)$ then*

$$\int_0^{\infty} \exp(-\alpha s) \cos\omega s\, \mu(s)\, ds > 0, \qquad \alpha \in \mathbb{R}^+, \quad \omega \in \mathbb{R}. \tag{13.12}$$

Proof. Observe that when $\alpha = 0$ the inequality (13.12) coincides with (13.6) and then holds. Let $\alpha > 0$ and consider the functions

$$\phi(s) = \mu(|s|),\ s \in \mathbb{R}, \qquad \psi(s) = \begin{cases} 0, & s \in \mathbb{R}^{--}, \\ \exp(-\alpha s), & s \in \mathbb{R}^+. \end{cases}$$

It follows at once that $\phi, \psi, \phi\psi \in L^1(\mathbb{R})$ and

$$\phi_F(\tau) = 2 \int_0^{\infty} \cos\tau s\, \mu(s)\, ds, \qquad \psi_F(\tau) = \frac{1}{\alpha + i\tau}.$$

Then

$$\Phi(\omega) = 2 \int_{-\infty}^{\infty} \frac{1}{\alpha + i(\omega - \tau)} \int_0^{\infty} \cos\tau s\, \mu(s)\, ds\, d\tau$$

is in $C(\mathbb{R})$. Thus Lemma 13.1 applies and we have

$$\int_{-\infty}^{\infty} \exp[-(\alpha + i\omega)s]\, \mu(s)\, ds = \frac{1}{\pi} \int_0^{\infty} \frac{1}{\alpha + i(\omega - \tau)} \int_0^{\infty} \cos\tau s\, \mu(s)\, ds\, d\tau.$$

The real part yields

$$\int_0^{\infty} \exp(-\alpha s) \cos\omega s\, \mu(s)\, ds = \frac{1}{\pi} \int_{-\infty}^{\infty} \frac{\alpha}{\alpha^2 + (\omega - \tau)^2} \int_0^{\infty} \cos\tau s\, \mu(s)\, ds\, d\tau.$$

The condition (13.6) provides the desired result (13.12). □

We are now in a position to establish the coerciveness of the bilinear form a.

Theorem 13.1. *If the relaxation function μ meets the conditions (13.5) and (13.6) then the bilinear form $a(\mathbf{v}, \mathbf{w}; p)$ is coercive for every $p \in \mathbb{C}^+$.*

Proof. Since $|a(\mathbf{v}, \mathbf{w}; p)| \geq \Re a(\mathbf{v}, \mathbf{w}; p)$, we need to show that

$$\Re a(\mathbf{v}, \mathbf{v}; p) \geq \nu(p) \|\mathbf{v}\|^2_{\overset{\circ}{1}(\Omega)}, \qquad \nu \in \mathbb{R}^{++}, \tag{13.13}$$

for any $p \in \mathbb{C}^+$. Letting $p = \alpha + i\beta$, $\alpha \in \mathbb{R}^+$, $\beta \in \mathbb{R}$, we have

$$\mu_L(p) = \int_0^\infty \exp(-\alpha s)\, \mu(s) \cos\beta s\, ds - i \int_0^\infty \exp(-\alpha s)\, \mu(s) \sin\beta s\, ds$$

whence

$$\Re a(\mathbf{v}, \mathbf{v}; p) = \int_0^\infty \exp(-\alpha s)\, \mu(s) \cos\beta s\, ds \int_\Omega |\nabla \mathbf{v}(\mathbf{x})|^2 dx + \int_\Omega |\mathbf{v}(\mathbf{x})|^2 dx.$$

Hence by Korn's inequality we obtain

$$\Re a(\mathbf{v}, \mathbf{v}; p) \geq c(\Omega) \Big[\int_0^\infty \exp(-\alpha s)\, \mu(s) \cos\beta s\, ds \Big] \, \|\mathbf{v}\|^2_{\overset{1}{0}(\Omega)}$$

where $c(\Omega)$ is a strictly positive constant which depends on the domain Ω. Recourse to Lemma 13.2, and then to the inequality (13.12), completes the proof. □

Now that, by Theorem 13.1, we are guaranteed about existence and uniqueness of the solution to the problem (13.6)-(13.8), we can investigate the properties of the solution $\mathbf{v}_L$. To this end we consider a representation in terms of the Green's tensor function such that

(13.14) $$\int_\Omega [\mu_L(p) \nabla'(\mathbf{x}, \mathbf{x}'; p) \nabla' \mathbf{v}(\mathbf{x}') + p\,(\mathbf{x}, \mathbf{x}'; p) \mathbf{v}(\mathbf{x}')] dx' = \int_\Omega \delta(\mathbf{x} - \mathbf{x}')\, \mathbf{v}(\mathbf{x}') dx', \quad \mathbf{v} \in_0^1 (\Omega),$$

δ denoting Dirac's delta function. In terms of the solution to (13.6)-(13.8) can be written as

(13.15) $$\mathbf{v}_L(\mathbf{x}, p) = \int_\Omega (\mathbf{x}, \mathbf{x}'; p) \tilde{\mathbf{b}}(\mathbf{x}', p)\, dx'.$$

The next results are based on the assumption that

(13.16) $$\mathbf{b} \in L^2(\mathbb{R}^+; H^{-1}(\Omega)), \qquad \mathbf{v}_0 \in_0^1 (\Omega),$$

and

$$\mathbf{V}_0 \in L^2(\mathbb{R}^+,^2 (\Omega)).$$

First we prove existence, uniqueness, and behaviour of with respect to p.

Lemma 13.6. *If (13.16) holds then there exists a unique solution to the problem (13.14) such that, for any* $p \in \mathbb{C}^+$,

(i) $(\mathbf{x}, \cdot; p) \in_0^1 (\Omega)$,

(ii) $(\mathbf{x}, \mathbf{x}'; \cdot)$ *is continuous,*

(iii) $\lim_{p \to \infty} p^{1-\varepsilon}(\mathbf{x}, \mathbf{x}'; p) = 0$, $\varepsilon \in \mathbb{R}^{++}$.

Proof. The proof of (i) is a consequence of Theorem 13.1 and of Lax-Milgram's theorem because $\delta(\mathbf{x} - \mathbf{x}')$, $\mathbf{x} \in \Omega$, is in $^{-1}(\Omega)$. To prove (ii) observe that the bilinear form $a(\mathbf{v}, \mathbf{w};\ )$

is continuous with respect to the third argument and then, by [22], Lemma 44.1, it follows the continuity of $(\mathbf{x},\mathbf{x}';\cdot)$. As regards (iii), by (13.14) we have

$$\int_\Omega p^{1-\varepsilon}(\mathbf{x},\mathbf{x}';p)[\mathbf{v}(\mathbf{x}')-\frac{\mu_L(p)}{p}\Delta'\mathbf{v}(\mathbf{x}')]dx' = p^{-\alpha}\mathbf{v}(\mathbf{x})$$

for any $\varepsilon \in \mathbb{R}^+$ and $\mathbf{v} \in C_0^\infty(\Omega)$. Taking the limit as $p \to \infty$ yields (iii). □

Further recourse to Green's function provides a representation for $\nabla \mathbf{v}_L(\mathbf{x},p)$. For convenience denote by $\nabla(\mathbf{x},\mathbf{x}';p)$ the (third-order) tensor function such that

$$\int_\Omega [\mu_L(p)\nabla'\,\nabla(\mathbf{x},\mathbf{x}';p))\nabla'\mathbf{w}(\mathbf{x}') + p(\nabla(\mathbf{x},\mathbf{x}';p))\mathbf{w}(\mathbf{x}')]\,dx' = \int_\Omega [\nabla\delta(\mathbf{x}-\mathbf{x}')]\mathbf{w}(\mathbf{x}')\,dx' \tag{13.17}$$

for every $\mathbf{w} \in_0^1 (\Omega)$. In terms of ∇ we have

$$\nabla\mathbf{v}_L(\mathbf{x};p) = \int_\Omega \nabla(\mathbf{x},\mathbf{x}';p)\,\tilde{\mathbf{b}}(\mathbf{x}';p)\,dx'.$$

By paralleling the proof of Lemma 13.3, and replacing with ∇ we obtain the following result.

Lemma 13.4. *If (13.16) holds then there exists a unique solution ∇' to the problem (13.17) such that, for any $p \in \mathbb{C}^+$,*

$\nabla(\mathbf{x},\cdot;p) \in^2 (\Omega)$, (Cf. [22], Lemma 26.2.)

$\nabla(\mathbf{x},\mathbf{x}';\cdot)$ *is continuous,*

$\lim_{p\to\infty} p^{1-\varepsilon}\,\nabla(\mathbf{x},\mathbf{x}';p) = 0,\ \ \varepsilon \in \mathbb{R}^{++}.$

These results enable us to state existence and uniqueness of the solution to the problem (13.1)-(13.4) as follows.

Theorem 13.2. *If (13.14) holds then there exists a unique solution, in $L^2(\mathbb{R}^+;_0^1(\Omega))$, to the problem (13.1)-(13.4).*

Proof. Since $\mathbf{b} \in L^2(\mathbb{R}^+;H^{-1}(\Omega))$ and $\mathbf{V}_0 \in L^2(\mathbb{R}^+;^2(\Omega))$ we have

$$\lim_{p\to\infty}\tilde{\mathbf{b}}(\mathbf{x},p) = \lim_{p\to\infty}\left[\mathbf{b}_L(\mathbf{x},p) + \mathbf{v}_0(\mathbf{x},0) + \int_0^\infty \exp(-ps)\mathbf{V}_0(\mathbf{x},s)\,ds\right] = \mathbf{v}_0(\mathbf{x},0).$$

Then by (13.15) and property (iii) of we obtain

$$\lim_{p\to\infty} p^{1-\varepsilon}\mathbf{v}_L(\mathbf{x},p) = \lim_{p\to\infty}\int_\Omega p^{1-\varepsilon}(\mathbf{x},\mathbf{x}';p)\,\mathbf{v}(\mathbf{x}',0) = 0. \tag{13.18}$$

Let $p = i\omega$, $\omega \in \mathbb{R}^+$, and $\epsilon \in (0,\frac{1}{2})$. Then (13.18) implies that $\mathbf{v}_L(\mathbf{x},i\omega)$ is in $L^2(\mathbb{R};L^2(\Omega)$ and we can view it as the Fourier transform of

$$\tilde{\mathbf{v}}(\mathbf{x},t) = \begin{cases} 0, & t<0,\\ \mathbf{v}(\mathbf{x},t), & t\geq 0.\end{cases}$$

Accordingly, Plancherel's theorem allows us to write

$$\int_{-\infty}^{\infty}\int_\Omega |\mathbf{v}_L(\mathbf{x},i\omega)|^2\,dx\,d\omega = \frac{1}{2\pi}\int_{-\infty}^{\infty}\int_\Omega |\tilde{\mathbf{v}}(\mathbf{x},t)|^2\,dx\,dt = \frac{1}{2\pi}\int_0^{\infty}\int_\Omega |\mathbf{v}(\mathbf{x},t)|^2\,dx\,dt. \tag{13.19}$$

By the same token we have

(13.20)
$$\int_{-\infty}^{\infty}\int_{\Omega}|\nabla \mathbf{v}_L(\mathbf{x},i\omega)|^2\,dx\,d\omega=\frac{1}{2\pi}\int_{-\infty}^{\infty}\int_{\Omega}|\nabla \tilde{\mathbf{v}}(\mathbf{x},t)|^2\,dx\,dt=\frac{1}{2\pi}\int_{0}^{\infty}\int_{\Omega}|\nabla \mathbf{v}(\mathbf{x},t)|^2\,dx\,dt.$$

This provides the sought result $\mathbf{v}\in L^2(\mathbb{R}^+;^1_0(\Omega))$. □

14. Counterexamples to uniqueness and stability

In this section two solutions two the problem (10.1)-(10.2) are considered which, as a by-product, provide a further support of the statement in §2.5 as being the right formulation of the second law of thermodynamics. Really, the more refined version of §2.5 is characterized by the inequality being strict for any non-constant history $\mathbf{E}^t$ and this is reflected by (2.4), instead of $\mathbf{G}_s'(\omega)\leq 0,\omega\in\mathbb{R}^+$, being the right restriction on the relaxation function $\mathbf{G}$. Specifically, we show by two examples that weakening (2.4) to $\mathbf{G}_s'(\omega)\leq 0,\omega\in\mathbb{R}^+$ provide counterexamples to uniqueness and asymptotic stability properties of the solutions.

Consider the dynamic problem (10.1)-(10.2) and let $\mathbf{b}$ be time-harmonic, namely $\mathbf{b}(\mathbf{x},t)=\boldsymbol{b}\exp(i\omega t)$. Then we look for solutions $\mathbf{u}$ in the form

$$\mathbf{u}(\mathbf{x},t)=\boldsymbol{u}(\mathbf{x})\exp(i\omega t).$$

The function $\boldsymbol{u}$ has to be a solution to the problem

(14.1)
$$\omega^2\boldsymbol{u}(\mathbf{x})+\nabla\cdot[(\mathbf{G}_0+\mathbf{G}_F(\omega))\nabla\boldsymbol{u}(\mathbf{x})]=-\boldsymbol{b}(\mathbf{x}),\qquad \mathbf{x}\in\Omega,$$

(14.2)
$$\boldsymbol{u}(\mathbf{x})=0\qquad \mathbf{x}\in\partial\Omega.$$

By Theorem 10.2, the validity of (2.4) and (6.1.1) ensures existence and uniqueness of the solution to the problem (14.1)-(14.2). Now, let (2.4) be replaced by $\mathbf{G}_s'(\omega)\leq 0,\omega\in\mathbb{R}^+$. Then there can exist at least one frequency, $\bar{\omega}$ say, such that

$$\mathbf{G}_s'(\bar{\omega})=0.$$

Now, to fix ideas and make the conclusion immediate, let $\Omega=\{x:x\in[0,l]\}$. Hence, as $\omega=\bar{\omega}$ the homogeneous problem becomes

(14.3)
$$\bar{\omega}^2u(x)+[G_0+G_c'(\bar{\omega})]u_{xx}=0,\qquad x\in(0,l),$$

$$u(x)=0,\qquad x=0,l.$$

Since

$$G_0+G_c'(\bar{\omega})=G_\infty+\int_0^\infty G'(s)\cos\bar{\omega}s-1)ds$$

and the integral is bounded because $G' \in L^1(\mathbb{R}^+)$, we can take it that G_∞ makes $G_0 + G'_c(\omega) > 0$. Then the quantity

$$\bar{\lambda} = \frac{\rho\bar{\omega}^2}{G_0 + G'_c(\bar{\omega})},$$

is strictly positive and we can find a value $\bar{l}$ of l such that $\bar{\lambda}$ is an eigenvalue of $-d^2/dx^2$. This proves the existence of non-zero solutions to the homogeneous problem (13.3)-(13.4) and then nonuniqueness of solutions to the nonhomogeneous problem.

As regards the stability properties, observe that the one-dimensional version of the problem (6.1)-(6.4) can be written as

(14.5) $$u_{tt}(x,t) - G_0 u_{xx}(x,t) - \int_0^t G'(s)\, u_{xx}(x,t-s)\, ds = B(x,t), \qquad \text{in} \qquad (0,l) \times \mathbb{R}^+,$$

(14.6) $$u(x,t) = u_0(x,t), \qquad \text{in} \qquad [0,l] \times \mathbb{R}^-,$$

(14.7) $$u(0,t) = u(l,t) = 0, \qquad t \in \mathbb{R}^+,$$

where

$$B(x,t) = b(x,t) + \int_t^\infty G'(s)\, u_{0xx}(x,t-s)\, ds \in L^2(\mathbb{R}^+; L^2((0,l))).$$

Then by Theorem 11.1 we know that (6.1.1) and (2.4) imply the decay of the solution as $t \to \infty$ in the sense that $u \in L^2(\mathbb{R}^+; H_0^1((0,l)))$. If, instead, $\mathbf{G}'_s(\omega) \le 0, \omega \in \mathbb{R}^+$ holds in place of (2.4) then asymptotic stability may cease to hold. Letting $\bar{\omega}, \bar{l}$ be as in the previous example, denote by $\bar{w} \in H^2((0,l))$ any non-zero solution to (14.3)-(14.4). The choice $b = 0$ makes $u(x,t) = \bar{w}\exp(i\bar{\omega}t)$, $x \in (0,l)$, $t \in \mathbb{R}^{++}$, with $u_0(x,t) = \bar{w}\exp(i\bar{\omega}t)$ in $[0,l] \times \mathbb{R}^-$, a solution to (14.5)-(14.7). Of course such $u(x,t)$ does not decay as $t \to \infty$.

Incidentally, if also $G_0 + G'_c(\bar{\omega}) = 0$ then by (14.3) even the existence of the solution ceases to hold.

Analogous results hold for fluids. Assume that an incompressible viscoelastic fluid occupies the region $\{(x,y,z):\ x \in [0,l],\ y,z \in \mathbb{R}\}$ and the velocity depends on x only and is directed along an orthogonal direction, y say. Then the corresponding one-dimensional problem for the velocity v has the form

(14.8) $$v_t(x,t) = \int_0^\infty u(s) v_{xx}(x,t-s)\, ds + b(x,t), \qquad x \in (0,l), \quad t \in \mathbb{R}^{++},$$

(14.9) $$v(0,t) = v(l,t) = 0, \qquad t \in \mathbb{R}^+,$$

(14.10) $$v(x,\tau) = v_0(x,\tau), \qquad x \in (0,l), \quad \tau \in \mathbb{R}^-.$$

We show that replacing (13.6) with the weaker condition $\mu_c(\omega) \geq 0$, $\omega \in \mathbb{R}^{++}$, implies that asymptotic stability may cease to hold. For technical reasons we let μ be monotonically decreasing on $\mathbb{R}^+$. Then we prove the following

Theorem 14.1 *Let $\mu_c(\bar{\omega}) = 0$ for some $\bar{\omega} \in \mathbb{R}^{++}$. Then there exists a critical length $\bar{l}$ such that for a given initial history $v_0(x,\tau) = \sin(\pi x/l)[c_1 \cos\bar{\omega}\tau + c_2 \sin\bar{\omega}\tau]$, $c_1, c_2 \in \mathbb{R}$, and vanishing body force, the problem (14.8)-(14.10) has a nontrivial solution.*

Proof. Let $v(x,t) = V(x)\exp(i\bar{\omega}t)$. By (14.8) and (14.9) we have

$$\bar{\omega}V + \mu_s(\bar{\omega})V_{xx} = 0, \qquad V(0) = V(l) = 0. \tag{14.11}$$

Because

$$\frac{1}{\bar{\omega}}\mu_s(\bar{\omega}) = \frac{1}{\bar{\omega}^2}\left[\mu_0 + \int_0^\infty \mu'(s)\cos\bar{\omega}s\, ds\right] = \frac{1}{\bar{\omega}^2}\int_0^\infty \mu'(s)[\cos\bar{\omega}s - 1]ds > 0 \tag{14.11}$$

we can define the length

$$\bar{l} = \pi\sqrt{\frac{\mu_s(\bar{\omega})}{\bar{\omega}}}.$$

Then the function $V(x) = c\sin(\pi x/\bar{l})$, $c \in \mathbb{R}$, is a solution to the problem (14.11). Now it is easy to see that, if $b = 0$,

$$v(x,t) = \sin\left(\frac{\pi x}{\bar{l}}\right)[c_1\cos\bar{\omega}t + c_2\sin\bar{\omega}t]$$

is the unique, nontrivial solution to the problem (14.8)-(14.10). □

Theorem (14.1) means that, with $b = 0$, asymptotic stability ceases to hold while the initial history $v_0(x,\tau)$ is bounded on $[0,\bar{l}] \times \mathbb{R}^-$. Incidentally, this proves also that uniqueness does not hold as $b \neq 0$.

The hypothesis of Theorem 14.1 is exemplified by considering the relaxation function

$$\mu(s) = \left(s^2 - \frac{\nu - 3}{\upsilon}s + \frac{\nu^2 - 8\nu + 4}{8\upsilon^2}\right)\exp(-\upsilon s)$$

where $\nu \in (2-\sqrt{2},\, 2+\sqrt{2})$, $\upsilon > 0$. It is easily seen that $\mu(s) > 0$, $\mu'(s) < 0$, $s \in [0,\infty)$. Moreover,

$$\mu_c(\omega) = \frac{\nu^2}{8\upsilon(\upsilon^2 + \omega^2)^3}\left(\omega^2 - \frac{8-\nu}{\nu}\upsilon^2\right)^2.$$

This means that $\mu_c \geq 0$ and $\mu_c = 0$ at $\omega = \sqrt{(8-\nu)/\nu}\,\upsilon$.

REFERENCES

[1] M. FABRIZIO, C. GIORGI, A. MORRO, *Free energies and dissipation properties for sistems with memory,*Arch. Rational Mech. Anal., 125 (1994), 341-373.

[2] M. FABRIZIO, A. MORRO, *Mathematical Problems in Linear Viscoelasticity,*SIAM, Philadelphia, 1992.

[3] V. VOLTERRA, *Sulle equazioni integro-differenziali della teoria dell'elasticità,* Atti Reale Accad. Lincei, 18 (1909), 295-301.

[4] V. VOLTERRA, *Equazioni integro-differenziali della elasticità nel caso della isotropia,* Atti Reale Accad. Lincei, 18 (1909), pp. 577-586.

[5] V. VOLTERRA, *Lesons sur les Fonctions des Lignes,* Gauthier-Villars, Paris, 1913.

[6] C. TRUESDELL AND W. NOLL, *The Classical Field Theories,* in Encyclopedia of Physics, S. Flügge, ed., vol. III/3, Springer, Berlin, 1965.

[7] V. Volterra, Theory of Functionals, Blackie & Son Limited, London and Glasgow, 1930.

[8] D. GRAFFI, *Sui problemi dell'ereditarietà lineare,* Nuovo Cimento, 5 (1928), pp. 53-71.

[9] Baggett, L., & W. Fulks Fourier Analysis. Api Anjou Press, Boulder, 1979.

[10] D. GRAFFI, *Sull'espressione dell'energia libera nei materiali visco-elastici lineari,* Ann. Mat. Pura Appl., 98 (1974), pp. 273-279.

[11] D. GRAFFI, *Sull'espressione analitica di alcune grandezze termodinamiche nei materiali con memoria,* Rend. Sem. Mat. Univ. Padova, 68 (1982), pp. 17-29.

[12] Day, A., Reversibility, recoverable work and free energy in linear viscoelasticity. Quart. J. Mech. Appl. Math. 23, 1-15 (1970)

[13] G. DUVAUT AND J. LIONS, *Les inéquations en Mécanique et en P hysique,* Dunod, Paris, 1972.

[14] C.M. DAFERMOS, *On abstract Volterra equations with applications to linear viscoelasticity,* J. Diff. Eq., 7 (1970), pp. 554-569.

[15] C.M. DAFERMOS, *Asymptotic stability in viscoelasticity,* Arch. Rational Mech. Anal., 37 (1970), pp. 297-308.

[16] G. FICHERA, *Avere una memoria tenace crea gravi problemi,* Arch. Rational Mech. Anal., 70 (1979), pp. 101-112.

[17] G. FICHERA, *Sul principio della memoria evanescente,* Rend. S em. Mat. Univ. Padova, 68 (1982), pp. 245-259.

[18] G. FICHERA, *Problemi Analitici Nuovi nella Fisica Matematica Cla ssica,* Scuola Tipo-Lito Istituto Anselmi, Marigliano (Napoli), 1985.

[19] M. FABRIZIO AND A. MORRO, *Thermodynamic restrictions on relaxati on functions in linear viscoelasticity,* Mech. Res. Comm., 12 (1985), pp. 101-105.

[20] C. GIORGI AND B. LAZZARI, *Uniqueness and stability in linear viscoelasticity: some counterexamples,* Proc. V Conf. Waves and Stability in Continuous Media, Sorrento, 1990.

[21] G. GRIOLI, *Continui con Memoria,* Accad. Naz. Lincei, Roma, 1 990.

[22] B. LAZZARI AND E. VUK, *Un teorema di esistenza e unicità per un problema dinamico in*

viscoelasticità lineare, Atti Sem. Mat. Fis. Univ. Modena, 33 (1985), pp. 267-290.

[23] F. TREVES, *Basic Linear Partial Differential Equations,* Aca demic Press, New York, 1975.

[24] G. FICHERA, *Existence Theorems in Elasticity,* in Encyclopedi a of Physics, C. Truesdell ed., vol. VIa/2, Springer, Heidelberg, 1972, pp. 347-389.

[25] A. MORRO AND M. FABRIZIO, *On uniqueness in linear viscoelastici ty: a family of counterexamples,* Quart. Appl. Math., 45 (1987), pp. 263-268.

[26] G. CAPRIZ, *Sulla impostazione di problemi dinamici in viscoelasticità,* Ref. [72], pp. 25-33.

[27] E.G. VIRGA AND G. CAPRIZ, *Un teorema di unicità in viscoelasticità lineare,* Rend. Sem. Mat. Univ. Padova, 79 (1988), pp. 15-24.

[28] G. CAPRIZ AND E.G. VIRGA, *Esempi di non-unicità in viscoelasticità lineare,* Atti Accad. Scienze Torino, 120 (1987), pp. 81-86.

[29] C. GIORGI, *Alcune conseguenze delle restrizioni termodinamiche per mezzi viscoelastici lineari,* Quaderno n. 6/89, Università, Dipartimento di Matematica, Brescia, 1989.

[30] M. FABRIZIO AND B. LAZZARI, *On the existence and the asymptotic stability of solutions for a linear viscoelastic solid system,* Arch. Rational Mech. Anal.,123, (1993) .

[31] M. FABRIZIO AND B. LAZZARI, *On asymptotic stability for linear viscoelastic fluids,* to appear.

[32] A. PAZY, *On the applicability of Lyapunov's theorem in Hilbert space,* SIAM J. Math. Anal., 3 (1972), pp. 291-294.

[33] M. SLEMROD, *A hereditary partial differential equation with applications in the theory of simple fluids,* Arch. Rational Mech. Anal., 62 (1976), pp. 303-321.

[34] M. SLEMROD, *An energy stability method for simple fluids,* Arch. Rational Mech. Anal., 68 (1978), pp. 1-18.

[35] R. TEMAM, *Navier-Stokes Equations,* North-Holland, Amsterdam, 1984.

GENERAL METHODS IN NON-INERTIAL VISCOELASTIC BOUNDARY VALUE PROBLEMS

J.M. Golden
Dublin Institute of Technology, Dublin, Ireland
and
G.A.C. Graham
Simon Fraser University, Burnaby, Canada

Chapter 1

INTRODUCTION

We will deal in these lecture notes with the behaviour of viscoelastic solids under the action of different boundary loading configurations. Strains will be assumed to be sufficiently small so that the linear approximation is valid. Also, inertial effects are generally assumed to be neglected, though occasional remarks are included on the effects of their inclusion.

Many materials of engineering interest exhibit viscoelastic behaviour, in particular those composed of polymers, both biological and man-made. Certain soils also exhibit this type of behaviour.

Viscoelastic materials characteristically possess internal friction or viscosity which causes a loss of mechanical energy into heat if the material is deformed. However, they also exhibit elastic recovery.

A fundamental distinction, not always easy to make in practice, is between viscoelastic solids and liquids. In these lectures, we will deal only with solids. The distinction between the two is discussed by Golden and Graham (1988) and many other authors.

We remark that this reference, Golden and Graham (1988), is the basic source for the material covered here, though various topics from papers published after 1988 are also discussed. The 1988 book will be referred to, for brevity, as [14].

Chapter 2

FUNDMENTAL RELATIONSHIPS

We summarize the basic relations of the theory of linear viscoelasticity in this chapter. Firstly, constitutive relations are discussed. Then the dynamical equation and the general boundary value problem are stated.

2.1 Linear Viscoelasticity

We refer briefly to the physical phenomena associated with linear viscoelasticity and then move on to discuss constitutive relations.

2.1.1 Characteristic Phenomena

The phenomenology of viscoelastic behaviour is described in a large number of texts both new and old. We will not dwell on the physical phenomena here except to mention the well-known property of stress relaxation and the quite distinct, though closely related phenomenon associated with strain, referred to as creep.

Consider, for simplicity, a viscoelastic material stressed in such a way that only the x, y components, σ and ϵ, of stress and strain can take non-zero values. We will refer to this as the one-dimensional case. Let the body be subjected to a sudden constant strain ϵ_0 at time t_1. Then the resulting stress σ is a time-dependent quantity, initially high and declining to a final value. This is stress relaxation, characterized mathematically by

$$\sigma(t) = G(t - t_1)\epsilon_0, \tag{2.1.1}$$

where $G(t)$ is a characteristic of the material, referred to as the relaxation function. A schematic representation of the typical form of $G(t)$ is shown on Figure 2.1.

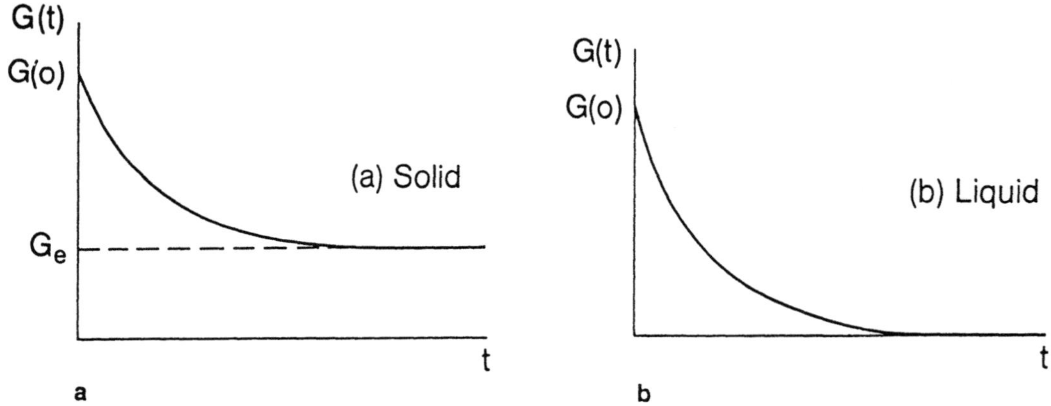

Figure 2.1: **a,b.** Schematic representation of the form of $G(t)$ for **a**) a viscoelastic solid (note that G_e may be zero, corresponding to $J(t)$ becoming infinite as t^p, $0 < p < 1$) and (**b**) a viscoelastic liquid.

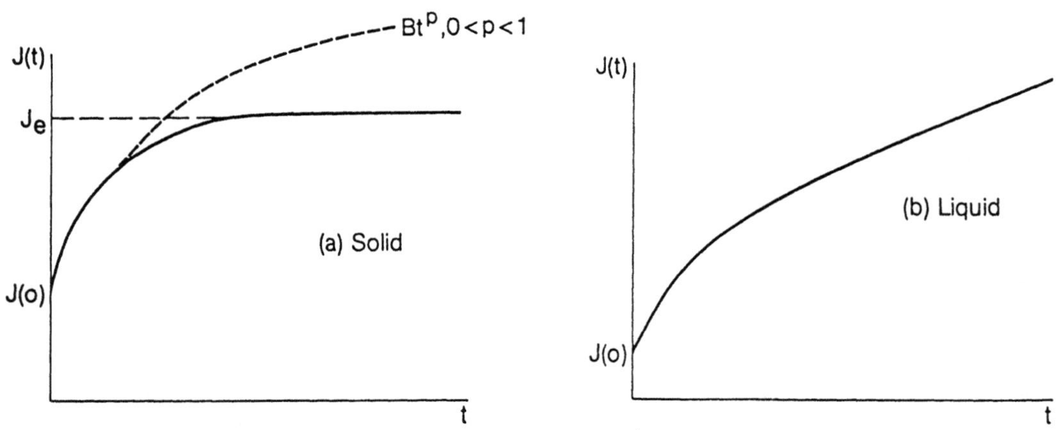

Figure 2.2: **a,b.** Schematic representation of the form of $J(t)$ for (**a**) a viscoelastic solid and (**b**) a viscoelastic liquid.

Similarly, if the material is subjected to a sudden, constant stress σ_0 at time t_1, this results in a certain instantaneous strain which subsequently increases either to some final value or indefinitely. This is the phenomenon of creep. In mathematical terms, we have

$$\epsilon(t) = J(t - t_1)\sigma_0, \tag{2.1.2}$$

where $J(t)$, the creep function, is a characteristic of the material, related, as we shall see, to the relaxation function. Its typical form is represented on Figure 2.2.

2.1.2 Hereditary Integrals

We postulate that the stress at time t is given as a linear functional of the history of the strain at all times up to and including t. This will involve hereditary integrals, a concept that will recur continuously throughout these lectures.

It is assumed that we are dealing with non-aging materials, namely those that have no intrinsic variation with time in their properties.

The constitutive relation in the one-dimensional case has the form

$$\sigma(t) = G(0)\epsilon(t) + \int_{-\infty}^{t} dt'\dot{G}(t-t')\epsilon(t') = G(\infty)\epsilon(-\infty) + \int_{-\infty}^{t} dt'G(t-t')\dot{\epsilon}(t'), \tag{2.1.3}$$

where the dot indicates time differentiation. We may rewrite this in the form

$$\sigma(t) = \int_{-\infty}^{t} dt'\mu(t-t')\epsilon(t') = [\mu * \epsilon](t),$$

$$\mu(t) = G(0)\delta(t) + \dot{G}(t)H(t) = \frac{d}{dt}[H(t)G(t)], \tag{2.1.4}$$

where $\delta(t)$ is the singular delta function and $H(t)$ is the Heaviside step function. Recall that $\dot{H}(t) = \delta(t)$. The rightmost expression of the first relation of (2.1.4) is the Faltung or convolution product notation. Note that in this relation, the upper limit of integration is t^+, the limit from above, so that the full contribution from the delta function is included.

The second form of (2.1.3) may be shown to imply (2.1.1). Note that $\epsilon(-\infty)$ in (2.1.3) will often be zero.

From the first relation of (2.1.3), we see that if $G(t)$ is a constant, then the material reduces to an elastic material with shear modulus equal to this constant. Also, let $\epsilon(t)$ vanish for all negative times and let t approach zero from above. Then we see that for an instant after deformation begins, the material acts like an elastic material with modulus $G(0)$, named for this reason the instantaneous elastic modulus.

From the second relation of (2.1.3), it may be seen that if constant deformation has existed in the material for a long time, it acts as an elastic material with modulus

$G(\infty)$. This is the modulus associated with a final equilibrium state which will exist for a solid. It is the natural generalization of the elastic modulus. We refer to it as the long time modulus.

Physical observations tell us that $G(t)$ is monotonically decreasing with time so that $G(0) > G(\infty)$ for a viscoelastic material.

Application of a well-known result from the theory of Volterra integral equations gives that if $G(0)$ is non-zero then (2.1.3) can be uniquely inverted to give $\epsilon(t)$ in terms of $\sigma(t)$. We write it in the form

$$\epsilon(t) = J(0)\sigma(t) + \int_{-\infty}^{t} dt' \dot{J}(t-t')\sigma(t') = J(\infty)\sigma(-\infty) + \int_{-\infty}^{t} dt' J(t-t')\dot{\sigma}(t'), \quad (2.1.5)$$

or, in convolution form, as

$$\epsilon(t) = \int_{-\infty}^{t} dt' \gamma(t-t')\sigma(t') = [\gamma * \sigma](t),$$

$$\gamma(t) = J(0)\delta(t) + \dot{J}(t)H(t) = \frac{d}{dt}[H(t)J(t)]. \quad (2.1.6)$$

We have

$$\int_{t_1}^{t_2} dt' \mu(t_2 - t')\gamma(t' - t_1) = \int_{t_1}^{t_2} dt' \gamma(t_2 - t')\mu(t' - t_1) = \delta(t_2 - t_1). \quad (2.1.7)$$

Putting $t_1 = 0$, $t_2 = t$ gives the more compact result

$$\int_0^t dt' \mu(t-t')\gamma(t') = \int_0^t dt' \gamma(t-t')\mu(t') = \delta(t), \quad (2.1.8)$$

from which the more general form (2.1.7) may be deduced. Also we have

$$\mu(t) = \gamma(t) = 0, \qquad \text{for all} \quad t < 0. \quad (2.1.9)$$

This is referred to as the Causality condition, since it imposes the causal constraint that $\sigma(t)$ cannot depend upon $\epsilon(t')$ where $t' > t$ and similarly for $\epsilon(t)$, $\sigma(t')$ [14].

Problem 2.1.1 *Show, using (2.1.8), that*

$$G(0)J(0) = 1,$$

$$\dot{G}(t)J(0) + G(0)\dot{J}(t) = -\int_0^t dt' \dot{G}(t-t')\dot{J}(t') = -\int_0^t dt' \dot{J}(t-t')\dot{G}(t'), \quad t \geq 0. \quad (2.1.10)$$

Show that the integrated form of this relationship

$$G(0)J(t) + \int_0^t dt' J(t-t')\dot{G}(t') = J(0)G(t) + \int_0^t dt' G(t-t')\dot{J}(t') = 1, \quad t \geq 0, \quad (2.1.11)$$

follows from (2.1.3) and (2.1.5). [Hint: if $\epsilon(t) = J(t)$, then $\sigma(t) = 1$, $t \geq 0$.] Hence deduce that if $\dot{G}(t) \to 0$ at large times

$$\lim_{t\to\infty} G(t)J(t) = 1. \tag{2.1.12}$$

It follows that if $G(\infty)$ is finite then $J(\infty)$ is also and

$$J(\infty)G(\infty) = 1. \tag{2.1.13}$$

Note that (2.1.5) implies (2.1.2), where $J(t)$ in the former relation is now identified as the creep function. Furthermore, from (2.1.10) for example, we see that the creep and relaxation functions are deeply related. In fact, since the inversion of the constitutive relation is unique, it must be that $\gamma(t)$ can be deduced from $\mu(t)$ and vice versa. Therefore $J(t)$ can be deduced from $G(t)$ and $G(t)$ from $J(t)$, so that creep behaviour can be deduced from relaxation behaviour and vice versa. We will see examples of this later.

We refer to the quantities $\mu(t)$ and $\gamma(t)$ as the singular relaxation and creep shear viscoelastic functions.

2.1.3 Complex Modulus

Define the Fourier transform of the causal function $f(t)$ (i.e. zero for negative values of t) as

$$\hat{f}(\omega) = \int_0^\infty dt f(t)e^{-i\omega t}. \tag{2.1.14}$$

The quantity $\hat{\mu}(\omega)$, given by

$$\begin{aligned}\hat{\mu}(\omega) &= \int_0^\infty dt\mu(t)e^{-i\omega t} = G(0) + \int_0^\infty dt\dot{G}(t)e^{-i\omega t}\\ &= G(\infty) + i\omega\int_0^\infty dt\{G(t) - G(\infty)\}e^{-i\omega t}\end{aligned} \tag{2.1.15}$$

is the shear complex modulus of the material. Note that in the last relation of (2.1.15). $G(\infty)$ is introduced for convergence purposes. The complex modulus is, as its name indicates, a complex quantity for a viscoelastic material:

$$\begin{aligned}\hat{\mu}(\omega) &= \hat{\mu}_1(\omega) + i\hat{\mu}_2(\omega),\\ \hat{\mu}_1(\omega) &= G(0) + \int_0^\infty dt\dot{G}(t)\cos\omega t = G(\infty) + \omega\int_0^\infty dt[G(t) - G(\infty)]\sin\omega t,\\ \hat{\mu}_2(\omega) &= -\int_0^\infty dt\dot{G}(t)\sin\omega t = \omega\int_0^\infty dt[G(t) - G(\infty)]\cos\omega t.\end{aligned} \tag{2.1.16}$$

If $G(t)$ is a constant the $\hat{\mu}_2(\omega)$ vanishes and $\hat{\mu}_1(\omega)$ reduces to $G(0)$. The quantity $\hat{\mu}_2(\omega)$ does not vanish in any other case.

It can be shown, using (2.1.15) and properties of the Fourier transform that [14]

$$\hat{\mu}(0) = G(\infty), \qquad \hat{\mu}(\infty) = G(0) > \hat{\mu}(0), \tag{2.1.17}$$

which limits are both real, so that $\hat{\mu}_2(\omega)$ vanishes for large and small ω. Further properties of $\hat{\mu}(\omega)$ are derived in [14] and various older texts.

For a sinusoidal strain history of the form

$$\epsilon(t) = \epsilon_0 \sin \omega t \tag{2.1.18}$$

equation (2.1.4) gives

$$\sigma(t) = \epsilon_0 |\hat{\mu}(\omega)| \sin(\omega t + \varphi), \tag{2.1.19}$$

where $\varphi(\omega)$ is the phase angle of $\hat{\mu}(\omega)$ defined by

$$\hat{\mu}(\omega) = |\hat{\mu}(\omega)| e^{i\varphi}. \tag{2.1.20}$$

We have that

$$\tan \varphi = \frac{\hat{\mu}_2(\omega)}{\hat{\mu}_1(\omega)}. \tag{2.1.21}$$

It may be shown that $\varphi(\omega)$ is a positive angle [14] which is chosen to be in $[0, \pi/2]$. This is the loss angle and $\tan \varphi$ is the loss tangent, a terminology connected with the fact that $\hat{\mu}_2(\omega)$ is related to the rate of dissipation of mechanical energy into heat per unit volume in the material [14]. For the same reason, $\hat{\mu}_2(\omega)$ is termed the loss modulus. The real part $\hat{\mu}_1(\omega)$, being related to the average level of stored energy per unit volume [14], is termed the storage modulus.

Equation (2.1.19) tells us that if the material is subjected to a long term sinusoidal strain of amplitude ϵ_0, the resulting stress is also sinusoidal, of amplitude $\epsilon_0|\hat{\mu}(\omega)|$. There is a phase difference between the two such that strain will always lag behind stress in phase, by an amount φ.

Let us take the Fourier transform of (2.1.4) and (2.1.6) and use a well-known property of the transform of convolution products, to obtain

$$\begin{aligned} \hat{\sigma}(\omega) &= \hat{\mu}(\omega)\hat{\epsilon}(\omega), \\ \hat{\epsilon}(\omega) &= \hat{\gamma}(\omega)\hat{\sigma}(\omega), \end{aligned} \tag{2.1.22}$$

so that

$$\hat{\mu}(\omega)\hat{\gamma}(\omega) = 1. \tag{2.1.23}$$

This important relation is in fact the Fourier transform of (2.1.8).

2.1.4 Types of Material

Various assumptions are made about the form of $G(t)$ to fit as completely as possible the observed behaviour of viscoelastic materials. The book by Ferry (1970) deals extensively with this question. There are perhaps three broad categories

1. Continuous and discrete spectrum models and degenerate forms of these;
2. Power law and associated models;
3. Forms deriving from molecular theories.

We will consider only discrete spectrum models in these lectures.

Older treatments, for example Ferry (1970), use mechanical (spring and dashpot) models of viscoelastic behaviour from which emerge essentially the discrete spectrum forms. It was common also to derive and use, in this context, differential forms of the constitutive relations. A consistent development of the theory is possible without these concepts.

The simplest discrete spectrum model is the standard linear solid, for which $G(t)$ has the form

$$G(t) = G_0 + G_1 e^{-t/\tau}, \tag{2.1.24}$$

where G_0 and G_1 are positive constants and τ is a positive decay time. Clearly

$$G(0) = G_0 + G_1, \quad G(\infty) = G_0 < G(0). \tag{2.1.25}$$

Also

$$\begin{gathered} \mu(t) = g_0\delta(t) + g_1 e^{-t/\tau} H(t), \\ g_0 = G_0 + G_1 = G(0), \quad g_1 = -G_1/\tau \ < \ 0, \end{gathered} \tag{2.1.26}$$

so that the complex modulus is given by

$$\begin{gathered} \hat{\mu}(\omega) = g_0 + \frac{g_1\tau}{1 + i\omega\tau} = G_0\frac{1 + i\omega\tau'}{1 + i\omega\tau}, \\ \tau' = \tau\frac{G_0 + G_1}{G_0} \ > \ \tau. \end{gathered} \tag{2.1.27}$$

The significance of τ' will emerge shortly. The storage and loss moduli have the form

$$\begin{gathered} \hat{\mu}_1(\omega) = G_0\frac{1 + \omega^2\tau\tau'}{1 + \omega^2\tau^2}, \\ \hat{\mu}_2(\omega) = \frac{G_1\omega\tau}{1 + \omega^2\tau^2}, \end{gathered} \tag{2.1.28}$$

and the loss tangent is given by

$$\tan\varphi = \frac{G_1\omega\tau}{G_0(1+\omega^2\tau\tau')}. \tag{2.1.29}$$

Equation (2.1.23) gives immediately that

$$\hat{\gamma}(\omega) = \frac{1}{G_0}\frac{1+i\omega\tau}{1+i\omega\tau'}. \tag{2.1.30}$$

Reversing the steps from (2.1.27) to (2.1.25) for $\hat{\gamma}(\omega)$ which has the same functional form as $\hat{\mu}(\omega)$, but with different parameters, we find that

$$J(t) = J_0 + J_1(1-e^{-t/\tau'}),$$

$$J_0 + J_1 = \frac{1}{G_0}; \quad \tau = \tau'\frac{J_0}{J_0+J_1}. \tag{2.1.31}$$

Comparing with the last equation of (2.1.27) we see that

$$\frac{G_0}{G_0+G_1} = \frac{J_0}{J_0+J_1}, \tag{2.1.32}$$

which, together with (2.1.31), gives

$$J_0 = \frac{1}{G_0+G_1}, \quad J_1 = \frac{G_1}{G_0(G_0+G_1)}. \tag{2.1.33}$$

From these relations, we see that J_0, J_1 and τ' are determined completely by G_0, G_1 and τ. In other words, noting that τ' is the decay time for the creep function, the creep parameters are determined completely by the relaxation parameters, and vice versa.

It follows from (2.1.33) that J_0 and J_1 are positive.

The well-known Maxwell and Voigt models are degenerate forms of the standard linear solid [14].

More general discrete spectrum models are characterized by sums of decaying exponentials with different decay times. Thus, for such materials

$$G(t) = G_0 + \sum_{i=1}^{N} G_i e^{-t/\tau_i}, \tag{2.1.34}$$

where G_0 and G_i, τ_i, $i = 1, N$ are positive. It follows from (2.1.4) that

$$\mu(t) = g_0\delta(t) + \sum_{i=1}^{N} g_i e^{-t/\tau_i},$$

$$g_0 = \sum_{i=0}^{N} G_i = G(0) > 0,$$

$$g_i = -\frac{G_i}{\tau_i} < 0, \quad i = 1, 2, \cdots, N. \tag{2.1.35}$$

The complex modulus has the form

$$\hat{\mu}(\omega) = g_0 + \sum_{i=1}^{N} \frac{g_i \tau_i}{1 + i\omega\tau_i} = G_0 + \sum_{i=1}^{N} \frac{iG_i\omega\tau_i}{1 + i\omega\tau_i}, \tag{2.1.36}$$

and the storage and loss moduli are given by

$$\hat{\mu}_1(\omega) = G_0 + \sum_{i=1}^{N} \frac{G_i\omega^2\tau_i^2}{1 + \omega^2\tau_i^2},$$

$$\hat{\mu}_2(\omega) = \sum_{i=1}^{N} \frac{G_i\omega\tau_i}{1 + \omega^2\tau_i^2}. \tag{2.1.37}$$

Equation (2.1.36) shows that $\hat{\mu}(\omega)$ is the ratio of two polynomials of degree N, so that $\hat{\gamma}(\omega) = 1/\hat{\mu}(\omega)$ is also, from which we conclude that $J(t)$ is a series of exponentials. On physical grounds we expect that these will be decaying exponentials with real decay times, a point which is discussed further below. In anticipation of this, we write

$$\gamma(t) = h_0\delta(t) + \sum_{i=1}^{N} h_i e^{-t/\tau_i'},$$

$$J(t) = J_0 + \sum_{i=1}^{N} J_i(1 - e^{-t/\tau_i'}),$$

$$h_0 = J(0) = J_0 > 0, \quad h_i = J_i/\tau_i' > 0, \tag{2.1.38}$$

where $J(t)$ is put in this form because it is an increasing function of t.

It is convenient to use a slightly different form of $\mu(t)$ and $\gamma(t)$ in order to discuss the relationship between their parameters. We put

$$\mu(t) = g_0\delta(t) + \sum_{i=1}^{N} g_i e^{-\alpha_i t},$$

$$\gamma(t) = h_0\delta(t) + \sum_{i=1}^{N} h_i e^{-\beta_i t},$$

$$\alpha_i = 1/\tau_i, \quad \beta_i = 1/\tau_i', \tag{2.1.39}$$

giving

$$\hat{\mu}(\omega) = g_0 + \sum_{i=1}^{N} \frac{g_i}{\alpha_i + i\omega},$$

$$\hat{\gamma}(\omega) = h_0 + \sum_{i=1}^{N} \frac{h_i}{\beta_i + i\omega}. \tag{2.1.40}$$

Consider the functions

$$\hat{\mu}(iz) = g_0 + \sum_{i=1}^{N} \frac{g_i}{\alpha_i - z},$$

$$\hat{\gamma}(iz) = h_0 + \sum_{i=1}^{N} \frac{h_i}{\beta_i - z}, \tag{2.1.41}$$

where z is a arbitrary complex number. These are analytic continuations of $\hat{\mu}(\omega)$ and $\hat{\gamma}(\omega)$ to the complex plane. From the properties of analytic continuation of functions defined on restricted portions of the complex plane, we deduce that

$$\hat{\gamma}(iz) = \frac{1}{\hat{\mu}(iz)}, \tag{2.1.42}$$

for all z, even at poles, since by inverting both sides, they become zeros. Since the poles of $\hat{\mu}(iz)$ are the zeros of $\hat{\gamma}(iz)$ and vice versa, we deduce that

$$g_0 + \sum_{i=1}^{N} \frac{g_i}{\alpha_i - \beta_j} = 0, \quad j = 1,\ 2, \cdots,\ N,$$

$$h_0 + \sum_{j=1}^{N} \frac{h_j}{\beta_j - \alpha_i} = 0, \quad i = 1,\ 2, \cdots,\ N, \tag{2.1.43}$$

which relations may also be derived by applying (2.1.8) to (2.1.39). From the latter it also follows that $g_0 h_0 = 1$. Also, as z approaches β_i, the left-hand side of (2.1.42) will have a pole and the right-hand side must have a zero in the denominator. Therefore,

$$h_i = \lim_{z \to \beta_i} (\beta_i - z)\hat{\gamma}(iz) = -\left\{\frac{d}{dz}\hat{\mu}(iz)\right\}_{z=\beta_i}^{-1} = \left\{-\sum_{j=1}^{N} \frac{g_j}{(\alpha_j - \beta_i)^2}\right\}^{-1},$$

$$i = 1,\ 2,\ \cdots,\ N. \tag{2.1.44}$$

If any of the α_j is equal to β_i, then h_i is zero. It follows that this cannot be the case.

The inverse decay constants β_j are solutions of the second relation of (2.1.43). On physical grounds their real parts must be positive. If they are complex, this will lead to oscillatory behaviour in $J(t)$ which would cause creep behaviour of a kind that is not observed. Thus, the range of allowed values of g_0, g_i, α_i, $i = 1,\ 2, \cdots,\ N$ must be such as to ensure that the β_i are real and positive.

Note that (2.1.44) gives a positive sign of h_i as proposed in (2.1.38).

Coincident decay times giving as a limiting process terms such as $te^{-\alpha_i t}$ are excluded from the relaxation function. We have also excluded them from the creep function which must be regarded, without more complete mathematical insight, as an assumption or a restriction on the range of the relaxation parameters.

The interrelationships between relaxation and creep function for discrete spectrum models were given by Gross (1953).

Problem 2.1.2 *Show that for a model with one decay time, i.e. a standard linear solid, these relations give*

$$h_0 = \frac{1}{g_0}, \qquad \beta = \alpha + \frac{g_1}{g_0},$$

$$h_1 = -\frac{(\alpha-\beta)^2}{g_1} = -\frac{g_1}{g_0^2}, \tag{2.1.45}$$

where

$$\mu(t) = g_0\delta(t) + g_1 e^{-\alpha t},$$

$$\gamma(t) = h_0\delta(t) + h_1 e^{-\alpha t}, \tag{2.1.46}$$

and that these are equivalent to (2.1.27) and (2.1.31) - (2.1.33).

2.2 Three-dimensional Constitutive and Dynamical Equations

We now move on to the three-dimensional case and write down the constitutive relations, dynamical equations and the general form of the boundary conditions.

2.2.1 Constitutive Relations for Isotropic Materials

The general form of the constitutive equations for a linear viscoelastic isotropic medium in the isothermal approximation may be written as [14]

$$\sigma_{ij}(\underline{r},t) = 2\int_{-\infty}^{t} dt' G(t-t')\dot{\epsilon}_{ij}(\underline{r},t') + \delta_{ij}\int_{-\infty}^{t} dt' M(t-t')\dot{\epsilon}(\underline{r},t'),$$

$$i,j = 1,2,3, \tag{2.2.1}$$

$$M(t) = K(t) - \frac{2}{3}G(t),$$

$$\epsilon(\underline{r},t) = \sum_{i=1}^{3} \epsilon_{ii}(\underline{r},t),$$

where $\sigma_{ij}(\underline{r},t)$ and $\epsilon_{ij}(\underline{r},t)$ are the components of the stress and strain tensors at position $\underline{r}$ and time t. The quantity $G(t)$ is the shear relaxation function, differing from the quantity used in the one-dimensional discussion by a factor 2, which affects no important property. The quantity $K(t)$ is the relaxation function for bulk deformation. We can rewrite (2.2.1) in terms of singular viscoelastic functions as follows.

$$\sigma_{ij}(\underline{r},t) = 2\int_{-\infty}^{t} dt'\mu(t-t')\epsilon_{ij}(\underline{r},t') + \delta_{ij}\int_{-\infty}^{t} dt'\lambda(t-t')\epsilon(\underline{r},t'), \tag{2.2.2}$$

$$\mu(t) = G(0)\delta(t) + \dot{G}(t)H(t),$$
$$\lambda(t) = M(0)\delta(t) + \dot{M}(t)H(t).$$

2.2.2 The General Boundary Value Problem

The dynamical equations are given by

$$\sum_{j=1}^{3} \frac{\partial}{\partial x_j}\sigma_{ij}(\underline{r},t) = \rho\frac{\partial^2}{\partial t^2}u_i(\underline{r},t), \qquad i = 1,2,3, \tag{2.2.3}$$

where ρ is the material density and $u_i(\underline{r},t)$ are the displacement components at the position $\underline{r}$ and time t.

In material with high internal friction losses, it often happens that inertial effects, namely those depending on ρ, may be neglected compared with viscous effects. This is analogous to the approximation in the theory of simple harmonic motion through a frictional medium where the acceleration term is neglected compared with the frictional resistance term. This approximation results in considerable simplification in both the particle mechanics case and in the continuum mechanics context. Throughout these lectures, this approximation will be assumed to hold.

We therefore neglect the right-hand side of (2.2.3). Inserting the expression for the stresses from (2.2.2) into (2.2.3) gives

$$\int_{-\infty}^{t} dt'\mu(t-t')\nabla^2 u_i(\underline{r},t') + \int_{-\infty}^{t} dt'[\mu(t-t')+\lambda(t-t')]\frac{\partial}{\partial x_i}\underline{\nabla}\cdot\underline{u}(\underline{r},t') = 0,$$
$$i = 1,\ 2,\ 3. \tag{2.2.4}$$

These equations are generalizations of those of static elasticity with a time-dependence added. In the elasticity context, the time-dependence is merely an extra parameter in the problem which does not effect the method of solution or the form of of the solution except in a trivial way. In the viscoelastic context, the time-dependence is an essential part of the problem. Without it, there would be no specifically viscoelastic effects.

The boundary conditions are stated in general as follows

$$u_i(\underline{r},t) = d_i(\underline{r},t), \quad \underline{r} \in B_u^{(i)}(t),$$

$$\sum_{j=1}^{3} \sigma_{ij}(\underline{r}, t) n_j(\underline{r}) = c_i(\underline{r}, t), \quad \underline{r} \in B_\sigma^{(i)}(t),$$

$$B_u^{(i)}(t) \bigcup B_\sigma^{(i)}(t) = B, \quad i = 1, 2, 3, \tag{2.2.5}$$

where B is the total boundary of the body, $n_j(\underline{r})$, the outward normal at $\underline{r}$ and $c_i(\underline{r}, t)$, $d_i(\underline{r}, t)$ are known functions.

2.2.3 Fourier Transformed Relationships

Taking the time Fourier transforms of (2.2.2) and (2.2.4) gives

$$\hat{\sigma}_{ij}(\underline{r}, \omega) = 2\hat{\mu}(\omega)\hat{\epsilon}_{ij}(\underline{r}, \omega) + \delta_{ij}\hat{\lambda}(\omega)\hat{\epsilon}(\underline{r}, \omega),$$

$$\hat{\mu}(\omega)\nabla^2 \hat{u}_i(\underline{r}, \omega) + [\hat{\mu}(\omega) + \hat{\lambda}(\omega)] \frac{\partial}{\partial x_i} \underline{\nabla} \cdot \underline{u}(\underline{r}, \omega) = 0, \tag{2.2.6}$$

which are identical to the equations of static elasticity with the complex moduli for shear and bulk deformation, $\hat{\mu}(\omega)$ and $\hat{\lambda}(\omega)$, replacing *Lamé*'s constants μ and λ.

The various moduli and other parameters that can be formed out of *Lamé*'s constants in elastic theory may be generalized to the viscoelastic case by simply substituting $\hat{\mu}(\omega)$ and $\hat{\lambda}(\omega)$ for μ and λ, and, if one wishes to return to the time representation, inverting the Fourier transform. One important example is $\hat{\nu}(\omega)$, give by

$$\hat{\nu}(\omega) = \frac{\hat{\lambda}(\omega)}{2(\hat{\lambda}(\omega) + \hat{\mu}(\omega))}, \tag{2.2.7}$$

which can be regarded as a generalization of Possion's ratio.

2.2.4 The Proportionality Assumption

Significant mathematical simplification results from an assumption that the relaxation function for bulk and shear are proportional to one another. This we will refer to as the proportionality assumption. It implies that $\hat{\nu}(\omega)$, defined by (2.2.7) is a constant ν and its inverse Fourier transform is $\nu\delta(t)$. The material therefore has a unique, time-independent Possion's ratio.

While this assumption is made essentially for reasons of mathematical simplicity, it is approximately justified for many materials [14]. It will be frequently invoked in these lecture notes.

2.3 Two-dimensional Viscoelastic Problems

Considerable simplification results if the number of dimensions in the problem reduces to two. One specific way in which this happens is when plane strain conditions prevail, which is the case for most problems considered in these lectures.

2.3.1 Plane Strain in Linear Viscoelasticity

Under plane strain conditions parallel to the xy plane, the displacement $u_z(\underline{r},t) = u_3(\underline{r},t)$ is constrained to be zero, while the other components of displacement are independent of z. This would occur for example if all the bodies in the problem were uniform and infinitely extensive in the z direction, and are acted upon by external stresses independent of z. It follows that

$$\epsilon_{xz} = \epsilon_{yz} = \epsilon_{zz} = 0. \tag{2.3.1}$$

The constitutive equations (2.2.2) give that

$$\sigma_{xz} = \sigma_{yz} = 0. \tag{2.3.2}$$

However, σ_{zz} is not zero. It is given by

$$\sigma_{zz}(\underline{r},t) = \int_{-\infty}^{t} dt'\lambda(t-t')\epsilon(\underline{r},t'),$$

$$\epsilon(\underline{r},t') = \epsilon_{xx}(\underline{r},t') + \epsilon_{yy}(\underline{r},t'), \quad \underline{r} = (x,y). \tag{2.3.3}$$

This normal stress must operate to maintain plane strain conditions.

2.3.2 Viscoelastic Kolosov-Muskhelishvili Equations

A generalization of the Papkovich-Neuber solution of the elastostatic equations, in terms of harmonic functions can be given for the non-inertial viscoelastic equations (2.2.4) [14] [Gurtin and Sternberg (1962)]. Under plane strain conditions, this solution reduces to a viscoelastic generalization of the Kolosov-Muskhelishvili equations of two-dimensional elastostatics [14] [Muskhelishvili (1963), Gladwell (1980)]. In formal terms, the steps are entirely analogous to those in the elastic case.

It will be adequate for our purposes to deal with a differentiated form of these equations which may be stated as follows:

$$\sigma_{11}(\underline{r},t) + \sigma_{22}(\underline{r},t) = 2[\varphi(z,t) + \bar{\varphi}(\bar{z},t)],$$

$$\Sigma(\underline{r},t) \equiv \sigma_{22}(\underline{r},t) - i\sigma_{12}(\underline{r},t) = \varphi(z,t) + \bar{\varphi}(\bar{z},t) + z\bar{\varphi}'(\bar{z},t) + \bar{\psi}(\bar{z},t),$$

$$2\int_{-\infty}^{t} dt'\mu(t-t')D'(\underline{r},t') = \int_{-\infty}^{t} dt'\kappa(t-t')\varphi(z,t') - \bar{\varphi}(\bar{z},t) - z\bar{\varphi}'(\bar{z},t) - \bar{\psi}(\bar{z},t), \quad (2.3.4)$$

$$D(\underline{r},t) = u_1(\underline{r},t) + iu_2(\underline{r},t),$$

$$D'(\underline{r},t) = \frac{\partial}{\partial x}D(\underline{r},t), \quad z = x + iy,$$

where $\kappa(t)$ which is zero for negative t, is defined by the relation

$$\hat{\kappa}(\omega) = 3 - 4\hat{\nu}(\omega), \quad (2.3.5)$$

in terms of $\hat{\nu}(\omega)$ given by (2.2.7). If the proportionality assumption applies, then

$$\kappa(t) = (3 - 4\nu)\delta(t). \quad (2.3.6)$$

The functions $\varphi(z,t)$ and $\psi(z,t)$, which will be referred to as the complex potentials, are derivatives of quantities related to the Papkovich-Neuber harmonic functions [14]. They are analytic with respect to z everywhere within the viscoelastic body under consideration.

The bar notation indicates complex conjugate. Over the function it implies taking the complex conjugate while leaving z untouched, while over z it effects only that quantity. Thus $\bar{\varphi}(\bar{z},t)$ is the complex conjugate, in the normal sense, of the value of $\varphi(z,t)$.

We will deal only with materials, in these lecture notes, that have an infinite boundary. At large distances from the origin in such materials [Green and Zerna 1968, for example].

$$\varphi(z,t) \sim -\frac{Q}{2\pi z}, \qquad \psi(z,t) \sim \frac{\bar{Q}}{2\pi z},$$

$$Q = X + iY \quad (2.3.7)$$

where Q is the resultant of the external forces acting on the finite boundaries, provided that Q is finite and that stresses and rotations at infinity are zero.

If Q is non-zero, as happens, for example, in the case of contact problems on the half-plane, then the displacements, which are given by integrals of $\varphi(z,t)$ and $\psi(z,t)$, are logarithmically divergent at infinity, and thus are indeterminate – though their derivatives are determinate. This is a flaw in the theory, indicating the idealized nature of such two-dimensional problems

If however Q vanishes, as generally happens for crack problems, then the displacements are determinate, because

$$\varphi(z,t) \sim \frac{1}{z^2}, \qquad \psi(z,t) \sim \frac{1}{z^2} \quad (2.3.8)$$

at large z.

Chapter 3

CORRESPONDENCE PRINCIPLES

Methods of solution of non-inertial viscoelastic boundary value problems are in all cases greatly dependent on solutions of the underlying or corresponding non-inertial elastic (effectively elastostatic, though as mentioned earlier, time can be introduced as a trivial external parameter) boundary value problems. However, in certain situations the relationship between the two is particularly close, sufficiently so that the viscoelastic solutions are given explicitly in terms of the elastic solutions. These cases are enshrined in certain 'correspondence principles' which we now discuss.

3.1 The Classical Correspondence Principle

It was observed in subsection 2.2.3 that the time Fourier transforms of the constitutive and dynamical equations are formally identical to the corresponding elastic equations with the elastic constants replaced by complex moduli. Let us further assume that the boundary regions $B_u^{(i)}$ and $B_\sigma^{(i)}$, $i = 1,\ 2,\ 3$ where the displacements and boundary stresses are specified, do not vary with time.

Taking the Fourier transform of (2.2.5) gives

$$\hat{u}_i(\underline{r},\omega) = \hat{d}_i(\underline{r},\omega), \qquad \underline{r} \in B_u^{(i)},$$

$$\sum_{j=1}^{3} \hat{\sigma}_{ij}(\underline{r},\omega) n_j(\underline{r}) = \hat{c}_i(\underline{r},\omega), \quad \underline{r} \in B_\sigma^{(i)}, \quad i = 1,\ 2,\ 3. \tag{3.1.1}$$

Since $B_u^{(i)}$ and $B_\sigma^{(i)}$ are time-independent, the quantities $\hat{c}(\underline{r},\omega)$, $\hat{d}(\underline{r},\omega)$ can be calculated without difficulty. Equations (2.2.6) and (3.1.1) are formally identical to the statement of the elastic boundary value problem with the same boundary regions, except that the elastic constants are replaced by the corresponding complex moduli. Thus, if the solution to the elastic problem is known, one can obtain the solution for a viscoelastic material by replacing moduli with complex moduli and calculating the inverse Fourier transforms. More specifically, let $\{u_i^{(e)}(\underline{r},\omega,t),\ \sigma_{ij}^{(e)}(\underline{r},\omega,t)\}$ be the elastic solutions of the boundary value problem specified by (2.2.5) but with moduli replaced by complex moduli – this being indicated by the frequency dependence. Then

$$\hat{u}_i(\underline{r},\omega) = \int_{-\infty}^{\infty} dt' u_i^{(e)}(\underline{r},\omega,t')e^{-i\omega t'},$$

$$\hat{\sigma}_{ij}(\underline{r},\omega) = \int_{-\infty}^{\infty} dt' \sigma_{ij}^{(e)}(\underline{r},\omega,t')e^{-i\omega t'}, \tag{3.1.2}$$

obey the transformed equations (2.2.6) and the boundary conditions (3.1.1) and thus (assuming that unique solutions exist) are the transformed viscoelastic solutions. Therefore the required solutions are given by

$$\begin{aligned} u_i(\underline{r},t) &= \frac{1}{2\pi}\int_{-\infty}^{\infty} d\omega e^{i\omega t}\int_{-\infty}^{\infty} dt' u_i^{(e)}(\underline{r},\omega,t')e^{-i\omega t'} \\ &= \frac{1}{2\pi}\int_{-\infty}^{\infty} dt'\int_{-\infty}^{\infty} d\omega e^{i\omega(t-t')}u_i^{(e)}(\underline{r},\omega,t'), \\ \sigma_{ij}(\underline{r},t) &= \frac{1}{2\pi}\int_{-\infty}^{\infty} dt'\int_{-\infty}^{\infty} d\omega e^{i\omega(t-t')}\sigma_{ij}^{(e)}(\underline{r},\omega,t'). \end{aligned} \tag{3.1.3}$$

It follows immediately from Fourier Integral Theorem that if $u_i^{(e)}(\underline{r},\omega,t')$ is independent of ω at any point $\underline{r}$ (i.e. if an elastic displacement component is independent of the moduli) then the viscoelastic component and the elastic component are equal at that point. Similarly for the stress components.

The above argument breaks down if the specified boundary functions $c_i(\underline{r},t)$ or $d_i(\underline{r},t)$ are allowed to depend upon the material moduli. This would generally not be the case, however.

Let us write the displacement relation in (3.1.3) in the form

$$u_i(\underline{r},t) = \int_{-\infty}^{\infty} dt' U_i^{(e)}(\underline{r},t-t',t')$$

$$U_i^{(e)}(\underline{r},t,t') = \frac{1}{2\pi}\int_{-\infty}^{\infty} d\omega e^{i\omega t}u_i^{(e)}(\underline{r},\omega,t'). \tag{3.1.4}$$

In the first equation, the $t-t'$ occurs in inverse transformed functions of the complex moduli, while the t' occurs in the boundary functions. Now, the state of the material at time t cannot by virtue of Causality depend on the values of the driving quantities, namely the boundary functions, at a later time. This implies that the integral over t' must stop at t or, equivalently, that $U_i^{(e)}(\underline{r},t,t')$ is zero for $t<0$. Thus, the requirement of Causality imposes some constraints on the form of the elastic Green's functions, which are discussed in [14]. By virtue of this property, we write

$$u_i(\underline{r},t) = \frac{1}{2\pi}\int_{-\infty}^{\infty} d\omega e^{i\omega t}\int_{-\infty}^{t} dt' u_i^{(e)}(\underline{r},\omega,t')e^{-i\omega t'}$$

$$\sigma_{ij}(\underline{r},t) = \frac{1}{2\pi}\int_{-\infty}^{\infty} d\omega e^{i\omega t}\int_{-\infty}^{t} dt' \sigma_{ij}^{(e)}(\underline{r},\omega,t')e^{-i\omega t'}. \tag{3.1.5}$$

Problems with fixed boundary regions are therefore solvable at least in principle, by application of the Classical Correspondence Principle, if the corresponding elastic solution is known. In this sense, such problems are trivial, and the focus must shift to problems where the boundary regions vary with time. Such problems are the central topic of these lecture notes.

3.2 Time-Dependent Boundary Regions

In sections 3.3 and 3.4, we shall consider two extensions of the Correspondence Principle, which apply, under limited conditions to problems with time-dependent boundary regions. We shall also in Chapter 5 sketch an approach to solving the general problem. It is convenient for illustration purposes to restrict the discussion to materials for which the proportionality assumption is a valid approximation. The practical effect of this restriction is that there is only one viscoelastic function in the equations.

There are in fact extremely important cases where this restriction is not necessary. These are problems where the dependence on the material parameters combines into one expression without the need for special assumptions. For example, it will be recalled that in elastic frictionless contact problems (essentially Hertz's theory) the material parameters combine into the quantity $\mu/(1-\nu)$, sometimes termed the compliance. The effect of this in the viscoelastic case is to reduce the number of viscoelastic functions to one, which is necessary for the methods outlined here, at least at our present level of understanding, to apply.

A fully satisfactory characterization of the class of problems where this collapse of the material parameters into a single expression occurs, has not been given. However, a particular characterization is given by Graham and Golden (1988b) which consists of problems where the shearing stress vanishes at all points in a plane - typically as a result of an assumption of frictionless contact.

In our detailed treatment of specific problems, it will always be obvious whether or not they are of this type, and if so, the correspondence principles and general solution technique derived under the proportionality assumption can be applied to them with minimal modification. However, at the present initial discussion of general techniques, it is a considerable simplification to ignore this special class.

Under the proportionality assumption

$$\lambda(t) = \frac{2\nu}{1-2\nu}\mu(t), \tag{3.2.1}$$

and equations (2.2.2) and (2.2.4) may be written in the form

$$\sigma_{ij}(\underline{r},t) = 2\mu_0[\eta_{ij}(\underline{r},t) + \frac{\nu}{1-2\nu}\delta_{ij}\eta(\underline{r},t)],$$

$$\nabla^2 v_i(\underline{r},t) + \frac{1}{1-2\nu}\frac{\partial}{\partial x_i}\nabla\cdot\underline{v}(\underline{r},t) = 0, \quad i,j = 1,2,3, \tag{3.2.2}$$

where μ_0 is an arbitrary constant and

$$v_i(\underline{r},t) = \frac{1}{\mu_0}\int_{-\infty}^{t} dt'\mu(t-t')u_i(\underline{r},t'),$$

$$\eta_{ij}(\underline{r},t) = \frac{1}{2}\left(\frac{\partial v_i(\underline{r},t)}{\partial x_j} + \frac{\partial v_j(\underline{r},t)}{\partial x_i}\right),$$

$$\eta(\underline{r},t) = \sum_{i=1}^{3}\eta_{ii}(\underline{r},t). \tag{3.2.3}$$

Alternatively, they may be written as

$$Q_{ij}(\underline{r},t) = 2\mu_0[\epsilon_{ij}(\underline{r},t) + \frac{\nu}{1-2\nu}\delta_{ij}\epsilon(\underline{r},t)],$$

$$\nabla^2 u_i(\underline{r},t) + \frac{1}{1-2\nu}\frac{\partial}{\partial x_i}\nabla\cdot\underline{u}(\underline{r},t) = 0, \tag{3.2.4}$$

where

$$Q_{ij}(\underline{r},t) = \mu_0\int_{-\infty}^{t} dt'\gamma(t-t')\sigma_{ij}(\underline{r},t'), \tag{3.2.5}$$

the quantity $\gamma(t)$ being the inverse of $\mu(t)$ in the sense of (2.1.8).

Equations (3.2.2) and (3.2.4) have the same form as the elastic equations with $v_i(\underline{r},t)$ replacing the displacements in the case of (3.2.2) and $Q_{ij}(\underline{r},t)$ replacing the stresses in the case of (3.2.4). We refer to these as the pseudo-displacements and the pseudo-stresses respectively.

The boundary relations (2.2.5) cannot however in general be put in elastic form if $B_u^{(i)}$ and $B_\sigma^{(i)}$ depend on time. If however, we could determine either the pseudo-displacements or the pseudo-stresses on the boundary regions, then the problem is reduced to an elastic problem. This will be our strategy in developing general methods of solution. Note that even at this stage, the problem has been reduced to that involving boundary quantities only, which is a considerable simplification.

Before attempting a general approach however, we consider certain special cases which amount to extensions of the Classical Correspondence Principle.

3.3 The Extended Correspondence Principle

Consider the case where $B_u^{(i)}(t)$, $i = 1,2,3$ is stationary or contracting, i.e. $B_u^{(i)}(t_2) \subseteq B_u^{(i)}(t_1)$ for all t_1, t_2 such that $t_2 \geq t_1$. Then if $\underline{r} \in B_u^{(i)}(t)$ it follows that $\underline{r} \in B_u^{(i)}(t')$ for all $t' \leq t$. Taking the convolution product of $\mu(t)$ with the displacement relation of the boundary conditions (2.2.5) gives

$$v_i(\underline{r},t) = \frac{1}{\mu_0}\int_{-\infty}^{t} dt'\mu(t-t')d_i(\underline{r},t'), \quad i = 1,2,3, \quad \underline{r} \in B_u^{(i)}(t), \tag{3.3.1}$$

where, by virtue of our assumption, the right-hand side is a known function.

The problem therefore has a form identical to the elastic problem, with the $v_i(\underline{r},t)$, $i = 1,2,3$ playing the part of displacements, as we see by considering (3.2.2) and the boundary conditions (2.2.5) modified as in (3.3.1). The solution is given in terms of the elastic Green's functions with specified pseudo-displacements given by the right-hand side of (3.3.1) and specified stresses by $c_i(\underline{r},t)$ in (2.2.5). The actual displacements can be obtained by taking the convolution product of $\gamma(t)$ with $v_i(\underline{r},t)$.

If the quantities $d_i(\underline{r},t)$ are zero for all times for $\underline{r} \in B_u^{(i)}(t)$ then the pseudo-displacements also vanish on $B_u^{(i)}(t)$ and the specified boundary functions for pseudo-displacements and stresses are identical to those for elastic displacements and stresses. It follows from the elastic uniqueness theorem that the stresses for the viscoelastic problem are identical to those for the elastic problem. This occurs typically in extending crack problems.

If on the other hand, $B_\sigma^{(i)}(t)$ is contracting or stationary, then the stress boundary condition in (2.2.5) gives, recalling (3.2.5),

$$\sum_{i=1}^{3} Q_{ij}(\underline{r},t)n_j(\underline{r}) = \mu_0\int_{-\infty}^{t} dt'\gamma(t-t')c_i(\underline{r},t'), \quad \underline{r} \in B_\sigma^{(i)}(t), \tag{3.3.2}$$

where, by an entirely analogous argument as previously, the right-hand side is a known function. Thus the displacements and the pseudo-stresses obey equations (3.2.4), the displacement condition in (2.2.5) and (3.3.2) which have the same form as the elastic equations, with boundary functions given by $d_i(\underline{r},t)$ and the right-hand side of (3.3.2). They are given in terms of elastic Green's functions for the problem. The stresses are then determined by taking the convolution product of $\mu(t)$ with the pseudo-stresses.

If the $c_i(\underline{r},t)$ vanish at all times for $\underline{r} \in B_\sigma^{(i)}(t)$ then the right-hand side of (3.3.2) also vanishes and the specified functions for displacements and pseudo-stresses in the viscoelastic problem are identical to those for the elastic problem. It follows that the displacements are identical to the elastic displacements. This happens typically for contact problems with expanding contact area.

These observations express the main content (at least for materials to which the proportionality assumption is applicable) of theorems extending the Classical Correspondence Principle, due to Graham (1968), generalized by Ting (1969) and further discussed by Graham and Sabin (1973). We refer to these results as the Extended Correspondence Principle.

3.4 The Generalized Partial Correspondence Principle

Let us assume that at a particular time t, $B_u^{(i)}(t) \subseteq B_u^{(i)}(t')$, $i = 1,2,3$ for all $t' \leq t$. We are only assuming that $B_u^{(i)}(t)$ is contained in the corresponding region for all previous time, not making any assumptions on the earlier history of the boundary regions. In particular, the behaviour of $B_u^{(i)}(t')$ need not be monotonic.

The displacement relation of the boundary conditions (2.2.5) then gives

$$v_i(\underline{r},t) = \frac{1}{\mu_0}\int_{-\infty}^{t} dt'\mu(t-t')d_i(\underline{r},t'), \quad \underline{r} \in B_u^{(i)}(t), \tag{3.4.1}$$

where, by virtue of our assumption, the right-hand side is a known function at this special time t.

The problems has therefore a form identical to the elastic problem with the pseudo-displacements playing the part of displacements, as observed in the previous section. The solution is given in terms of the elastic Green's functions with specified pseudo-displacements given by the right-hand side of (3.4.1) and specified stresses by $c_i(\underline{r},t)$ in (2.2.5). However, the actual displacements cannot in general be obtained by operating with $\gamma(t)$ because the earlier history of the pseudo-displacements is not known. The usefulness of the observation lies in the fact that the stresses are known everywhere at time t if the elastic Green's functions are known.

If the specified displacements are always zero for $\underline{r} \in B_u^{(i)}(t')$, $t' \leq t$, then the right-hand side of (3.4.1) vanishes also, and the stresses are everywhere identical to the elastic stresses. This happens typically for crack problems.

If on other hand we assume that, at time t, $B_\sigma^{(i)}(t)$ has the property that $B_\sigma^{(i)}(t) \subseteq B_\sigma^{(i)}(t')$ for all $t' \leq t$, with no further assumption on the history of the boundary regions, then the stress boundary condition in (2.2.5) gives

$$\sum Q_{ij}(\underline{r},t)n_j(\underline{r}) = \mu_0\int_{-\infty}^{t} dt'\gamma(t-t')c_i(\underline{r},t'), \quad \underline{r} \in B_\sigma(t), \tag{3.4.2}$$

where the right-hand side is a known function. The pseudo-stresses $Q_{ij}(\underline{r},t)$ are defined by (3.2.5). By virtue of an argument entirely analogous to that above, the pseudo-stresses and the displacements are given in terms of the elastic Green's functions with specified functions $d_i(\underline{r},t)$ and the right-hand side of (3.4.2). Thus, the displacements are known but, in general, the stresses cannot be determined, at this special time t.

If the $c_i(\underline{r},t)$ vanish at all times for $\underline{r} \in B_\sigma(t')$, $t' \leq t$, then the right-hand side of (3.4.2) also vanishes and the displacements at time t are identical to the elastic displacements. This happens typically for contact problems.

These observations express the main content (at least for materials to which the proportionality assumption is applicable) of the Generalized Partial Correspondence Principle [Graham and Golden (1988b)]. In that reference, this principle is applied to derive a number of interesting results for the normal contact problem, which is discussed in detail in chapter 7.

Chapter 4

DECOMPOSITION OF HEREDITARY INTEGRALS

Central to the solution of problems with time-dependent boundary regions is a fundamental decomposition of hereditary integrals which we shall now discuss. The reason for its importance will become clear in later chapters as it is shown to be the key to the solution of widely different types of problems.

4.1 Detailed form

Let us first phrase the problem in fairly abstract terms. Let $u(t)$ and $v(t)$ be two functions related by

$$v(t) = \int_{-\infty}^{t} dt' l(t-t') u(t'),$$

$$u(t) = \int_{-\infty}^{t} dt' k(t-t') v(t'), \tag{4.1.1}$$

where $k(t)$ and $l(t)$ are causal functions i.e. zero for negative times, which are inverses with respect to the convolution product. In other words, they are related in the same manner as $\mu(t)$ and $\gamma(t)$ as given by (2.1.8). Let $\theta(t)$ be the set of the present and all past times $(-\infty, t]$, which we decompose into two disjoint sets $W_u(t)$ and $W_v(t)$, i.e.

$$\theta(t) = W_u(t) \bigcup W_v(t), \tag{4.1.2}$$

where $u(t')$ is given for $t' \in W_u(t)$ and $v(t')$ is given (or can be usefully represented; we shall see what is meant by this phrase in the next chapter) for $t' \in W_v(t)$. If we can decompose $v(t)$, for example, as follows:

$$v(t) = \int_{W_u(t)} dt' \Pi_u(t,t') u(t') + \int_{W_v(t)} dt' \Pi_v(t,t') v(t'), \tag{4.1.3}$$

then everything on the right-hand side is known, or can be usefully represented, thus giving an expression for $v(t)$. The problem is therefore to decompose a hereditary integral in this manner.

We first define the sets $W_u(t)$ and $W_v(t)$ in more detail. Let $t \in W_u(t)$; if it were in $W_v(t)$ we would in some sense know $v(t)$ to begin with and there would be no need for further discussion. Let $t_1, t_2, t_3, \cdots$ be a monotone decreasing sequence of times marking when t' changes from $W_u(t)$ to $W_v(t)$ and vica versa. We refer to them as transition times. The sequence may be finite or infinite. Let us write $v(t)$ as

$$\begin{aligned} v(t) &= \int_{t_1}^{t} dt' l(t-t')u(t') + \int_{-\infty}^{t_1} dt' l(t-t')u(t') \\ &= \int_{t_1}^{t} dt' l(t-t')u(t') + \int_{-\infty}^{t_1} dt' T_1(t,t')v(t'), \\ T_1(t,t') &= \int_{t'}^{t_1} dt'' l(t-t'')k(t''-t'). \end{aligned} \tag{4.1.4}$$

This last step involved substituting for $u(t')$ from (4.1.1) and an interchange of integrations. The procedure can be repeated. We obtain

$$\begin{aligned} v(t) &= \int_{t_1}^{t} dt' l(t-t')u(t') + \int_{t_2}^{t_1} dt' T_1(t,t')v(t') \\ &\quad + \int_{-\infty}^{t_2} dt' T_2(t,t')u(t'), \\ T_2(t,t') &= \int_{t'}^{t_2} dt'' T_1(t,t'')l(t''-t'). \end{aligned} \tag{4.1.5}$$

Continuing in this way, we obtain a decomposition of the form indicated by (4.1.3) where

$$\begin{aligned} \Pi_u(t,t') &= T_0(t,t')R(t';t_1,t) \\ &\quad + T_2(t,t')R(t';t_3,t_2) \\ &\quad + T_4(t,t')R(t';t_5,t_4) \\ &\quad + \cdots, \\[1em] \Pi_v(t,t') &= T_1(t,t')R(t';t_2,t_1) \\ &\quad + T_3(t,t')R(t';t_4,t_3) \\ &\quad + T_5(t,t')R(t';t_6,t_5) \\ &\quad + \cdots, \end{aligned} \tag{4.1.6}$$

where the function $R(t';t_2,t_1)$ projects onto the interval $[t_2,t_1]$ in the sense that

$$R(t';t_2,t_1) = \begin{cases} 1, & t' \in [t_2,t_1], \\ 0, & \text{otherwise}, \end{cases} \tag{4.1.7}$$

for all t_2, t_1, t'. Also:

$$T_0(t,t') = l(t-t'),$$

$$T_j(t,t') = \begin{cases} \int_{t'}^{t_j} dt'' T_{j-1}(t,t'') l(t''-t'), & j \text{ even}, \\ \int_{t'}^{t_j} dt'' T_{j-1}(t,t'') k(t''-t'), & j \text{ odd}. \end{cases} \tag{4.1.8}$$

Similarly, we wish to decompose $u(t)$ given by (4.1.1) in the form

$$u(t) = \int_{W_u(t)} dt' \Gamma_u(t,t') u(t') + \int_{W_v(t)} dt' \Gamma_v(t,t') v(t'). \tag{4.1.9}$$

In this case, the non-trivial assumption is to take $t \in W_v(t)$. Going through a similar argument, we find that

$$\begin{aligned} \Gamma_v(t,t') &= N_0(t,t') R(t';t_1,t) \\ &+ N_2(t,t') R(t';t_3,t_2) \\ &+ N_4(t,t') R(t';t_5,t_4) \\ &+ \cdots, \end{aligned} \tag{4.1.10}$$

$$\begin{aligned} \Gamma_u(t,t') &= N_1(t,t') R(t';t_2,t_1) \\ &+ N_3(t,t') R(t';t_4,t_3) \\ &+ N_5(t,t') R(t';t_6,t_5) \\ &+ \cdots, \end{aligned} \tag{4.1.11}$$

where the function $N_j(t,t')$ are given by

$$N_0(t,t') = k(t-t'),$$

$$N_j(t,t') = \begin{cases} \int_{t'}^{t_j} dt'' N_{j-1}(t,t'') l(t''-t'), & j \text{ odd}, \\ \int_{t'}^{t_j} dt'' N_{j-1}(t,t'') k(t''-t'), & j \text{ even}. \end{cases} \tag{4.1.12}$$

It is clear that the difference between Γ_u, Γ_v and Π_u, Π_v amounts to an interchange of the role of the functions $l(t)$, $k(t)$.

These formulae will find application essentially unchanged in form in later chapters. We will also need a decomposition of $v_i(\underline{r},t)$ as defined in (3.2.3) and this involves a straightforward generalization of the formulae in this section.

4.2 Point-Based Decomposition

We seek a decomposition of $v_i(\underline{r},t)$ into a term involving $u_i(\underline{r},t')$ if $\underline{r} \in B_u^{(i)}(t')$ - on which $u_i(\underline{r},t')$ is given - and a term involving $v_i(\underline{r},t')$ if $\underline{r} \in B_\sigma^{(i)}(t')$. The quantity

$v_i(\underline{r}, t')$ is not necessarily given on $B_\sigma^{(i)}(t')$ but we will see that it can be usefully represented.

The decomposition now derived will depend on the point $\underline{r}$ under consideration and thus will be referred to as the point-based decomposition. The set of all past times and the current time $\theta(t)$ is the union of two disjoint sets $W_u^{(i)}(\underline{r}, t)$ and $W_\sigma^{(i)}(\underline{r}, t)$ where $W_u^{(i)}(\underline{r}, t)$ is the set of all those times $t' \leq t$ such that $\underline{r} \in B_u^{(i)}(t')$ and $W_\sigma^{(i)}(\underline{r}, t)$ is all those times $t' \leq t$ such that $\underline{r} \in B_\sigma^{(i)}(t')$. In these expressions, $\underline{r}$ is a point on the boundary of the material. Note that the subscript 'σ' has replaced 'v' on the set of times $W_\sigma^{(i)}(\underline{r}, t)$.

Applying the same techniques as in section 4.1, we find that the required decomposition is given by

$$v_i(\underline{r}, t) = \int_{W_u^{(i)}(\underline{r},t)} dt' \Pi_u^{(i)}(t, t'; \underline{r}) u_i(\underline{r}, t') + \int_{W_\sigma^{(i)}(\underline{r},t)} dt' \Pi_\sigma^{(i)}(t, t'; \underline{r}) v_i(\underline{r}, t'),$$

$$\underline{r} \in B_u^{(i)}(t), \tag{4.2.1}$$

where

$$\Pi_u^{(i)}(t, t'; \underline{r}) = \sum_{j=0,2,4,\cdots} T_j^{(i)}(t, t'; \underline{r}) R(t'; t_{j+1}(\underline{r}), t_j(\underline{r})),$$

$$\Pi_\sigma^{(i)}(t, t'; \underline{r}) = \sum_{j=1,3,5,\cdots} T_j^{(i)}(t, t'; \underline{r}) R(t'; t_{j+1}(\underline{r}), t_j(\underline{r})), \tag{4.2.2}$$

where $t_0(\underline{r})$ is understood here to be the current time t and

$$T_0^{(i)}(t, t'; \underline{r}) = T_0(t, t') = l(t - t'),$$

$$T_j^{(i)}(t, t'; \underline{r}) = \begin{cases} \int_{t'}^{t_j^i(\underline{r})} dt'' T_{j-1}^{(i)}(t, t''; \underline{r}) l(t'' - t'), & j \text{ even}, \\ \int_{t'}^{t_j^i(\underline{r})} dt'' T_{j-1}^{(i)}(t, t''; \underline{r}) k(t'' - t'), & j \text{ odd}, \end{cases} \tag{4.2.3}$$

the times $t_j^i(\underline{r})$ being times of transition of $\underline{r}$ between $B_u^{(i)}(t)$ and $B_\sigma^{(i)}(t)$. It should be noted that these transition times $t_j^i(\underline{r})$ (and the t_j considered earlier) depend, in a discontinuous manner, on the current value of t. However, to simplify notation, this has not been shown explicitly.

A similar decomposition is possible for $u_i(\underline{r}, t)$, $\underline{r} \in B_\sigma^{(i)}(t)$ analagous to (4.1.9). For future reference, we write it out:

$$u_i(\underline{r}, t) = \int_{W_u^{(i)}(\underline{r},t)} dt' \Gamma_u^{(i)}(t, t'; \underline{r}) u_i(\underline{r}, t') + \int_{W_\sigma^{(i)}(\underline{r},t)} dt' \Gamma_\sigma^{(i)}(t, t'; \underline{r}) v_i(\underline{r}, t'),$$

$$\underline{r} \in B_\sigma^{(i)}(t), \tag{4.2.4}$$

where

$$\Gamma_u^{(i)}(t, t'; \underline{r}) = \sum_{j=1,3,5,\cdots} N_j^{(i)}(t, t'; \underline{r}) R(t'; t_{j+1}(\underline{r}), t_j(\underline{r})),$$

$$\Gamma_{\sigma}^{(i)}(t,t';\underline{r}) = \sum_{j=0,2,4,\cdots} N_j^{(i)}(t,t';\underline{r})R(t';t_{j+1}(\underline{r}),t_j(\underline{r})), \qquad (4.2.5)$$

where again $t_0(\underline{r})$ is understood to be the current time t and

$$N_0^{(i)}(t,t';\underline{r}) = N_0(t,t') = k(t-t'),$$

$$N_j(t,t';\underline{r}) = \begin{cases} \int_{t'}^{t_j(\underline{r})} dt'' N_{j-1}^{(i)}(t,t'';\underline{r})l(t''-t'), & j \text{ odd}, \\ \int_{t'}^{t_j(\underline{r})} dt'' N_{j-1}^{(i)}(t,t'';\underline{r})k(t''-t'), & j \text{ even}. \end{cases} \qquad (4.2.6)$$

The decomposition developed in these two sections is an alternative form of that given by Graham (1967), Ting (1968) and later, for aging case, by Sabin and Graham (1980), in the context of normal contact problems. The form developed here (and in [14]) is more convenient in some respects than those given previously.

4.3 Steady-State Case

A frequent assumption is that the boundary regions vary cyclically with time with period Δ, usually generated by periodic applied loads. After a sufficient number of cycles, transient effects will die away and all physical quantities will become periodic functions with period Δ. We assume that such a situation prevails throughout all past history.

Consider (4.1.3) where $u(t)$ and $v(t)$ obey

$$u(t+\Delta) = u(t), \qquad v(t+\Delta) = v(t). \qquad (4.3.1)$$

Now t_1 is the time of transition from $W_u(t)$ to $W_v(t)$, t_2 is the time of transition from $W_v(t)$ to $W_u(t)$ and so on. We claim that

$$t_{j+2} = t_j - \Delta. \qquad (4.3.2)$$

This is clearly true if the sets of times W_u and W_v (or W_σ) are defined as in section 4.2, given that $B_u(t)$ and $B_\sigma(t)$ are periodic. However, the other, generally simpler contexts in which we will be applying the results of section 4.1 are in fact special cases of the formalism developed in section 4.2, so that (4.3.2) will always hold. Therefore (4.1.3) can be written as

$$v(t) = \int_{t_1}^{t} dt' T_0(t,t')u(t') + \int_{t_1}^{t_2+\Delta} dt'\Pi_0(t,t')u(t') + \int_{t_2}^{t_1} dt'\Pi_1(t,t')v(t'), \qquad (4.3.3)$$

where

$$\Pi_0(t,t') = \sum_{j=1}^{\infty} T_{2j}(t,t'-j\Delta),$$

$$\Pi_1(t,t') = \sum_{j=0}^{\infty} T_{2j+1}(t,t'-j\Delta). \tag{4.3.4}$$

We have separated out the first integral in (4.3.3) because the current time t is usually within the range $[t_1, t_2+\Delta]$ and so the integral is over part of a period rather that a complete period.

Similarly (4.1.9) can be written as

$$u(t) = \int_{t_1}^{t} dt' N_0(t,t')v(t') + \int_{t_2}^{t_1} dt'\Gamma_1(t,t')u(t') + \int_{t_1}^{t_2+\Delta} dt'\Gamma_0(t,t')v(t'), \tag{4.3.5}$$

where

$$\Gamma_1(t,t') = \sum_{j=0}^{\infty} N_{2j+1}(t,t'-j\Delta),$$

$$\Gamma_0(t,t') = \sum_{j=1}^{\infty} N_{2j}(t,t'-j\Delta). \tag{4.3.6}$$

We shall now show that the summations Π_i, Γ_i, $i = 1,2$ obey integral equations which if they can be solved directly (i.e. without iteration) allow us to express the summations in closed form. From (4.1.8), it follows that

$$T_n(t,t') = \int_{t'}^{t_n} dt''' \int_{t'''}^{t_{n-1}} dt'' T_{n-2}(t,t'')l(t''-t''')k(t'''-t'), \qquad \text{n odd},$$

$$T_n(t,t') = \int_{t'}^{t_n} dt''' \int_{t'''}^{t_{n-1}} dt'' T_{n-2}(t,t'')k(t''-t''')l(t'''-t'), \qquad \text{n even}. \tag{4.3.7}$$

The integral over t'' can in fact be extended to t' at the lower limit since $l(t''-t''')$ or $k(t''-t''')$ vanish over this interval. This allows the integrations to be interchanged without difficulty and one has

$$T_n(t,t') = \int_{t'}^{t_{n-1}} dt'' T_{n-2}(t,t'')G_n(t'',t') \tag{4.3.8}$$

$$G_n(t'',t') = \begin{cases} \int_{t'}^{t_n} dt''' l(t''-t''')k(t'''-t'), & n \text{ odd}, \\ \int_{t'}^{t_n} dt''' k(t''-t''')l(t'''-t'), & n \text{ even}. \end{cases}$$

Noting that

$$G_n(t'',t') = \delta(t''-t'), \quad t' < t_n \tag{4.3.9}$$

we have

$$T_n(t,t') = T_{n-2}(t,t') + \int_{t_n}^{t_{n-1}} dt'' T_{n-2}(t,t'')G_n(t'',t'), \qquad t' < t_n. \tag{4.3.10}$$

Consider for definiteness the case where n is odd. Let us take $t' < t_1$, and rewrite (4.3.10) as

$$T_{2n+1}(t,t'-n\Delta) = T_{2n-1}(t,t'-n\Delta) + \int_{t_{2n+1}}^{t_{2n}} dt'' T_{2n-1}(t,t'')G_{2n+1}(t'',t'-n\Delta), \tag{4.3.11}$$

where

$$G_{2n+1}(t'',t'-n\Delta)=\int_{t'-n\Delta}^{t_{2n+1}} dt''' l(t''-t''')k(t'''-t'+n\Delta)$$
$$=\int_{t'}^{t_1} du\, l(t''-u+n\Delta)k(u-t')=G(t''+n\Delta,t') \tag{4.3.12}$$

on transforming the variable of integration according to $u=t'''+n\Delta$. The function $G(t'',t')$ is given by

$$G(t'',t')=\int_{t'}^{t_1} du\, l(t''-u)k(u-t'), \tag{4.3.13}$$

which has the same functional form as $T_1(t,t')$ with t'' replacing t but t_1, which may depend on t, left untouched. Using (4.3.12) in (4.3.11) gives, after a further change of variable ($u=t''+(n-1)\Delta$)

$$T_{2n+1}(t,t'-n\Delta)=T_{2n-1}(t,t'-n\Delta)+\int_{t_3}^{t_2} du T_{2n-1}(t,u-(n-1)\Delta)G(u+\Delta,t'), \tag{4.3.14}$$

so that $\Pi_1(t,t')$, given by (4.3.4) obeys the equation

$$\Pi_1(t,t')=T_1(t,t')+\Pi_1(t,t'-\Delta)+\int_{t_3}^{t_2} du\Pi_1(t,u)G(u+\Delta,t'). \tag{4.3.15}$$

It follows that

$$\Pi_1(t,t'-\Delta)=T_1(t,t'-\Delta)+\Pi_1(t,t'-2\Delta)+\int_{t_3}^{t_2} du\Pi_1(t,u)G(u+\Delta,t'-\Delta), \tag{4.3.16}$$

and so on. Repeated substitution of (4.3.16) and its successors into (4.3.15), together with the assumption, which will be justified later, that

$$\lim_{n\to\infty}\Pi_1(t,t'-n\Delta)=0, \tag{4.3.17}$$

finally gives an integral equation for $\Pi(t,t')$ of the form

$$\Pi_1(t,t')=K(t,t')+\int_{t_3}^{t_2} du\Pi_1(t,u)K(u+\Delta,t'), \tag{4.3.18}$$

where

$$K(u,t')=\sum_{n=0}^{\infty} G(u,t'-n\Delta). \tag{4.3.19}$$

In a similar maner, it can be shown that $\Pi_0(t,t')$ obeys the integral equation

$$\Pi_0(t,t')=K''(t,t')+\int_{t_2}^{t_1} du\Pi_0(t,u)K'(u,t'),$$

$$K'(u,t')=\sum_{n=0}^{\infty} G'(u+\Delta,t'-n\Delta),$$

$$G'(t'',t') = \int_{t'}^{t_2} du k(t''-u) l(u-t'), \tag{4.3.20}$$

$$K''(t,t') = \int_{t_2}^{t_1} du l(t-u) K'(u,t') + \sum_{n=1}^{\infty} l(t-t'+n\Delta),$$

provided that

$$\lim_{n\to\infty} \Pi_0(t, t'-n\Delta) = 0. \tag{4.3.21}$$

Equations for $\Gamma_0(t,t')$ and $\Gamma_1(t,t')$ can be written down by interchanging the roles of $l(t)$ and $k(t)$ in the above equations. Thus $\Gamma_1(t,t')$ obeys

$$\Gamma_1(t,t') = L(t,t') + \int_{t_3}^{t_2} du \Gamma_1(t,u) L(u+\Delta, t'), \tag{4.3.22}$$

$$L(u,t') = \sum_{n=0}^{\infty} H(u, t'-n\Delta),$$

$$H(t'',t') = \int_{t'}^{t_1} du k(t''-u) l(u-t'),$$

and $\Gamma_0(t,t')$ obeys

$$\Gamma_0(t,t') = L''(t,t') + \int_{t_2}^{t_1} du \Gamma_0(t,u) L'(u,t'), \tag{4.3.23}$$

$$L'(u,t') = \sum_{n=0}^{\infty} H'(u+\Delta, t'-n\Delta),$$

$$H'(t'',t') = \int_{t'}^{t_2} du l(t''-u) k(u-t'),$$

$$L''(t,t') = \int_{t_2}^{t_1} du k(t-u) L'(u,t') + \sum_{n=1}^{\infty} k(t-t'+n\Delta),$$

provided that

$$\lim_{n\to\infty} \Gamma_i(t, t'-n\Delta) = 0, \quad i = 0, 1. \tag{4.3.24}$$

These equations provide a convenient method of determining the four quantities for discrete spectrum materials discussed in section 2.1. In fact, for the standard linear model, they are not needed; the summations can be carried out directly. For more that one decay time however, they are required. Note that a prerequisite of usefulness is that the kernels and inhomogeneous terms can be calculated easily in simple, closed form, and that the equations can be solved in closed form, without iteration. This turns out to be the case for spectrum models.

Chapter 5

GENERAL METHOD OF SOLUTION

We derive a general integral equation in this chapter for solving viscoelastic problems where a complete elastic solution is known. Also, where the time variation of the boundary regions is of a special kind, a technique is developed leading to results generally much simpler and more explicit that the general integral equation method. Again, however, the corresponding elastic solution must be known explicitly.

It proves necessary to restrict the coordinate dependence of the boundary regions $B_u^{(i)}(t)$, $B_\sigma^{(i)}(t)$, $i = 1,2,3$ in a manner discussed in section 5.1. for progress to be made.

5.1 General Integral Equation

We write the first relation of (3.2.3) and its inverse in the form

$$v_i(\underline{r},t) = \int_{-\infty}^{t} dt' l(t-t') u_i(\underline{r},t'),$$

$$u_i(\underline{r},t) = \int_{-\infty}^{t} dt' k(t-t') v_i(\underline{r},t'), \qquad (5.1.1)$$

$$l(t) = \frac{\mu(t)}{\mu_0}; \qquad k(t) = \mu_0 \gamma(t),$$

where μ_0 is an arbitrary constant. Let $\underline{r} \in B_u^{(i)}(t)$. We decompose $v_i(\underline{r},t)$ according to the point-based formula (4.2.1). The quantity $v_i(\underline{r},t')$ in the second integral on the right is always evaluated in $\underline{r} \in B_\sigma^{(i)}(t')$ by definition of $W_\sigma^{(i)}(\underline{r},t)$. If the quantities $v_i(\underline{r},t)$ were known on $B_u^{(i)}(t)$ (which they are not; our central goal is to determine them) then, as discussed in section 3.2, we are dealing with an elastic problem, with the pseudo-displacements $v_i(\underline{r},t)$ playing the part of displacements. As stated earlier, it is assumed that the elastic Green's functions for the mixed boundary value problem are known. Therefore the pseudo-displacements $v_i(\underline{r},t)$ for $\underline{r}$ anywhere in the material, and in particular on $B_\sigma^{(i)}(t)$, are given by

$$v_i(\underline{r},t) = \sum_{j=1}^{3} \int_{B_u^{(j)}(t)} ds' H_{ij}(\underline{r},\underline{r}',t) v_j(\underline{r}',t) + \sum_{j=1}^{3} \int_{B_\sigma^{(j)}(t)} ds' L_{ij}(\underline{r},\underline{r}',t) c_j(\underline{r}',t), \quad (5.1.2)$$

where ds' is the surface element, the $c_j(\underline{r},t)$ are the known applied stresses on $B_\sigma^{(j)}(t)$ and $H_{ij}(\underline{r},\underline{r}',t)$, $L_{ij}(\underline{r},\underline{r}',t)$ are the known elastic Green's functions, for the mixed boundary value problem. Their time dependence results from the fact that the boundary regions are time-dependent.

In order to proceed to the next stage, it is necessary to restrict the coordinate dependence of the boundary regions. We shall assume either (a) that $B_u^{(i)}(t)$ and $B_\sigma^{(i)}(t)$ are the same for each value of i; or (b) that they are time-dependent only for one coordinate direction, taken to be $i = 3$ or the z direction. In case (b) $v_1(\underline{r},t)$ and $v_2(\underline{r},t)$ are therefore known on $B_u^{(1)}$ and $B_u^{(2)}$ respectively.

Let us consider case (a) first. On the right-hand side of (4.2.1), we substitute $v_i(\underline{r},t)$ given by (5.1.2), to obtain the integral equations

$$v_i(\underline{r},t) = \sum_{j=1}^{3} \int_{W_\sigma(\underline{r},t)} dt' \int_{B_u(t')} ds' K_{ij}(\underline{r},\underline{r}';t,t') v_j(\underline{r}',t') + I_i(\underline{r},t), \quad \underline{r} \in B_u(t), \quad (5.1.3)$$

$$K_{ij}(\underline{r},\underline{r}';t,t') = \Pi_\sigma(t,t';\underline{r}) H_{ij}(\underline{r},\underline{r}',t'),$$

$$I_i(\underline{r},t) = \int_{W_u(\underline{r},t)} dt' \Pi_u(t,t';\underline{r}) u_i(\underline{r},t') + \sum_{j=1}^{3} \int_{W_\sigma(\underline{r},t)} dt' \Pi_\sigma(t,t';\underline{r}) \int_{B_\sigma(t')} ds' L_{ij}(\underline{r},\underline{r}',t') c_j(\underline{r}',t').$$

This is a set of coupled integral equations for the pseudo-displacements on $B_u(t)$. If the sets $W_u(\underline{r},t)$ and $W_\sigma(\underline{r},t)$ are known, the inhomogeneous terms $I_i(\underline{r},t)$ are known. The system of equations involves integrals over both space and time variables in a coupled manner, which suggests that, in general, analytic solutions are unlikely to be obtained. In fact, under a special assumption (steady-state behaviour) considerable analytic progress can be made, as we shall see. However, in general, numerical techniques must be resorted to at a fairly early stage.

In case (b), the only unknown quantity on $B_u^{(3)}(t) (= B_u(t))$ is $v_3(\underline{r},t)$, which we denote as $v(\underline{r},t)$. Proceeding in a similar manner to case (a), we derive an integral equation for $v(\underline{r},t)$ of the form

$$v(\underline{r},t) = \int_{W_\sigma(\underline{r},t)} dt' \int_{B_u(t')} ds' K(\underline{r},\underline{r}';t,t') v(\underline{r}',t') + I(\underline{r},t), \quad (5.1.4)$$

$$K(\underline{r},\underline{r}';t,t') = K_{33}(\underline{r},\underline{r}';t,t'),$$

$$I(\underline{r},t) = I_3(\underline{r},t) + \sum_{j=1}^{2} \int_{W_\sigma(\underline{r},t)} dt' \int_{B_u^{(j)}} ds' K_{3j}(\underline{r},\underline{r}';t,t') v_j(\underline{r}',t'),$$

$$W_\sigma(\underline{r},t) = W_\sigma^{(3)}(\underline{r},t),$$

in terms of quantities already defined in (5.1.3), but where $\Pi_u(t,t';\underline{r})$ and $\Pi_\sigma(t,t';\underline{r})$ now indicate $\Pi_u^{(3)}$ and $\Pi_\sigma^{(3)}$ respectively. Also, the space integrals in $I_3(\underline{r},t)$ are over $B_\sigma^{(j)}(t')$, two of which are time-independent. Again, if $W_u(\underline{r},t)(= W_u^{(3)}(\underline{r},t))$ and $W_\sigma(\underline{r},t)$ are known, the inhomogeneous term $I(\underline{r},t)$ is known.

The problems discussed in later chapters will in all cases be examples of case (b) where in fact $B_u^{(1)}$ and $B_u^{(2)}$ are the empty set. We have included mention of case (a) (which has not been given before) here for completeness. The sets $W_u(\underline{r},t)$ and $W_\sigma(\underline{r},t)$ are in general not known in advance but must be determined as part of the problem, with the aid of extra physical conditions, the nature of which is generally fairly obvious. This means however, that the integral equation must be solved subject to extra constraints which of course complicate numerical iteration procedures.

Because of the definition of $W_\sigma(\underline{r},t)$ in (5.1.3) and (5.1.4) the singularity in the kernel (essentially the singularity in the Green's function) is not integrated over, and so the integral equation is non-singular. This may be perceived more clearly later when we discuss special cases. It is the property of non-singularity that may break down in the case of (5.1.3) if the boundary regions depend on coordinate direction.

5.2 Repetitive Expansion and Contraction of Boundary Regions

For a particular kind of time behaviour of the boundary regions, a special and much simpler solution method is applicable, which we now explore.

Once again, the coordinate dependence of the boundary regions must be restricted, as described for cases (a) and (b) in section 5.1. We shall restrict our attention to case (b) for brevity, since this is the only case that will be used later. The general argument developed goes through however for case (a).The coordinate index is dropped and all quantities and sets are assumed to be those for the z-coordinate.

Let the region $B_u(t)$ be entirely characterized by a single parameter $a(t)$ which contains all the time-dependence. Thus, $B_u(t) \equiv A(a(t))$ where $A(a)$ is a time-independent region for fixed a. This must then also be true for $B_\sigma(t)$, so that $B_\sigma(t) \equiv B(a(t))$. Furthermore, let $A(a)$ have the property that

$$A(a_1) \subseteq A(a_2), \quad \text{for all } a_1 \leq a_2. \tag{5.2.1}$$

It follows that

$$B(a_1) \supseteq B(a_2), \quad \text{for all } a_1 \leq a_2. \tag{5.2.2}$$

Thus, if $a(t)$ fluctuates in value, $B_u(t)$ and $B_\sigma(t)$ constitute a one parameter sequence of regions, each contained in or containing all other regions. We refer to this as repetitive expansion and contraction. A standard example is a smooth indentor on a half-space under a fluctuating normal load; which will be considered in Chapter 7.

Let $\theta(t) = (-\infty, t]$ be the union of the disjoint sets $W_g(t)$ and $W_l(t)$ where $W_g(t)$ is all those times such that $a(t') \geq a(t)$ and $W_l(t)$ is all times such that $a(t') < a(t)$. Then for all $\underline{r} \in B_u(t)$, it follows that $\underline{r} \in B_u(t')$ if $t' \in W_g(t)$ and for all $\underline{r} \in B_\sigma(t)$ it follows that $\underline{r} \in B_\sigma(t')$ if $t' \in W_l(t)$.

Consider the times $t_j(t), j = 1, 2, \cdots$ such that $a(t_j(t)) = a(t)$. These are the transition times between $W_g(t)$ and $W_l(t)$. If $B_u(t)$ is contracting at time t, then $(t_1(t), t) \subset W_g(t)$, while if it is expanding $[t_1(t), t] \subset W_l(t)$.

Consider the boundary conditions (2.2.5) at a time when $B_u(t)$ is contracting. We take the convolution product of $l(t)$ given by (5.1.1), with both sides of the z component of the displacement condition, and decompose the right-hand side, according to (4.1.3). The set $W_u(t)$ is identified as $W_g(t)$, namely the times such that if $\underline{r} \in B_u(t)$ then $\underline{r} \in B_u(t')$, $t' \in W_g(t)$. It follows that $W_v(t)$ is identified as $W_l(t)$. The expansion (4.1.3) rather than (4.1.9) must be chosen since $t \in W_g(t)$. Thus we obtain

$$v(\underline{r}, t) = \int_{W_g(t)} dt' \Pi_u(t, t') d(\underline{r}, t') + \int_{W_l(t)} dt' \Pi_v(t, t') v(\underline{r}, t'), \tag{5.2.3}$$

where $d(\underline{r}, t')$ is the known form of $u_3(\underline{r}, t')$, $\underline{r} \in B_u(t')$. The other pseudo-displacements $v_1(\underline{r}, t)$ and $v_2(\underline{r}, t)$ are known on $B_u^{(1)}$ and $B_u^{(2)}$ since these are time-independent. Let us define

$$v_i^{(1)}(\underline{r}, t) = v_i(\underline{r}, t) - \int_{W_l(t)} dt' \Pi_v(t, t') v_i(\underline{r}, t'), \quad i = 1, 2, 3. \tag{5.2.4}$$

The right-hand side is a linear operator acting only on the time coordinate of $v_i(\underline{r}, t)$. Thus the $v^{(1)}(\underline{r}, t)$ obey the equations in (3.2.2), and if substituted into the constitutive equations, give, instead of $\sigma_{ij}(\underline{r}, t)$, the quantities

$$\sigma_{ij}^{(1)}(\underline{r}, t) = \sigma_{ij}(\underline{r}, t) - \int_{W_l(t)} dt' \Pi_v(t, t') \sigma_{ij}(\underline{r}, t'). \tag{5.2.5}$$

All three of the quantities $v_i^{(1)}(\underline{r}, t)$ are known on their respective $B_u^{(i)}$, including $v_3^{(1)}(\underline{r}, t)$ on $B_u(t)$ by virtue of (5.2.3). They are known if $W_g(t)$ is known. The transition times determining this set and its complement are in general determined by extra constraints, which can be ignored for purpose of the present discussion.

But what about the stress boundary conditions? In terms of the $\sigma_{ij}^{(1)}$, these take the form

$$\sum_{j=1}^{3} \sigma_{ij}^{(1)}(\underline{r}, t) n_j(\underline{r}) = c_i(\underline{r}, t) - \int_{W_l(t)} dt' \Pi_v(t, t') c_i(\underline{r}, t'), \quad \underline{r} \in B_\sigma^{(i)}(t). \tag{5.2.6}$$

For $i = 1, 2$ the quantities $c_i(\underline{r}, t')$ are known for all $t' \leq t$, since the boundary regions are constant. Thus for these coordinate directions, the boundary conditions have elastic form. However, this is also true for the z direction since by definition of $W_l(t)$, if $\underline{r} \in B_\sigma(t)$ then $\underline{r} \in B_\sigma(t')$ for $t' \in W_l(t)$.

Therefore, the problem, rephrased in terms of $v_i^{(1)}(\underline{r}, t)$ and $\sigma_{ij}^{(1)}(\underline{r}, t)$, has the form of an elastic problem. The quantity

$$v_e(\underline{r}, t) = \int_{W_g(t)} dt' \Pi_u(t, t') d(\underline{r}, t'), \quad \underline{r} \in B_u(t), \tag{5.2.7}$$

plays the part of the known elastic displacement in the z direction.

There is however the further stage of determining the stresses and pseudo- displacements from $\sigma_{ij}^{(1)}$ and $v_i^{(1)}$. For simple histories of time-dependence of the boundary regions, this can be straight forward, but for more complicated histories, involves solving an integral equation in the time variable.

If $B_u(t)$ is expanding at time t, then we decompose $u(\underline{r}, t)$ in the boundary condition (2.2.5) according to (4.1.9). In this case we define

$$u_i^{(1)}(\underline{r}, t) = \int_{W_l(t)} dt' \Gamma_v(t, t') v_i(\underline{r}, t'),$$

$$\mathcal{Q}_{ij}^{(1)}(\underline{r}, t) = \int_{W_l(t)} dt' \Gamma_v(t, t') \sigma_{ij}(\underline{r}, t'). \tag{5.2.8}$$

It can be shown as before that these quantities obey the dynamical equations and boundary conditions of an elastic boundary value problem. Also

$$u_e(\underline{r}, t) = d(\underline{r}, t) - \int_{W_g(t)} dt' \Gamma_u(t, t') d(\underline{r}, t'), \quad \underline{r} \in B_u(t), \tag{5.2.9}$$

plays the part of the known elastic displacement in the z direction.

In two of the examples discussed later (chapters 8, 9) where repetitive expansion and contraction prevails, it is of a degenerate form that deserves special mention. This is where the time-dependence of $B_u(t)$ and $B_\sigma(t)$ consists of abrupt changes at discrete times. The cases of interest can be characterized as follows. The region $B_u(t)$ is either B_G or B_S, $B_S \subset B_G$ and transitions between these states is instantaneous. Let

$$\begin{aligned} B_u(t') &= B_G, \quad t' \in W_G(t), \\ &= B_S, \quad t' \in W_S(t), \\ t' \in (-\infty, t] &= W_G \bigcup W_S, \end{aligned} \tag{5.2.10}$$

where the dependence of W_G and W_S on t consists of discontinuous jumps at discrete times.

If $B_u(t) = B_G$ then $t \in W_G(t)$ and we use the decomposition (5.2.3) but with $W_l(t)$, $W_g(t)$ replaced by $W_S(t)$, $W_G(t)$. The argument from (5.2.4) to (5.2.7) goes through unchanged. If $B_u(t) = B_S$ then the decomposition (4.1.9) must be used, giving (5.2.8) as the quantities obeying the equations of an elastic boundary value problem.

In specific applications, it is convenient to define the quantities $v_i^{(1)}(\underline{r}, t)$, $\sigma_{ij}^{(1)}(\underline{r}, t)$ and so on, as above, but with an extra time-dependent factor, which does not change the argument as given. We shall see examples of this later.

The general derivation, based on the fundamental equations, presented in this section, has not been given before. Previous treatments start from displacement-traction relations [14] (see also chapter 7) or are derived starting from the viscoelastic Kolosov-Muskhelishvili relations in the context of relatively simple problems [14] (see also chapters 8, 9).

We see that in the general case discussed in section 5.1, it is necessary to trace the history of each point on the boundary as it moves between $B_u(t)$ and $B_\sigma(t)$. We term solution methods based on the integral equations (5.1.3) or (5.1.4) point-based methods.

If on the other hand, the special time behaviour of $B_u(t)$ and $B_\sigma(t)$ described in this section prevails, then we need not consider the history of each boundary point, but rather the behaviour of $B_u(t')$, $t' \leq t$ as it changes from containing $B_u(t)$ to being contained by it. Thus, the object of attention is $B_u(t')$ itself. Solution methods in this special case will be termed area-based methods.

A feature of viscoelastic boundary value problems, as we shall see, is that the determination of the boundary regions is part of the problem. The loading is generally specified and this, together with the geometry of the problem, determines the qualitative nature of the boundary regions. The quantitative aspects are usually only determined as part of the solution process.

Chapter 6

ADAPTION OF THE VISCOELASTIC KOLOSOV-MUSKHELISHVILI EQUATIONS FOR CONTACT AND CRACK PROBLEMS

In this chapter, we adapt equations (2.3.4) to forms suitable for solving contact problems on a viscoelastic half-plane and crack problems in an infinite medium. It must be assumed of course that plane strain conditions prevail. This will be so if the indentors on the half-plane and the cracks are uniformly infinite in the z−direction, and uniformly loaded with respect to that variable.

6.1 Equations for the Half-Plane

Our object is to apply equations (2.3.4) to problems involving loads on viscoelastic half-spaces. It is desirable to re-express these equations in an alternative form [Muskhelishvili (1963)] which facilitates reduction of such problems to a Hilbert problem. Another approach is that of Galin (1961) which leads to problems of the Riemann-Hilbert type. These however are somewhat more difficult to deal with, from a mathematical point of view.

Let the material occupy the upper half-plane $y > 0$ so that $\varphi(z,t)$ and $\psi(z,t)$ are analytic in this region. We extend the analyticity of $\varphi(z,t)$ to the lower half-plane according to

$$\varphi(z,t) = -\bar{\varphi}(z,t) - z\bar{\varphi}'(z,t) - \bar{\psi}(z,t), \quad \text{Im}(z) < 0, \tag{6.1.1}$$

recalling that if $f(z)$ is analytic for $\mathrm{Im}(z) > 0$ then $\bar{f}(z)$ is analytic for $\mathrm{Im}(z) < 0$. This form is chosen because, as we shall see below, it leads to discontinuities in $\varphi(z,t)$ on the x-axis wherever the boundary is loaded. Equation (6.1.1) enables us to eliminate $\psi(z,t)$ from the equations since for $\mathrm{Im}(z) > 0$, it may be rewritten as

$$\bar{\psi}(\bar{z},t) = -\varphi(\bar{z},t) - \bar{\varphi}(\bar{z},t) - \bar{z}\bar{\varphi}'(\bar{z},t), \tag{6.1.2}$$

which may be substituted into (2.3.4) to give for $\mathrm{Im}(z) > 0$

$$\begin{aligned} (\sigma_{11}+\sigma_{22})(\underline{r},t) &= 2[\varphi(z,t)+\bar{\varphi}(\bar{z},t)] = 4\mathrm{Re}[\varphi(z,t)], \\ \sum(\underline{r},t) &= (\sigma_{22}-i\sigma_{12})(\underline{r},t) = \varphi(z,t)-\varphi(\bar{z},t)+(z-\bar{z})\bar{\varphi}'(\bar{z},t), \\ 2\int_{-\infty}^{t} dt'\mu(t-t')D'(\underline{r},t') &= \kappa\varphi(z,t)+\varphi(\bar{z},t)+(\bar{z}-z)\bar{\varphi}'(\bar{z},t), \\ \kappa &= 3-4\nu. \end{aligned} \tag{6.1.3}$$

The function $\varphi(\bar{z},t)$ is evaluated in the lower half-plane so that as z approaches the real axis from within the material, $\varphi(z,t)$ and $\varphi(\bar{z},t)$ approach $\varphi^{+}(x,t)$ and $\varphi^{-}(x,t)$ respectively, which are the limits of this complex function from above and below. We see therefore, from the second equation of (6.1.3), that at points on the real axis where the stresses are non-zero, $\varphi(z,t)$ is discontinuous, as noted earlier.

From (2.3.7) it follows that, if there are no stresses or rotations at infinity, and if the stresses on the half-plane fall off as $1/x^2$ or faster at large distances from the origin, then $\varphi(z,t)$ behaves as $1/z$ at large $|z|$.

We will sometimes refer to $\sum(\underline{r},t)$ and $D'(\underline{r},t)$ as the complex stress and the complex displacement derivative.

For simplicity, we continue to restrict the discussion to materials for which the proportionality assumption is a good approximation.

As z approaches the real axis, (6.1.3) gives relations between $\varphi^{\pm}(x,t)$ and the boundary complex stresses and displacement derivatives $\sum(\underline{r},t)$ and $D'(\underline{r},t)$. In the problem that we wish to deal with (Chapter 10) it is the case that $D'(\underline{r},t)$ is not known completely at any point, however. What is known, in the contact region, is the imaginary part of $D'(\underline{r},t)$, which is the normal displacement derivative $u_2'(x,0,t)$. This means that we must take real and imaginary parts of the displacement equation, which introduces another function, namely $\bar{\varphi}(z,t)$. We therefore seek a simple relationship between $\varphi(z,t)$ and $\bar{\varphi}(z,t)$ [Muskhelishvili (1963)], for the case of moving indentors on the boundary, with limiting friction between the indentors and the half-space. We take the motion to be in the negative x direction. This choice yields a minor advantage of symmetry between the signs of space and time quantities at a later stage. The assumption of limiting friction gives that

$$\sigma_{12}(\underline{r},t) = -f\sigma_{22}(\underline{r},t), \tag{6.1.4}$$

for $\underline{r}$ on the x-axis, where f is the coefficient of friction. We therefore deduce from the second equation of (6.1.3) that

$$\begin{aligned}(1+if)\sigma_{22}(x,t) &= \varphi^+(x,t)-\varphi^-(x,t),\\ (1-if)\sigma_{22}(x,t) &= \bar{\varphi}^-(x,t)-\bar{\varphi}^+(x,t).\end{aligned} \tag{6.1.5}$$

It follows that

$$\begin{gathered}\bar{\varphi}^+(x,t)-\alpha\varphi^+(x,t)=\bar{\varphi}^-(x,t)-\alpha\varphi^-(x,t),\\ \alpha=-\frac{1-if}{1+if},\end{gathered} \tag{6.1.6}$$

for all x. Since $\varphi(z,t)$ is analytic in both the upper and lower half-plane, except on parts of the real axis, it follows that $\bar{\varphi}(z,t)$ is also. What (6.1.6) tells us is that the combination $\bar{\varphi}(z,t)-\alpha\varphi(z,t)$ has no discontinuity across the real axis. Since $\varphi(z,t)$ and therefore $\bar{\varphi}(z,t)$ vanish as $1/z$ at infinity, it follows from Liouville's Theorem that

$$\bar{\varphi}(z,t)=\alpha\varphi(z,t), \tag{6.1.7}$$

for all z. This is the desired relationship. Its derivation did not require the proportionality assumption. The complex conjugate of the displacement equation in (6.1.3) therefore reads

$$2\int_{-\infty}^{t} dt'\mu(t-t')\bar{D}'(\underline{r},t)=\alpha\kappa\varphi(\bar{z},t)+\alpha\varphi(z,t)+(z-\bar{z})\varphi'(z,t), \tag{6.1.8}$$

giving that on the boundary

$$2\int_{-\infty}^{t} dt'\mu(t-t')u_2'(x,0,t')=\frac{1}{2i}[(1-\alpha\kappa)\varphi^-(x,t)+(\kappa-\alpha)\varphi^+(x,t)]. \tag{6.1.9}$$

This relation applies to the case where the material is in the upper half-plane since we are taking the surface displacements to be the limits from above.

If there is no boundary friction, then $\alpha=-1$, giving

$$\bar{\varphi}(z,t)=-\varphi(z,t) \tag{6.1.10}$$

and (6.1.9) reduces to

$$\begin{gathered}i\int_{-\infty}^{t} dt'l(t-t')u_2'(x,0,t')=\varphi^+(x,t)+\varphi^-(x,t),\\ l(t)=\frac{4\mu(t)}{1+\kappa}=\frac{\mu(t)}{1-\nu}.\end{gathered} \tag{6.1.11}$$

In fact (6.1.11) can be obtained without the use of the proportionality assumption [14], where $l(t)$ is in general defined by the fact that its Fourier transform is given by $\hat{\mu}(w)/(1-\hat{\nu}(w))$ where $\hat{\nu}(w)$ is given by (2.2.7). This is an example of one of the important special cases where the proportionality assumption is not necessary, which were remarked upon in section 3.2.

The first boundary value problem, where the stresses are specified everywhere on the boundary, and the second boundary value problem where the displacement derivatives are known everywhere on the boundary, are discussed in [14]. We deal only with the mixed boundary value problem here.

The problem under consideration is a series of rigid indentors (finally only one) pressed into a viscoelastic half-space ($y > 0$) and moving across it. If plane strain conditions are to hold, the indentors must be infinitely long in one direction, taken to be the z direction, and of uniform cross-section. Also, the loading distribution must be uniform along each punch. We consider a typical cross-section of this configuration.

Let us denote the contact region by $C(t)$ and its complement on the x-axis by $C'(t)$. The displacement derivatives $u_2'(x,0,t)$ are known on $C(t)$ and the stresses are known to be zero on $C'(t)$. In the frictionless case, the shear stress is known to be zero on $C(t)$ also. This case is an example of the general boundary condition (2.2.5) where $B_u^{(1)}$ is empty while $B_u^{(2)}$, $B_\sigma^{(2)}$ are $C(t)$ and $C'(t)$ respectively. Where limiting friction is present, the shear stress in $C(t)$ is proportional to the normal stress. This case does not fit neatly into the general scheme (2.2.5). However, we shall see that it can be processed satisfactorily. Let

$$\sigma_{22}(x,0,t) = -p(x,t), \tag{6.1.12}$$

so that $p(x,t)$ is the pressure exerted by the indentors on the boundary. Then, for limiting friction, the complex stress on the boundary is given by

$$\sum(x,t) = -(1+if)p(x,t). \tag{6.1.13}$$

From the second relation of (6.1.3) and (6.1.9) we deduce the boundary equations

$$\begin{aligned}
&\varphi^+(x,t) - \varphi^-(x,t) = 0, \qquad x \in C'(t),\\
&\varphi^+(x,t) - \eta\varphi^-(x,t) = iv(x,t), \quad x \in C(t),\\
&v(x,t) = \int_{-\infty}^{t} dt' l(t-t') u_2'(x,0,t),\\
&l(t) = \frac{4\mu(t)}{\kappa - \alpha},\\
&\eta = -\frac{1-\alpha\kappa}{\kappa-\alpha} = \frac{f+ih}{f-ih}, \quad h = \frac{\kappa+1}{\kappa-1} = \frac{2(1-\nu)}{1-2\nu}.
\end{aligned} \tag{6.1.14}$$

It is convenient to put

$$\eta = e^{2\pi i\theta} \quad \text{or} \quad \theta = \frac{\ln\eta}{2\pi i} = \frac{1}{\pi}\tan^{-1}\left(\frac{h}{f}\right). \tag{6.1.15}$$

The quantity θ is in the interval $[0, 1/2]$. In the frictionless case $f = 0$ and $\theta = 1/2$.

Relations (6.1.14) constitute a Hilbert problem [Muskhelishvili (1963)] for $\varphi(z,t)$ if $v(x,t)$ is known. This however is not generally the case for problems with varying boundary regions, which is the source of the added difficulty of such non-inertial viscoelastic problems over elastic problems, or viscoelastic problems with fixed boundary regions. Nevertheless, we proceed for the moment as if $v(x,t)$ were known.

Let $C(t)$ consist of n intervals $[a_i(t), b_i(t)]$, $i = 1, 2, \ldots, n$. We assume that the indentors are smooth so that there are no singular stresses acting on the boundary. Therefore $\varphi(z,t)$ has no singularity on the boundary. Under these circumstances, there is a solution to the Hilbert problem only if $v(x,t)$ obeys certain conditions [Muskhelishvili (1963)] [14]. We have

$$\varphi(z,t) = \frac{X(z,t)}{2\pi} \int_{C(t)} dx' \frac{v(x',t)}{(x'-z)X^+(x',t)}, \tag{6.1.16}$$

$$X(z,t) = \prod_{i=1}^{n} [z - a_i(t)]^{1-\theta} [z - b_i(t)]^{\theta},$$

where $v(x,t)$ is subject to the n constraints

$$\int_{C(t)} dx \frac{x^r v(x,t)}{X^+(x,t)} = 0, \quad r = 0, 1, 2, \ldots, n-1. \tag{6.1.17}$$

The function $X(z,t)$ has branch points at each $a_i(t)$ and $b_i(t)$ and is analytic on the plane cut over $C(t)$. The quantity $X^+(x,t)$ is the limit of $X(z,t)$ on $C(t)$ from the upper half-plane.

We will require an expression for $v(x,t)$ when $x \in C'(t)$, in other words, on the boundary, outside the contact region. There is no discontinuity in $\varphi(z,t)$ at such points and we deduce from (6.1.14) and (6.1.15) that

$$v(x,t) = -2\sin(\pi\theta) e^{i\pi\theta} \varphi(x,t), \quad x \in C'(t), \tag{6.1.18}$$

or, more explicitly, letting $X(x,t)$ denote the unique limits of $X(z,t)$,

$$v(x,t) = \frac{-\sin(\pi\theta) e^{i\pi\theta} X(x,t)}{\pi} \int_{C(t)} dx' \frac{v(x',t)}{(x'-x)X^+(x',t)}, \quad x \in C'(t). \tag{6.1.19}$$

This is essentially the Green's function relation (5.1.2) for case (b) described in section 5.1, with the minor difference that displacement derivative, rather that displacement, is now in use.

The pressure function $p(x,t)$ can be determined form (6.1.5):

$$p(x,t) = q\left[\frac{X^+(x,t)\sin(\pi\theta)}{\pi} \int_{C(t)} dx' \frac{v(x',t)}{(x'-x)X^+(x',t)} + v(x,t)\cos(\pi\theta)\right], \tag{6.1.20}$$

$$q = \frac{-ie^{-i\pi\theta}}{1+if}.$$

The contact intervals $[a_i(t), b_i(t)]$ are usually unknown before the problem is solved, while the overall motion of the punches (say the horizontal position of the point of deepest penetration) is generally specified. In order to determine $a_i(t), b_i(t), i = 1, 2, \ldots, n$, we require $2n$ conditions. Equation (6.1.17) supplies n conditions. The remaining relations can be supplied in a number of ways, the simplest being to specify the loads W_k per unit length on each punch. This gives n conditions of the form

$$W_k = \int_{a_k(t)}^{b_k(t)} dx p(x,t) = \int_{C(t)} dx J_k(x,t) v(x,t) + q\cos(\pi\theta) \int_{a_k(t)}^{b_k(t)} dx v(x,t), \quad (6.1.21)$$

from (6.1.20), where

$$J_k(x,t) = q\left(\frac{\sin(\pi\theta)}{\pi X^+(x,t)} \int_{a_k(t)}^{b_k(t)} dx' \frac{X^+(x',t)}{x-x'}\right). \quad (6.1.22)$$

It is clear from these relationships that once the quantity $v(x,t)$, $x \in C(t)$, and the extent of $C(t)$ itself are known, all other quantities of interest can be determined easily.

6.1.1 Single Contact Interval

We will now discuss the case where there is only one indentor and contact region, which will be denoted by $[a(t), b(t)]$, the subscripts being now omitted. The quantity n is equal to unity, and $X(z,t)$ has the form

$$X(z,t) = [z - a(t)]^{1-\theta}[z - b(t)]^{\theta}, \quad (6.1.23)$$

or more specifically, the branch of this function that tends to z at large $|z|$. One deduces that, on the real axis, outside the contact interval:

$$X(x,t) = \begin{cases} n(x,t), & x > b(t), \\ -n(x,t), & x < a(t), \end{cases} \quad (6.1.24)$$

$$n(x,t) = |x - a(t)|^{1-\theta}|x - b(t)|^{\theta}.$$

Also, on the contact interval:

$$X^+(x,t) = m(x,t)e^{i\pi\theta},$$

$$m(x,t) = [b(t) - x]^{\theta}[x - a(t)]^{1-\theta}. \quad (6.1.25)$$

Equation (6.1.19) gives

$$v(x,t) = \mp \sin(\pi\theta)\frac{n(x,t)}{\pi}\int_{a(t)}^{b(t)} dx' \frac{v(x',t)}{(x'-x)m(x',t)}, \quad x \notin [a(t),b(t)], \tag{6.1.26}$$

the upper sign referring to $x > b(t)$, the lower sign to $x < a(t)$. Similarly, (6.1.20) gives

$$p(x,t) = q\left[\frac{\sin(\pi\theta)m(x,t)}{\pi}\int_{a(t)}^{b(t)} dx' \frac{v(x',t)}{(x'-x)m(x',t)} + v(x,t)\cos(\pi\theta)\right], \tag{6.1.27}$$

while (6.1.17) reduces to one relationship:

$$\int_{a(t)}^{b(t)} dx \frac{v(x,t)}{m(x,t)} = 0 \tag{6.1.28}$$

as does (6.1.21), giving with the aid of (6.1.22):

$$W = \int_{a(t)}^{b(t)} dx J(x,t)v(x,t) + q\cos(\pi\theta)\int_{a(t)}^{b(t)} dx v(x,t), \tag{6.1.29}$$

where

$$J(x,t) = q\left(\frac{\sin(\pi\theta)}{\pi m(x,t)}\int_{a(t)}^{b(t)} dx' \frac{m(x',t)}{x-x'}\right) = \frac{q}{m(x,t)}[x - a(1-\theta) - b\theta], \tag{6.1.30}$$

with the aid of integrals in [14]. With the help of (6.1.28), condition (6.1.29) becomes

$$W = q\int_{a(t)}^{b(t)} dx \frac{v(x,t)x}{m(x,t)}. \tag{6.1.31}$$

6.1.2 Frictionless Contact

In the frictionless case, $\theta = \frac{1}{2}$, giving

$$X(z,t) = [z - a(t)]^{1/2}[z - b(t)]^{1/2}. \tag{6.1.32}$$

Equation (6.1.26) reduces to

$$v(x,t) = \mp\frac{n(x,t)}{\pi}\int_{a(t)}^{b(t)} dx' \frac{v(x',t)}{(x'-x)m(x',t)}, \quad x \notin [a(t),b(t)],$$

$$n(x,t) = |x - a(t)|^{1/2}|x - b(t)|^{1/2}, \quad m(x,t) = [b(t) - x]^{1/2}[x - a(t)]^{1/2}. \tag{6.1.33}$$

The pressure $p(x,t)$, given by (6.1.27), reduces to

$$p(x,t) = -\frac{m(x,t)}{\pi}\int_{a(t)}^{b(t)} dx' \frac{v(x',t)}{(x'-x)m(x',t)}, \tag{6.1.34}$$

since, q, given by (6.1.20) tends to -1 in this limit. Finally, (6.1.31) reduces to

$$W = -\int_{a(t)}^{b(t)} dx \frac{v(x,t)x}{m(x,t)}. \tag{6.1.35}$$

Recall that in the frictionless case, there is no need to assume proportionality, as noted after (6.1.11).

6.1.3 The General Integral Equations

Let us decompose $v(x,t)$, $x \in C(t)$, as described in section 5.1 with the aim of deriving an integral equation for this quantity that is a particular case of the equation given by (5.1.4). We have

$$v(x,t) = \int_{-\infty}^{t} dt' l(t-t') u'(x,t')$$

$$= \int_{W_u(x,t)} dt' \Pi_u(t,t';x) u'(x,t') + \int_{W_\sigma(x,t)} dt' \Pi_\sigma(t,t';x) v(x,t'), \quad x \in C(t), \qquad (6.1.36)$$

where Π_u, Π_σ are defined by (4.2.2), with $\underline{r}$ replaced by x, and coordinate dependence dropped while $W_u(x,t)$, $W_\sigma(x,t)$ are defined in the previous paragraph, where $B_u^{(2)}(t)$, $B_\sigma^{(2)}(t)$ are replaced by $C(t)$, $C'(t)$.

The quantity $v(x,t')$, occurring in the second integral on the right of (6.1.36), is always evaluated outside of $C(t')$, by definition of $W_\sigma(x,t)$. We substitute (6.1.19) for it, to obtain the integral equation

$$v(x,t) = \int_{W_\sigma(x,t)} dt' \int_{C(t')} dx' K(x,x';t,t') v(x',t') + I(x,t), \quad x \in C(t), \qquad (6.1.37)$$

where

$$K(x,x';t,t') = -\frac{\sin(\pi\theta) e^{i\pi\theta} \Pi_\sigma(t,t';x) X(x,t')}{\pi X^+(x',t')(x'-x)},$$

$$I(x,t) = \int_{W_u(x,t)} dt' \Pi_u(t,t';x) u'(x,t'). \qquad (6.1.38)$$

Once $v(x,t)$ is known over the contact region $C(t)$, the extent of which has also to be determined, everything else can be calculated by elementary means, according to the formulae derived earlier in this section.

It was pointed out, without detailed discussion, in section 5.1 that the kernel of (6.1.37) possesses no non-integrable singularities. We can see this in a somewhat more detailed manner in the present special case. From the definition of Π_σ, one perceives that the only situations where singularities might arise are at times t' equal to one of the transition points $t_j(x)$. At such times x', at one of the endpoints of $C(t')$, could become equal to x. However, the singularity will be integrable because, at such times, the function $X(x,t')$ is zero. There is another source of difficulty, though. If the factor $X^+(x',t')$ in the denominator of $K(x,x';t,t')$ is not cancelled, it causes a singularity in the kernel, which may not be square-integrable, thus casting doubt upon the applicability of Fredholm theory. However, the singularity can be transformed away by a change of variable. An explicit example is discussed in section 10.2.

6.2 Two-dimensional Crack Problems

We consider a viscoelastic material occupying all of space with cracks, uniformly infinite in the z-direction, lying along the x-axis. When we come to discuss specific problems, only the case of a single crack will be considered. The loading is assumed to be independent of the z-coordinate so that plane strain conditions prevail.

When modeling real problems, it is generally appropriate to take the loading as being applied at large distance from the cracks. From a mathematical point of view, it is usually simpler to consider the applied stresses as acting on the crack faces. One can easily modify the original problem, by subtracting the applied stresses from the stress tensor, to obtain a related problem where the stresses are acting on the crack faces. The connection between these two problems is discussed by Golden and Graham (1994a). It will be assumed henceforth that the problem so modified is the one under consideration.

Let us first write down the boundary conditions. All stresses are zero at infinity (once the subtraction has been carried out). Off the crack face, the displacements are continuous everywhere, in particular along the x-axis. Let the region of the x-axis on the crack face be $F(t)$, made up of $O(t)$ and $C(t)$, two disjoint sets, $O(t)$ being the region on which the crack face is open and $C(t)$ being the region on which it is closed. On $O(t)$, we have

$$\begin{gathered} u_2(x,0^+,t) - u_2(x,0^-,t) > 0, \\ \sigma_{22}(x,0^+,t) = \sigma_{22}(x,0^-,t) = -p(x,t), \\ \sigma_{12}(x,0^+,t) = \sigma_{12}(x,0^-,t) = -s(x,t), \\ x \in O(t), \end{gathered} \tag{6.2.1}$$

where $0^\pm$ denotes y approaching zero from above $(+)$ and below $(-)$. The quantities $p(x,t)$, $s(x,t)$ are the specified pressure and shear on the crack face. On $C(t)$, frictional forces between the faces will be neglected. We have therefore

$$\begin{gathered} u_2(x,0^+,t) - u_2(x,0^-,t) = 0, \\ \sigma_{22}(x,0^+,t) = \sigma_{22}(x,0^-,t) < -p(x,t), \\ \sigma_{12}(x,0^+,t) = \sigma_{12}(x,0^-,t) = -s(x,t), \\ x \in C(t). \end{gathered} \tag{6.2.2}$$

In the elastic case, if $p(x,t)$ has the same sign everywhere, it is usually positive. Tensile stresses on the crack face would simply cause instant closure. In the viscoelastic case, closure would not always be instant, and applied stresses which change sign over time are of some interest.

Note that the stresses are continuous everywhere on the x-axis, even across the open crack. It follows from (2.3.4) that

$$\varphi^+(x,t)+\bar{\varphi}^-(x,t)+x\bar{\varphi}'^-(x,t)+\bar{\psi}^-(x,t)=\varphi^-(x,t)+\bar{\varphi}^+(x,t)+x\bar{\varphi}'^+(x,t)+\bar{\psi}^+(x,t), \tag{6.2.3}$$

at every point on the x-axis, where $\varphi^\pm(x,t)$ and $\bar{\psi}^\pm(x,t)$ are the limits of $\varphi(z,t)$ and $\bar{\psi}(z,t)$ from above and below the real axis. We write (6.2.3) as

$$\varphi^+(x,t)-\bar{\varphi}^+(x,t)-x\bar{\varphi}'^+(x,t)-\bar{\psi}^+(x,t)=\varphi^-(x,t)-\bar{\varphi}^-(x,t)-x\bar{\varphi}'^-(x,t)-\bar{\psi}^-(x,t), \tag{6.2.4}$$

so that the function $[\varphi(z,t)-\bar{\varphi}(z,t)-z\bar{\varphi}'(z,t)-\bar{\psi}(z,t)]$ is analytic everywhere. It is also zero at infinity, by virtue of (2.3.8) (since Q is zero in this case). It follows from Liouville's Theorem that it is zero everywhere so that

$$\bar{\psi}(\bar{z},t)=\varphi(\bar{z},t)-\bar{\varphi}(\bar{z},t)-\bar{z}\bar{\varphi}'(\bar{z},t), \tag{6.2.5}$$

which should be compared with (6.1.2). We can therefore write (2.3.4) as

$$\begin{aligned}
(\sigma_{11}+\sigma_{22})(\underset{\sim}{r},t)&=2[\varphi(z,t)+\bar{\varphi}(\bar{z},t)]=4\mathrm{Re}[\varphi(z,t)],\\
\sum(\underset{\sim}{r},t)=(\sigma_{22}-i\sigma_{12})(\underset{\sim}{r},t)&=\varphi(z,t)+\varphi(\bar{z},t)+(z-\bar{z})\bar{\varphi}'(\bar{z},t),\\
2d(\underset{\sim}{r},t)&=[\kappa*\varphi](z,t')-\varphi(\bar{z},t)-(z-\bar{z})\bar{\varphi}'(\bar{z},t),\\
d(\underset{\sim}{r},t)&=\int_{-\infty}^{t}dt'\mu(t-t')D'(\underset{\sim}{r},t')=[\mu*D'](\underset{\sim}{r},t).
\end{aligned} \tag{6.2.6}$$

Note that the proportionality assumption has not been imposed in the displacement equation. We have used the convolution product as defined in (2.1.6). Approaching the x-axis from above and below in the displacement equation gives

$$2[d(x,0^+,t)-d(x,0^-,t)]=[(1+\kappa)*(\varphi^+-\varphi^-)](x,t), \tag{6.2.7}$$

or

$$\frac{1}{2}\int_{-\infty}^{t}dt'l(t-t')\Delta'(x,t')=\varphi^+(x,t)-\varphi^-(x,t), \tag{6.2.8}$$

where

$$\Delta'(x,t)=D'(x,0^+,t)-D'(x,0^-,t), \tag{6.2.9}$$

which is the x derivative of the complex displacement across the gap. Also, $l(t)$ is defined by the Fourier transformed relation

$$\hat{l}(w)=\frac{4\hat{\mu}(w)}{1+\hat{\kappa}(w)}=\frac{\hat{\mu}(w)}{1-\hat{\nu}(w)}, \tag{6.2.10}$$

which should be compared with (6.1.11). In this case it was possible to remove the dependence on $\kappa(t)$ from the right-hand side without invoking proportionality.

If the left-hand side of (6.2.8) is zero then $\varphi(z,t)$ is continuous over the x-axis at that point. This will be true if $\Delta'(x,t)$ is zero for all $t' \leq t$. In particular it is true off the crack face.

The region of the x-axis off the crack face may be denoted by $F'(t)$, the complement of $F(t)$ on the x-axis.

In the problems considered in later chapters, the shear stress on the crack face will in general be taken to be zero, so we will deal exclusively with mode I displacements [Sih and Liebowitz (1968), for example].

Consider initially the case of a series of n cracks along the x-axis that are fully open and either stationary or growing. In this case $C(t)$ is empty and $F(t)$ is expanding or stationary. This is an example of the type of problem covered by the Extended Correspondence Principle discussed in section 3.3, so that we expect to obtain solutions closely related to the corresponding elastic solutions. From the second relation of (6.2.6) we have

$$\sum(x,t) = -p(x,t) + is(x,t) = \varphi^{+}(x,t) + \varphi^{-}(x,t), \quad x \in F(t), \tag{6.2.11}$$

Also, for $x \in F'(t)$, off the crack faces, the left-hand side of (6.2.8) is known to be zero for all t since if $x \in F'(t)$ it follows that $x \in F'(t')$, $t' \leq t$ and the complex displacement $\Delta(x,t')$ is zero for all $t' \leq t$. Therefore, the function $\varphi(z,t)$ is continuous on $F'(t)$. Equations (6.2.8) for $x \in F'(t)$ and (6.2.11) constitute a Hilbert problem [Muskhelishvili (1963), [14], for example] the solution of which is

$$\varphi(z,t) = \frac{X(z,t)}{2\pi i}\int_{F(t)} dx' \frac{\Sigma(x't)}{X^{+}(x',t)(x'-z)} + P(z,t)X(z,t), \tag{6.2.12}$$

where $P(z,t)$ is a polynomial, as yet undetermined, and $X(z,t)$ is given by

$$X(z,t) = \left\{\prod_{i=1}^{n}(z-a_i(t))(z-b_i(t))\right\}^{-1/2}, \tag{6.2.13}$$

$$F(t) = \bigcup_{i=1}^{n}[a_i(t), b_i(t)].$$

We choose the branch of $X(z,t)$ such that $z^n X(z,t) \to 1$ as $|z| \to \infty$. Note that we have allowed singularities at the end points, because it is not clear a priori whether they can be excluded.

We will now focus on the case of a single crack, since this is the only configuration which will be needed later. In this case $n = 1$ and the correct behaviour at infinity is obtained by choosing $P(z,t) = 0$. Note that for $n > 1$, this is not true. Let us put

$$X(z,t) = [(z-a(t))(z-b(t))]^{-1/2}. \tag{6.2.14}$$

On the real axis, for $x \in [a(t), b(t)]$

$$X^+(x,t) = -X^-(x,t) = \frac{1}{im(x,t)}, \tag{6.2.15}$$

$$m(x,t) = [(x - a(t))(b(t) - x)]^{1/2}$$

and for $x \notin [a(t), b(t)]$

$$X(x,t) = \pm\frac{1}{n(x,t)}, \tag{6.2.16}$$

$$n(x,t) = [(x - a(t))(x - b(t))]^{1/2}.$$

It is positive for $x > b(t)$ and negative for $x < a(t)$, as a consequence of choosing the branch of $X(z,t)$ as specified above. We write $\varphi(z,t)$ as

$$\varphi(z,t) = \frac{X(z,t)}{2\pi}\int_{a(t)}^{b(t)} dx'\frac{\sum(x',t)m(x',t)}{x'-z}. \tag{6.2.17}$$

Note that it is singular at the end points. This implies that the stresses will also be singular there.

Consider now the expression for the derivative of the gap. Applying a Plemelj formulae [14] to (6.2.17), we find that

$$\varphi^+(x,t) - \varphi^-(x,t) = \frac{1}{\pi i m(x,t)}\int_{a(t)}^{b(t)} dx'\frac{\sum(x',t)m(x',t)}{x'-x}, \quad x \in [a(t), b(t)], \tag{6.2.18}$$

which is a principal value integral. Using (6.2.8), we obtain

$$\Delta'(x,t) = \frac{2}{\pi i}\int_{t_1(x)}^{t} dt'\frac{k(t-t')}{m(x,t')}\int_{a(t')}^{b(t')} dx'\frac{\sum(x',t')m(x',t')}{x'-x}, \tag{6.2.19}$$

where $k(t)$ is the inverse of $l(t)$ under the convolution product, obeying a relation similar to (2.1.8), and $t_1(x)$ is the time at which x crossed the crack tip. Actually, we could replace $t_1(x)$ by $(-\infty)$ because, before this time $\varphi(z,t)$ is continuous and the space integral must be zero.

Applying the other Plemelj formulae [14], we deduce that the complex stress off the crack face is given by

$$\sum(x,t) = \pm\frac{1}{\pi n(x,t)}\int_{a(t)}^{b(t)} dx'\frac{\sum(x',t)m(x',t)}{x'-x}, \quad x \notin [a(t), b(t)], \tag{6.2.20}$$

the upper sign referring to $x > b(t)$ and the lower to $x < a(t)$. This is the same as in elastic theory. It is easily checked that the predictions of the Extended Correspondence Principle agree with these results.

We can determine the complex stress intensities from (6.2.20):

$$K_1(b) - iK_2(t) = \lim_{x\to b}[2(x-b)]^{1/2}\sum(x,t)$$

$$= -\frac{1}{\pi\sqrt{c(t)}}\int_{a(t)}^{b(t)} dx'\sum(x',t)\left(\frac{x'-a(t)}{b(t)-x'}\right)^{1/2}, \qquad (6.2.21)$$

$$K_1(a) - iK_2(a) = -\frac{1}{\pi\sqrt{c(t)}}\int_{a(t)}^{b(t)} dx'\sum(x',t)\left(\frac{b(t)-x'}{x'-a(t)}\right)^{1/2},$$

$$c(t) = \frac{b(t)-a(t)}{2}.$$

Problem 6.2.1 *If the middle point of the crack face is chosen as the origin and $\sum(x,t)$ is even in x over the crack face, show that these formulae give the same result at both ends, namely, choosing $b(t) = -a(t) = c(t)$*

$$K_1 - iK_2 = -\frac{\sqrt{c(t)}}{\pi}\int_{-c(t)}^{c(t)} dx'\frac{\sum(x',t)}{(c^2(t)-x'^2)^{1/2}}. \qquad (6.2.22)$$

A special case required in Chapter 8 is where $\sum(x,t)$ is independent of x on the crack face. We write it as $\sum(t)$. Then (6.2.19) gives with the aid of an integral to be found in [14]

$$\Delta'(x,t) = -2i\int_{t_1(x)}^{t} dt'\frac{k(t-t')}{m(x,t')}\left(\frac{b(t')+a(t')}{2} - x\right)\sum(t'), \qquad (6.2.23)$$

which can be integrated to give

$$\Delta(x,t) = -2i\int_{t_1(x)}^{t} dt' k(t-t')\sum(t')m(x,t'). \qquad (6.2.24)$$

The complex stress $\sum(x,t)$, given by (6.2.20), becomes (using another integral from the same reference)

$$\sum(x,t) = \sum(t)\left\{1 \pm \frac{\frac{b(t)+a(t)}{2} - x}{n(x,t)}\right\}, \qquad x \notin [a(t), b(t)], \qquad (6.2.25)$$

One deduces that the complex stress intensities are given by

$$K_1(b) - iK_2(b) = K_1(a) - iK_2(a) = -c^{1/2}(t)\sum(t). \qquad (6.2.26)$$

In particular, for purely normal stresses acting on the crack face, $\sum(t) = -p(t)$ and

$$K_1(a) = K_1(b) = c^{1/2}(t)p(t). \tag{6.2.27}$$

We will show in chapter 8 that a stationary crack may be open even when $p(t) < 0$ which does not happen in elastic theory. It follows that, in viscoelastic theory, the stress intensity factor K_1 may be zero or negative.

We shall consider certain fixed length crack problems in detail in Chapter 8. Other results, including viscoelastic extension criteria, are discussed in [14].

The remaining five chapters deal with specific problems. Chapters 7-9 consider examples of area-based methods while chapter 10 and 11 consider examples of point-based methods. While there is a need to keep the algebra as simple as possible in a context such as this, there is also an argument for including recent material, such as the method for summing the steady-state decomposition series introduced in section 4.3. These two somewhat conflicting requirements have been balanced in choosing the problems to be dealt with.

Chapter 7

THE THREE DIMENSIONAL STEADY-STATE INDENTATION PROBLEM FOR A GENERAL VISCOELASTIC MATERIAL

We consider in this chapter the normal time-dependent loading of a smooth indentor on a viscoelastic half-space. This problem has a long history [14] but we consider some recent developments here. Steady-state solutions under periodic loading are obtained for materials with more than one decay time.

Plane strain conditions are not present in this problem so that the laboriously prepared methods developed in chapter 6 are not applicable - though they will be in the remaining chapters. The corresponding plane strain problem could equally well be considered [14] but, from the remarks in section 2.3, we see that the extent of indentation of the indentor is not determinate for that case.

This problem, being in the area-based category, is reducible to elastic form, as shown in section 5.2. We in fact start with the displacement-pressure relation on the surface, reduce it to elastic form and use the well-known Hertzian solution.

The treatment presented here is based on a recent paper by Golden, Graham and Lan (1994).

7.1 General Formulation

Let a rigid indentor with axisymmetric form $S(r)$ be pressed into a lubricated viscoelastic half-space, occupying the region $z > 0$, by a time-dependent load $W(t)$. Here, r and z are the usual cylindrical coordinates. Because of the axisymmetry of the problem, we can assume that contact area $C(t)$ is a circle. Let $a(t)$ be the radius of this circle and $u(r,t)$ the normal surface displacement at position $\underline{r}$ and the time t. Let $p(r,t)$ be the normal pressure on the surface at $\underline{r}$ and t. These two quantities are related by

$$v(r,t) = \frac{1}{2\pi}\int_{C(t)} ds' \frac{p(r',t)}{|\underline{r}' - \underline{r}|}, \quad r \le a(t),$$

$$v(r,t) = \int_{-\infty}^{t} dt' l(t-t') u(r,t'), \tag{7.1.1}$$

where $l(t)$ is defined by (6.2.10). It is zero for $t < 0$. The quantity ds' is the surface element at $\underline{r}'$. The point $\underline{r}$ is understood to be on the surface. Equation (7.1.1) is the viscoelastic generalization of the well-known Boussinesq relation and may be deduced from the elastic form by applying the Classical Correspondence Principle. Note that the proportionality assumption is not used. The relation between $u(\underline{r},t)$ and $v(\underline{r},t)$ may be inverted to give

$$u(r,t) = \int_{-\infty}^{t} dt' k(t-t') v(r,t'), \tag{7.1.2}$$

where $k(t)$ is the inverse of $l(t)$ under the convolution product. They obey a relation similar to (2.1.8).

Let $D(t)$ be the indentation of the tip of the indentor into the half-space. Then the surface displacement $u(r,t)$ in the contact region takes the form

$$u(r,t) = D(t) - S(r), \qquad S(0) = 0. \tag{7.1.3}$$

Thus we take the indentor to be centered on the origin.

The applied load $W(t)$ is related to $p(r,t)$ by

$$W(t) = \int_{C(t)} ds p(r,t). \tag{7.1.4}$$

We consider the case where the applied load is oscillating in magnitude so that the contact area radius $a(t)$ passes through a series of maxima and minima before the current time t. This is an example of repetitive expansion and contraction of the boundary regions as defined in section 5.2. The method developed in that section is applied here. However, we start not with the basic equations but with (7.1.1). This is recast in elastic form and the elastic solutions are used. In fact, we could simply invoke the general results of section 5.2. However, it is interesting to restate the argument in a different way.

First we consider the case where the contact area is shrinking at the current time t. The quantity $v(r,t)$ can be decomposed according to (5.2.3), where $d(\underline{r},t')$ is $u(r,t')$ given by (7.1.3). Also, from (7.1.1)

$$\int_{W_l(t)} dt' \Pi_v(t,t') v(r,t') = \frac{1}{2\pi}\int_{C(t)} ds' \frac{q_c(r',t)}{|\underline{r}' - \underline{r}|}, \quad r \le a(t),$$

$$q_c(r,t) = \int_{W_l(t)} dt' \Pi_v(t,t') p(r,t'). \tag{7.1.5}$$

In other words, the space and time integrals can be interchanged. This is because $W_l(t)$ consists of all those times such that $a(t') < a(t)$, $t' < t$ and $p(r,t')$ vanishes outside of $C(t')$. We conclude that (5.2.3), combined with (7.1.1) and (7.1.3) gives

$$v(r,t) = v_c(r,t) + \frac{1}{2\pi}\int_{C(t)} ds' \frac{q_c(r',t)}{|\underline{r}' - \underline{r}|}, \quad r \leq a(t), \tag{7.1.6}$$

where

$$v_c(r,t) = D_c(t) - S(r)\Pi(t),$$

$$D_c(t) = \int_{W_g(t)} dt' \Pi_u(t,t') D(t'),$$

$$\Pi(t) = \int_{W_g(t)} dt' \Pi_u(t,t'). \tag{7.1.7}$$

Using (7.1.1) again, we rewrite (7.1.6) in the form

$$v_c(r,t) = \frac{1}{2\pi}\int_{C(t)} ds' \frac{p(r',t) - q_c(r',t)}{|\underline{r}' - \underline{r}|}, \quad r \leq a(t). \tag{7.1.8}$$

If we define a quantity $D_e(t)$ such that

$$D_c(t) = \Pi(t) D_e(t), \tag{7.1.9}$$

then

$$v_c(r,t) = \Pi(t)(D_e(t) - S(r)),$$

and (7.1.8) has elastic form where

$$p_e(r,t) = \frac{p(r,t) - q_c(r,t)}{k_e \Pi(t)}, \tag{7.1.10}$$

where k_e is an arbitrarily chosen elastic constant. The quantity $k_e p_e(r,t)$ is known from the well-known solution of the elastic contact problem and we have

$$p(r,t) = q_c(r,t) + \Pi(t) k_e p_e(r,t), \tag{7.1.11}$$

which is essentially an integral equation in time for $p(r,t)$. Equation (7.1.9) is an integral equation for $D(t)$ in terms of the elastic indentation $D_e(t)$. We will choose k_e to be the long time modulus

$$k_e = \frac{1}{l_e} = \int_0^\infty dt k(t). \tag{7.1.12}$$

If $C(t)$ is expanding at time t we decompose $u(r,t)$ according to (4.1.9) and go through similar manipulations to obtain

$$\int_{W_l(t)} dt'\Gamma_v(t,t')p(r,t') = k_e(1-\Gamma(t))p_e(r,t),$$

$$D(t) = \int_{W_g(t)} dt'\Gamma_u(t,t')D(t') + D_e(t)(1-\Gamma(t)),$$

$$\Gamma(t) = \int_{W_g(t)} dt'\Gamma_u(t,t'), \tag{7.1.13}$$

which are integral equations for the pressure function and indentation in terms of the elastic pressure and indentation.

It is interesting to compare (7.1.10) with (5.2.5) and (7.1.7), (7.1.9) with (5.2.7); also (7.1.13) with (5.2.8) and (5.2.9). The time-dependent multiplying factors mentioned after (5.2.9) are easily identified.

Integrating the pressure equations over the contact area gives

$$W(t) = \int_{C(t)} ds q_c(r,t) + k_e\Pi(t)W_e(t), \quad (C(t) \text{ contracting}),$$

$$\int_{W_l(t)} dt'\Gamma_v(t,t')W(t') = k_e(1-\Gamma(t))W_e(t), \quad (C(t) \text{ expanding}),$$

$$W_e(t) = \int_{C(t)} ds p_e(r,t). \tag{7.1.14}$$

For a spherical indentor of large radius R

$$S(r) = \frac{r^2}{2R}, \tag{7.1.15}$$

and the elastic solution is given by

$$p_e(r,t) = \frac{4}{\pi R k_e}(a^2(t) - r^2)^{1/2},$$

$$W_e(t) = \frac{8a^3(t)}{3k_e R},$$

$$D_e(t) = \frac{a^2(t)}{R}. \tag{7.1.16}$$

These results can be used to deduce solutions for simple time histories, in particular for monotonically changing $C(t)$ (which may anyway be determined by applying the Extended Correspondence Principle) and a history with one maximum. In the latter case, the results of Hunter (1960) and Graham (1965) are reproduced. This solution forms the basis of the analysis of an impact problem [Hunter (1960), [14]]. We will focus here on the steady-state problem under an oscillating load, using the results of section 4.3.

Three cases are considered in this context by Golden and Graham (1990), Golden, Graham and Lan (1994): the stress-controlled case where $W(t)$ is specified, the strain-controlled case where $D(t)$ is specified and, of less physical interest, the area controlled case where $C(t)$ is specified. We will consider only the stress-controlled case here.

7.2 Steady-State Limit

Let $W(t)$ be a known, periodic function of t with period Δ. It is assumed that sufficient time has passed so that steady-state conditions have been established. This means in particular that all physical quatities are periodic functions with period Δ. We consider one particular period $[\Delta_1, \Delta_2]$ such that $\Delta = \Delta_2 - \Delta_1$. At times Δ_1 and Δ_2, the contact region $C(t)$ is maximum. We assume that the behaviour of $W(t)$ is such that there is only one time in $[\Delta_1, \Delta_2]$ denoted t_0 at which $C(t)$ is minimum. The transition times used in the decomposition of hereditary integrals in the present context are the times of transition between $W_l(t)$ and $W_g(t)$. We will denote them by $\theta_l(t)$, $l = 1, 2, \ldots$. They are defined as solutions to the equation

$$a(\theta_l(t)) = a(t), \qquad \theta_l(t) \leq t. \tag{7.2.1}$$

For the contracting phase $t \in [\Delta_1, t_0]$, we define $t_1(t)$ to be the solution of the equation $a(t_1(t)) = a(t)$ in $[t_0, \Delta_2]$. The transition times can be given in terms of this quantity (recall (4.3.2)):

$$\begin{aligned} \theta_1(t) &= t_1(t) - \Delta, \\ \theta_2(t) &= t - \Delta, \\ \theta_3(t) &= t_1(t) - 2\Delta, \quad \text{and so on.} \end{aligned} \tag{7.2.2}$$

Following the argument leading to (4.3.3), we can write (7.1.9), (7.1.11) and the first equation of (7.1.14) as

$$\int_{t_1(t)-\Delta}^{t} dt' \Pi_u^{(p)}(t,t') D(t') = \Pi(t) D_e(t), \tag{7.2.3}$$

and

$$\begin{aligned} p(r,t) &= \int_t^{t_1(t)} dt' \Pi_v^{(p)}(t,t') p(r,t') + k_e \Pi(t) p_e(r,t), \\ W(t) &= \int_t^{t_1(t)} dt' \Pi_v^{(p)}(t,t') W(t') + k_e \Pi(t) W_e(t), \end{aligned} \tag{7.2.4}$$

where

$$\Pi_u^{(p)}(t,t') = T_0(t,t') + \Pi_0(t,t') = l(t-t') + \Pi_0(t,t'),$$

$$\Pi_v^{(p)}(t,t') = \Pi_1(t,t'-\Delta),$$
$$\Pi(t) = \int_{t_1(t)-\Delta}^{t} dt' \Pi_u^{(p)}(t,t'), \tag{7.2.5}$$

in term of $\Pi_0(t,t')$, $\Pi_1(t,t')$ defined by (4.3.4). From (4.3.20), it follows that $\Pi_u^{(p)}(t,t')$ obeys the integral equation

$$\Pi_u^{(p)}(t,t') = S(t,t') + \int_{t-\Delta}^{t_1(t)-\Delta} du \Pi_u^{(p)}(t,u) K'(u,t'),$$

$$S(t,t') = \sum_{n=0}^{\infty} l(t-t'+n\Delta), \tag{7.2.6}$$

where $K'(u,t')$ is defined by (4.3.20). Similarly $\Pi_v^{(p)}(t,t')$ obeys

$$\Pi_v^{(p)}(t,t') = K(t,t'-\Delta) + \int_{t_1(t)-\Delta}^{t} du \Pi_v^{(p)}(t,u) K(u,t'-\Delta) \tag{7.2.7}$$

by virtue of (4.3.18) and a shift in the range of integration.

For the expanding phase, i.e. $t \in [t_0, \Delta_2]$ we define $t_1(t) \in [\Delta_1, t_0]$ such that $a(t_1(t)) = a(t)$. In this case the transition times are given by

$$\theta_1(t) = t_1(t),$$
$$\theta_2(t) = t - \Delta,$$
$$\theta_3(t) = t_1(t) - \Delta, \quad \text{and so on.} \tag{7.2.8}$$

From (7.1.13) and the second equation of (7.1.14) we obtain

$$\int_{t_1(t)}^{t} dt' \Gamma_v^{(p)}(t,t') p(r,t') = k_e(1-\Gamma(t)) p_e(r,t),$$

$$\int_{t_1(t)}^{t} dt' \Gamma_v^{(p)}(t,t') W(t') = k_e(1-\Gamma(t)) W_e(t), \tag{7.2.9}$$

and

$$D(t) = \int_{t-\Delta}^{t_1(t)} dt' \Gamma_u^{(p)}(t,t') D(t') + D_e(t)(1-\Gamma(t)), \tag{7.2.10}$$

where

$$\Gamma_u^{(p)}(t,t') = \Gamma_1(t,t'),$$
$$\Gamma_v^{(p)}(t,t') = k(t-t') + \Gamma_0(t,t'),$$
$$\Gamma(t) = \int_{t-\Delta}^{t_1(t)} dt' \Gamma_u^{(p)}(t,t'), \tag{7.2.11}$$

in terms of $\Gamma_0(t,t')$ and $\Gamma_1(t,t')$ defined by (4.3.6). From (4.3.22) it follows that $\Gamma_u^{(p)}(t,t')$ obeys the integral equation

$$\Gamma_u^{(p)}(t,t') = L(t,t') + \int_{t_1(t)-\Delta}^{t-\Delta} du \Gamma_u^{(p)}(t,u) L(u+\Delta,t'), \tag{7.2.12}$$

where $L(t,t')$ is defined by (4.3.22). Also, from (4.3.23) $\Gamma_v^{(p)}(t,t')$ obeys the integral equation

$$\Gamma_v^{(p)}(t,t') = S'(t,t') + \int_{t-\Delta}^{t_1(t)} du \Gamma_v^{(p)}(t,u) L'(u,t'),$$

$$S'(t,t') = \sum_{n=0}^{\infty} k(t-t'+n\Delta), \tag{7.2.13}$$

where $L'(u,t')$ is defined by (4.3.23).

7.3 Summation of Series for Discrete Spectrum Models

If the integral equations derived in the last section can be solved in closed form, rather than by iteration for example, then we have effectively carried out the infinite summations involved in the quantities $\Pi_u^{(p)}$, $\Pi_v^{(p)}$, $\Gamma_u^{(p)}$, $\Gamma_v^{(p)}$. We shall show in this section that this is possible for discrete spectrum materials introduced in subsection (2.1.4). The method does not in general work for continuous spectra because it simply leads to another integral equation.

The proportionality assumption is adopted so that $l(t)$ is given by (6.1.11) and $k(t) = (1-\nu)\gamma(t)$. Utilizing (2.1.39), (2.1.43) and (2.1.44), we write

$$l(t) = l_0\delta(t) + \sum_{i=1}^{N} l_i e^{-\alpha_i t},$$

$$k(t) = k_0\delta(t) + \sum_{i=1}^{N} k_i e^{-\beta_i t}, \tag{7.3.1}$$

where the parameters on the right are related by

$$l_0 k_0 = 1;$$

$$l_0 + \sum_{i=1}^{N} \frac{l_i}{\alpha_i - \beta_j} = 0, \quad j = 1,2,\ldots,N;$$

$$k_0 + \sum_{j=1}^{N} \frac{k_j}{\beta_j - \alpha_i} = 0, \quad i = 1,2,\ldots,N; \tag{7.3.2}$$

$$k_i = -\{\sum_{j=1}^{N} \frac{l_j}{(\alpha_j - \beta_i)^2}\}^{-1}, \quad i = 1,2,\ldots,N;$$

$$l_i = \frac{g_i}{1-\nu}, \quad i = 0,1,2,\ldots,N;$$

$$k_i = (1-\nu)h_i, \quad i = 0,1,2,\ldots,N.$$

The quantities l_i, $i = 1,2,\ldots,N$ are negative and k_i, $i = 1,2,\ldots,N$ positive. It can be shown that $G(t,t')$ defined by (4.3.13) has the following form for discrete spectrum models:

$$G(t'',t') = \sum_{i,j=1}^{N} \frac{l_i k_j}{\alpha_i - \beta_j} e^{-\alpha_i(t''-\theta_1)-\beta_j(\theta_1-t')}, \quad t'' > \theta_1(t). \tag{7.3.3}$$

Then $K(t, t'-\Delta)$ in (7.2.7) defined by (4.3.19) can be summed to give

$$K(u, t'-\Delta) = \sum_{i,j=1}^{N} K_{ij} e^{-\alpha_i(u-\theta_1)-\beta_j(\theta_1-t')},$$

where

$$K_{ij} = \frac{l_i k_j}{\alpha_i - \beta_j} \frac{e^{-\beta_j \Delta}}{1 - e^{-\beta_j \Delta}}. \tag{7.3.4}$$

To solve (7.2.7), we make the ansatz for $\Pi_v{}^{(p)}(t,t')$ of the form

$$\Pi_v{}^{(p)}(t,t') = \sum_{i,j=1}^{N} P_{ij}(t) e^{\beta_j t'}, \tag{7.3.5}$$

which clearly obeys (4.3.17). Substitution into (7.2.7) gives

$$\begin{aligned} \sum_{i=1}^{N} P_{ij} &= \sum_{i=1}^{N} K_{ij} e^{-\alpha_i(t-\theta_1)-\beta_j\theta_1} \\ &+ \sum_{i,m,n=1}^{N} \frac{P_{im}K_{nj}}{\beta_m - \alpha_n} \{e^{\beta_m t - \alpha_n(t-\theta_1)-\beta_j\theta_1} - e^{(\beta_m-\beta_j)\theta_1}\}. \end{aligned} \tag{7.3.6}$$

This algebraic equation will certainly be satisfied if a stronger condition is imposed that cancellation takes place term by term in the variable i, a condition which gives the matrix equation

$$P = K_1 + PAK_2, \tag{7.3.7}$$

where P is a square matrix formed by P_{ij} while

$$\begin{aligned} (K_1)_{ij} &= K_{ij} e^{-\alpha_i(t-\theta_1)-\beta_j\theta_1}, \\ (K_2)_{nj} &= K_{nj} e^{-\beta_j\theta_1}, \\ A_{mn} &= \frac{e^{-\alpha_n(t-\theta_1)+\beta_m t} - e^{\beta_m\theta_1}}{\beta_m - \alpha_n}. \end{aligned} \tag{7.3.8}$$

The formal solution of (7.3.7) is

$$P = K_1(I - AK_2)^{-1}. \tag{7.3.9}$$

Using a similar approach on (7.2.6), one obtains

$$\Pi_u{}^{(p)}(t,t') = l_0\delta(t-t') + \sum_{i,j=1}^{N} Q_{ij}(t)e^{\alpha_j t'}, \tag{7.3.10}$$

where $Q_{ij}(t)$ are the components of a square matrix Q given by

$$Q = L_1(I - BL_2)^{-1}, \tag{7.3.11}$$

with

$$\begin{aligned}(L_1)_{ij} &= \frac{\delta_{ij} l_j e^{-\alpha_j t}}{1 - e^{-\alpha_j \Delta}},\\ (L_2)_{ij} &= \frac{k_i l_j e^{-\alpha_j \theta_2}}{\beta_i - \alpha_j}\frac{e^{-\alpha_j \Delta}}{1 - e^{-\alpha_j \Delta}},\end{aligned} \tag{7.3.12}$$

and

$$B_{ij} = \frac{e^{(\alpha_i - \beta_j)\theta_1 + \beta_j \theta_2} - e^{\alpha_i \theta_2}}{\alpha_i - \beta_j}. \tag{7.3.13}$$

By comparing the definition of $\Pi_u{}^{(p)}(t,t')$ and $\Gamma_v{}^{(p)}(t,t')$, we find that $\Gamma_v{}^{(p)}(t,t')$ may be obtained from $\Pi_u{}^{(p)}(t,t')$ by interchanging the roles of $l(t)$ and $k(t)$. Furthermore we see that $\Gamma_u{}^{(p)}(t,t'-\Delta)$ is the same as $\Pi_v{}^{(p)}(t,t')$ if the roles of $l(t)$ and $k(t)$ are interchanged. A similar observation allowed us to write (4.3.22) and (4.3.23). Therefore, for time $t \in [t_0, \Delta_2]$, we have

$$\Gamma_v{}^{(p)}(t,t') = k_0\delta(t-t') + \sum_{i,j=1}^{N} \hat{Q}_{ij}(t)e^{\beta_j t'}, \tag{7.3.14}$$

and

$$\Gamma_u{}^{(p)}(t,t') = \sum_{i,j=1}^{N} \hat{P}_{ij}(t)e^{\alpha_j (t'+\Delta)}, \tag{7.3.15}$$

where $\hat{P}$, $\hat{Q}$, are obtained from (7.3.9) and (7.3.11), respectively, by interchanging the role of $l(t)$ and $k(t)$.

Putting expression (7.3.10) into the third relation of (7.2.5) and (7.3.15) into the third relation of (7.2.11) gives

$$\Pi(t) = l_0 + \sum_{i,j=1}^{N} Q_{ij}(t)\frac{e^{\alpha_j t} - e^{\alpha_j \theta_1}}{\alpha_j}, \tag{7.3.16}$$

and

$$\Gamma(t) = \sum_{i,j=1}^{N} \hat{P}_{ij}(t) \frac{e^{\alpha_j(\theta_1+\Delta)} - e^{\alpha_j(\theta_2+\Delta)}}{\alpha_j}. \tag{7.3.17}$$

Substituting all the kernels into the periodic solution equations (7.2.3), (7.2.4) and (7.2.9), (7.2.10) finally gives

$$p(r,t) = \sum_{i,j=1}^{N} P_{ij}(t) \int_{t}^{t_1(t)} dt' e^{\beta_j t'} p(r,t') + k_e\{l_0 + \sum_{i,j=1}^{N} Q_{ij}(t) \frac{e^{\alpha_j t} - e^{\alpha_j(t_1-\Delta)}}{\alpha_j}\} p_e(r,t), \quad for\ t \in [\Delta_1, t_0], \tag{7.3.18}$$

$$k_0 p(r,t) + \sum_{i,j=1}^{N} \hat{Q}_{ij}(t) \int_{t_1(t)}^{t} dt' e^{\beta_j t'} p(r,t') = k_e\{1 - \sum_{i,j=1}^{N} \hat{P}_{ij}(t) \frac{e^{\alpha_j(\Delta+t_1)} - e^{\alpha_j t}}{\alpha_j}\} p_e(r,t), \quad for\ t \in [t_0, \Delta_2], \tag{7.3.19}$$

$$W(t) = \sum_{i,j=1}^{N} P_{ij}(t) \int_{t}^{t_1(t)} dt' e^{\beta_j t'} W(t') + k_e\{l_0 + \sum_{i,j=1}^{N} Q_{ij}(t) \frac{e^{\alpha_j t} - e^{\alpha_j(t_1-\Delta)}}{\alpha_j}\} W_e(t), \quad for\ t \in [\Delta_1, t_0], \tag{7.3.20}$$

$$k_0 W(t) + \sum_{i,j=1}^{N} \hat{Q}_{ij}(t) \int_{t_1(t)}^{t} dt' e^{\beta_j t'} W(t') = k_e\{1 - \sum_{i,j=1}^{N} \hat{P}_{ij}(t) \frac{e^{\alpha_j(\Delta+t_1)} - e^{\alpha_j t}}{\alpha_j}\} W_e(t), \quad for\ t \in [t_0, \Delta_2], \tag{7.3.21}$$

$$l_0 D(t) + \sum_{i,j=1}^{N} Q_{ij}(t) \int_{t_1-\Delta}^{t} dt' e^{\alpha_j t'} D(t') = \{l_0 + \sum_{i,j=1}^{N} Q_{ij}(t) \frac{e^{\alpha_j t} - e^{\alpha_j(t_1-\Delta)}}{\alpha_j}\} D_e(t), \quad for\ t \in [\Delta_1, t_0], \tag{7.3.22}$$

$$D(t) = \sum_{i,j=1}^{N} \hat{P}_{ij}(t) e^{\alpha_j \Delta} \int_{t-\Delta}^{t_1} dt' e^{\alpha_j t'} D(t') + \{1 - \sum_{i,j=1}^{N} \hat{P}_{ij}(t) \frac{e^{\alpha_j(t_1+\Delta)} - e^{\alpha_j t}}{\alpha_j}\} D_e(t), \quad for\ t \in [t_0, \Delta_2]. \tag{7.3.23}$$

It can be shown [Golden, Graham and Lan (1994)] that $p(r,t)$, $W(t)$, $D(t)$ satisfying the above equations are continuous at time t_0, and equal at Δ_1 and Δ_2.

Also, one can show that kernels for the case $N = 1$ are the same as the ones obtained by Golden and Graham (1987b), [14] and Graham and Golden (1988a) by a direct summation.

7.4 Numerical Results

In this section, numerical solutions to integral equations (7.3.18)-(7.3.23) by the quadrature method are given for the stress controlled mode.

We consider the spherical indentor of large radius R, for which the elastic solutions are given by (7.1.16). All the numerical calculations are carried out for the case $N = 2$. The dimensionless quantities $ca(t)$, $cD(t)$, $c^2 k_0 W(t)$, $k_0 p(r,t)$ and the dimensionless parameters $t' = \omega t$, $\beta_i' = \beta_i/\omega$, $k_i' = k_i/(\omega k_0)$, $\alpha_i' = \alpha_i/\omega$, $l_i' = l_i k_0/\omega$, $i = 1, 2$, and $k = k_e/k_0$ are used, where $c = 1/(2R)$. This method can be applied to the cases when $N > 2$ without any difficulty. The applied load is assumed to have the simple sinusoidal form

$$W(t) = K(d - cos(\omega t)),\ d \geq 1, \Delta = \frac{2\pi}{\omega}. \tag{7.4.1}$$

To solve for $a(t), p(r,t), D(t)$, we need to know t_0 and Δ_1, Δ_2. The Generalized Partial Correspondence Principle [Graham and Golden (1988b)] discussed in section 3.4, applied to the normal loading problem for a loading of the form (7.4.1) gives that $t_0 = 0$ and Δ_2, obeys

$$\tan(\omega\Delta_2) = \tan(\varphi(\omega)) = \left\{\sum_{i=1}^{N} \frac{k_i\omega}{\omega^2 + \beta_i^2}\right\} / \left\{k_0 + \sum_{i=1}^{N} \frac{k_i\beta_i}{\omega^2 + \beta_i^2}\right\}, \tag{7.4.2}$$

where $\varphi(\omega)$ is the loss angle and (2.1.21), (2.1.40) together with the last two relations of (7.3.2) have been used. We choose the solution $\omega\Delta_2 = \pi + \varphi(w)$. Also $\Delta_1 = \Delta_2 - \Delta$.

Once t_0 and Δ_1, Δ_2 are known, Newton's iteration method is employed to solve the functional equation, obtained by eliminating $W_e(t) = W_e(t_1(t))$ from (7.3.20) and (7.3.21), for $t_1(t)$. Then $W_e(t)$ can be obtained from one of these equations. If $W_e(t)$ is known, the contact radius $a(t)$, elastic indentation $D_e(t)$ and elastic pressure $p_e(r,t)$ can be found by using the relations (7.1.16). Therefore we can get $D(t)$ and $p(r,t)$. In particular, the indentation $D(t)$ is obtained by solving equation (7.3.22) starting at $t = \Delta_1$ and equation (7.3.23) beginning with $t = \Delta_2$, using the marching method. Considering that all the kernels are smooth functions, we use the trapezoidal rule for the numerical computations. Results are presented in Fig. 7.1-Fig. 7.3. From Fig. 7.1 and Fig. 7.2, we can see that $ca(t)$ and $cD(t)$ increases with the value of k and there is very little variation with β_i and k_i if k is close to 1. Note that, from Fig. 7.1 the minimum contact area occurs at $t = 0$ and is only dependent on the value of k and independent of the individual values of α_i' and l_i' or β_i' and k_i'. Fig. 7.2 also shows that the indentation achieves minimum a little later than the load. Contact area and indentation reach their maximum simultaneously and after the load [Graham and Golden (1988b)].

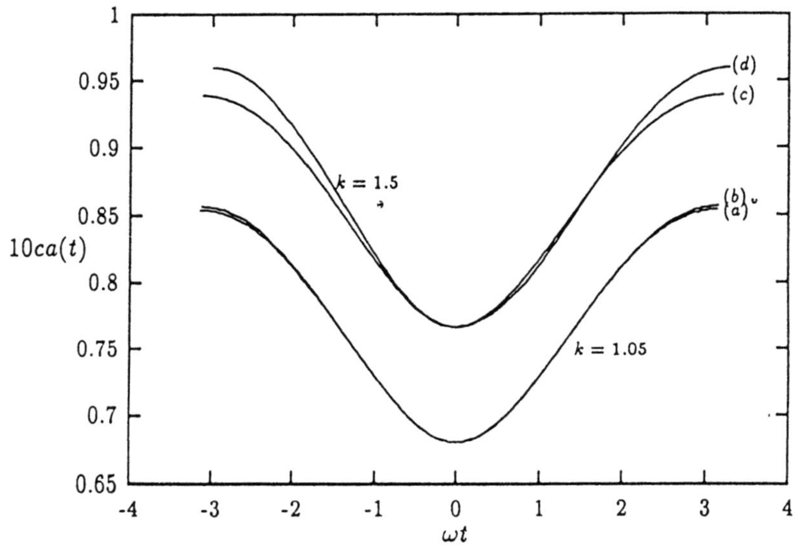

Figure 7.1: Contact area radius for the stress-controlled mode. This gives the dimensionless contact area radius $ca(t)$ over a complete cycle under stress-controlled condition for the following cases (a): $k_1' = 0.0096$, $k_2' = 0.003$, $\beta_1' = 0.2$, $\beta_2' = 1.5$, $k = 1.05$; (b): $k_1' = 0.09$, $k_2' = 0.03$, $\beta_1' = 2$, $\beta_2' = 6$, $k = 1.05$;(c): $k_1' = 1.26$, $k_2' = 0.12$, $\beta_1' = 0.9$, $\beta_2' = 1.2$, $k = 2.5$ *and* (d): $k_1' = 0.52$, $k_2' = 0.12$, $\beta_1' = 0.4$, $\beta_2' = 0.6$, $k = 2.5$. Also $c^2 K k_0 = 0.0008$, $d = 3$.

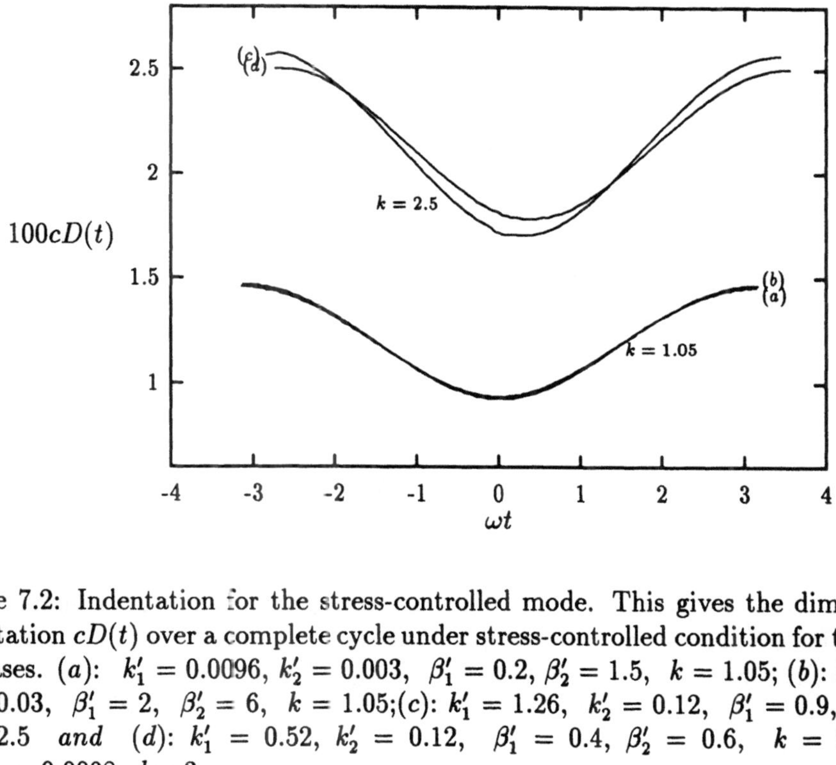

Figure 7.2: Indentation for the stress-controlled mode. This gives the dimensionless indentation $cD(t)$ over a complete cycle under stress-controlled condition for the following cases. (a): $k_1' = 0.0096$, $k_2' = 0.003$, $\beta_1' = 0.2$, $\beta_2' = 1.5$, $k = 1.05$; (b): $k_1' = 0.09$, $k_2' = 0.03$, $\beta_1' = 2$, $\beta_2' = 6$, $k = 1.05$;(c): $k_1' = 1.26$, $k_2' = 0.12$, $\beta_1' = 0.9$, $\beta_2' = 1.2$, $k = 2.5$ *and* (d): $k_1' = 0.52$, $k_2' = 0.12$, $\beta_1' = 0.4$, $\beta_2' = 0.6$, $k = 2.5$. Also $c^2 K k_0 = 0.0008$, $d = 3$.

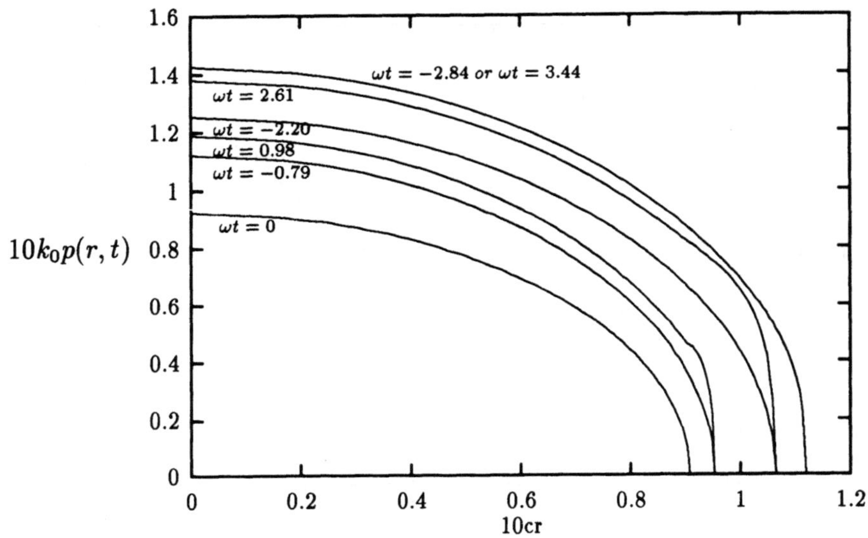

Figure 7.3: Pressure distribution for the stress-controlled mode. This picture gives the dimensionless pressure distribution $p(r,t)$ at various times during the cycle for the stress-controlled mode with $k_1' = 0.52$, $k_2' = 0.12$, $\beta_1' = 0.4$, $\beta_2' = 0.6$, $k = 2.5$ and $c^2Kk_0 = 0.0008$. The lines on the top and bottom are for the times when the contact area is maximum and minimum, respectively. The pairs of lines which meet on the horizontal axis are for times when the contact area radii are same.

Chapter 8

A FIXED LENGTH CRACK IN A SINUSOIDALLY LOADED GENERAL VISCOELASTIC MEDIUM

8.1 Introduction

The problem of a fixed length or stationary crack in a viscoelastic medium where the crack has always been open, is relatively trivial in that it is covered by the Classical Correspondence Principle. This means that its solution is closely related to the corresponding elastic solution. However, if crack closure can occur, the problem is not so trivial. This case has been considered by Graham (1976) and Graham and Sabin (1981) who present a method of solution and detailed numerical results for Maxwell and Voigt materials as well as for the more general standard linear model.

The problem of a fixed length crack under a fluctuating normal load, which causes closure, is considered here using a different, more general approach, which is applicable to materials with viscoelastic response more general than that of a standard linear model [Golden and Graham (1989)]. Specifically, we consider discrete spectrum models and, in the steady-state limit, use the summation method outlined in section 4.3 and in the previous chapter. This is another, rather simpler, example of a problem to which the area based method is applicable. The contact region has only two states - the full crack length or no contact. Because of the trivial nature of this time variation and the fact that it is discontinuous, the general formalism developed in section 5.2 and used (though not directly) in chapter 7, is not necessary. The degenerate case applies here.

8.2 General Relationships

Consider a viscoelastic body occupying all space. Let a uniformly infinite crack lie along the z-axis and let the body be subject to loads independent of z. Thus, plane strain conditions prevail and we focus on the particular cross-section lying in the xy plane. In this plane, let the crack opening lie on the x-axis, occupying the fixed interval $[-c, c]$. Let the body be subjected to a time-dependent loading parallel to the y-axis, which, in the standard manner we transfer to the crack face, for convenience, as noted in section 6.2. When the crack is open, this results in a normal stress $p(t)$ on the crack face, which is assumed to be continuous. Equation (6.2.24) gives that $\Delta(x,t)$ is purely imaginary. We denote it by $ig(x,t)$ where $g(x,t)$ is the normal gap. Then from (6.2.24) we obtain

$$g(x,t) = 2m(x)q(t), \qquad |x| < c, \tag{8.2.1}$$

where

$$q(t) = \int_{-\infty}^{t} dt' k(t-t')p(t'); \quad m(x) = (c^2 - x^2)^{1/2}, \tag{8.2.2}$$

provided the crack has been open up to time t.

It is clear from (8.2.1) that the crack will remain completely open while $q(t)$ remains positive. When $q(t)$ vanishes, the crack is closed completely and the pressure $p(t)$ on the crack faces is no longer known. Using (6.2.1) and (6.2.2), we put

$$p(t) = \begin{cases} p_o(t), & \text{crack open,} \\ p_c(t) > p_o(t), & \text{crack closed,} \end{cases} \tag{8.2.3}$$

where $p_o(t)$ is known and $p_c(t)$ must be determined. After an arbitrary number of closures, (8.2.1) holds provided we substitute the appropriate quantities for $p(t)$, as given by (8.2.3).

Inverting the first relation of (8.2.2) gives

$$p(t) = \int_{-\infty}^{t} dt' l(t-t')q(t'). \tag{8.2.4}$$

Let the crack be open for portion of $(-\infty, t]$ denoted by $W_o(t)$ and closed during the complementary set of times $W_c(t)$. For $t' \in W_o(t)$, $p(t')$ is equal to $p_o(t')$. For $t' \in W_c(t)$, $q(t')$ is known; in fact it is zero.

Consider first the case where the crack is closed at time t. Our object is to find $p_c(t)$ where, from (8.2.3), (8.2.4)

$$p_c(t) = \int_{-\infty}^{t} dt' l(t-t')q(t'). \tag{8.2.5}$$

We decompose according to (4.1.3) the hereditary integral on the right into integrals over $W_o(t)$ $\{= W_v(t)\}$ involving only $p_o(t')$ and over $W_c(t)$ $\{= W_u(t)\}$ involving only

$q(t')$, thus obtaining an expression for $p_c(t)$. By virtue of the fact that $q(t')$ is zero on $W_c(t)$, we obtain

$$p_c(t) = \int_{W_o(t)} dt' \Pi_v(t,t') p_o(t'), \tag{8.2.6}$$

where $\Pi_v(t,t')$ is given by (4.1.6) with the transition times equal to the times of opening and closing which we denoted by θ_r. The sets $W_o(t)$, $W_c(t)$ are defined in terms of these by

$$W_o(t) = (\theta_2, \theta_1) \bigcup (\theta_4, \theta_3) \bigcup \ldots,$$
$$W_c(t) = [\theta_1, t] \bigcup [\theta_3, \theta_2] \bigcup \ldots. \tag{8.2.7}$$

Thus, once the times of opening and closing are known, (8.2.6) is an expression for $p_c(t)$ in terms of known quantities. The time at which the crack next opens is the solution θ_s of the equation

$$p_c(\theta_s) = p_o(\theta_s). \tag{8.2.8}$$

Equation (8.2.6) is in fact the manifeslation of (5.2.3) (in the degenerate case discussed after (5.2.9)) for the present problem. This can be seen by defining $v(x,t) = \int_{-\infty}^{t} dt' l(t-t') g(x,t')$. Equation (5.2.4) gives that $v^{(1)}(x,t)$ vanishes on the closed crack face. There is no stress boundary condition in this case.

Now consider a time t at which the crack is open. In order to calculate the gap $g(x,t)$ according to (8.2.1), we need to know $q(t)$. Decomposing the first relation of (8.2.2) according to (4.1.9) gives

$$q(t) = \int_{W_o(t)} dt' \Gamma_v(t,t') p_o(t'), \tag{8.2.9}$$

where

$$W_o(t) = (\theta_1, t) \bigcup (\theta_3, \theta_2) \bigcup \ldots,$$
$$W_c(t) = [\theta_2, \theta_1] \bigcup [\theta_4, \theta_3] \bigcup \ldots. \tag{8.2.10}$$

The phase where the crack is open corresponds to (5.2.8), for the degenerate case discussed after (5.2.9). Equation (8.2.9) can be regarded as determining $Q_{22}^{(1)}(\underline{r},t)$ on the crack face to be $q(t)$, and providing a means of calculating this quantity.

Note that in (8.2.7), odd values of the subscript correspond to times of closing and even values to times of opening. In (8.2.10), that situation is reversed. The time at which the crack closes again, θ_c, is characterized by the condition that

$$q(\theta_c) = 0. \tag{8.2.11}$$

It remains zero until the time of next opening. It is clear that if the number of openings and closings is finite, one can establish the times θ_r of opening and closing and, form these, the complete solution, by starting at the beginning of the deformation history and working through the various phases, with the aid of above formulae. This program is carried out to a limited extent in the next section.

It can be proved that the pressure is continuous at times of crack closure and displacement at times of crack opening [Golden and Graham (1989)].

8.3 Early Opening and Closing Times

Let us assume that the deformation begins at $t = 0$ and that the applied load is sinusoidal, initially positive, giving

$$p_o(t) = p_m \sin(\omega t), \quad t \geq 0; \quad p_o(t) = 0, \quad t \leq 0. \tag{8.3.1}$$

We will write down explicit expressions for the time of first closing and reopening. These results, which generalize those given by Graham (1976) and Graham and Sabin (1981), will provide insight that will prove useful in the later discussion of the more difficult steady-state problem.

The time of first closing, that is after $t = 0$, follows from (8.2.9), (8.2.11) with θ_1 put equal to zero. We denote it by $\theta_c' = (\pi + \epsilon)/\omega$. Before this time

$$q(t) = p_m \int_0^t dt' k(t-t') \sin(\omega t') = p_m \left\{ \mathrm{Im}[e^{i\omega t}\hat{k}(\omega)] + Q(t) \right\}, \tag{8.3.2}$$

where

$$\hat{k}(\omega) = \int_0^\infty dt' e^{-i\omega t'} k(t'); \quad Q(t) = -\mathrm{Im}\left[e^{i\omega t} \int_t^\infty dt' k(t') e^{-i\omega t'} \right]. \tag{8.3.3}$$

Under the proportionality assumption

$$\hat{k}(\omega) = 1/\hat{l}(\omega) = (1-\nu)/\hat{\mu}(\omega), \tag{8.3.4}$$

by virtue of (6.2.10). Thus we can write $q(t)$ in the form

$$q(t) = p_m[|\hat{k}(\omega)| \sin(\omega t - \varphi) + Q(t)], \tag{8.3.5}$$

where φ is the loss angle. The condition (8.2.11) gives

$$|\hat{k}(\omega)| \sin(\epsilon - \varphi) = Q(\theta_c'). \tag{8.3.6}$$

After time θ_c' and before reopening, we have from (8.2.6), (4.1.6) and (4.1.8) that

$$p_c(t) = p_m \int_0^{\theta_c'} dt' T_1(t,t') \sin(\omega t'); \quad T_1(t,t') = \int_{t'}^{\theta_c'} dt'' l(t-t'') k(t''-t'). \tag{8.3.7}$$

The viscoelastic response (7.3.1) is now adopted. Equations (7.3.2) enable us to determine l_0, l_i, α_i, $i = 1, 2, \ldots, N$ in terms of k_0, k_i, β_i, $i = 1, 2, \ldots, N$. For example, if $N = 2$ we obtain

$$A\alpha^2 + B\alpha + C = 0;$$

$$A = k_0; \quad B = -(k_1 + k_2 + (\beta_1 + \beta_2)k_0); \quad C = k_0\beta_1\beta_2 + k_1\beta_2 + k_2\beta_1, \tag{8.3.8}$$

the two solutions of which are α_1 and α_2 respectively. Note that the values of α_1, α_2 depend on β_1, β_2 and the ratios k_1/k_0, k_2/k_0 only. The values of l_1, l_2 are then determined from the fourth relation of (7.3.2), with l_j, α_j and k_j, β_j exchanged.

The quantity $\hat{k}(\omega)$ is given by

$$\hat{k}(\omega) = k_0 + \sum_{i=1}^{N} \frac{k_i}{\beta_i + i\omega}, \tag{8.3.9}$$

while $Q(t)$ has the form

$$Q(t) = \sum_{i=1}^{N} \frac{k_i\omega}{\beta_i^2 + \omega^2} e^{-\beta_i t}, \tag{8.3.10}$$

which is a positive quantity. Thus, from (8.3.10) we see that $\epsilon > \varphi$. Since the loss angle is positive, it follows that ϵ is positive so that closure takes place after the pressure becomes negative.

From (8.3.7), on interchanging integrals

$$p_c(t) = \int_0^{\theta_c'} dt'' l_1(t - t'')q(t''), \tag{8.3.11}$$

$$l(t) = l_0\delta(t) + l_1(t),$$

where $q(t'')$, given by (8.3.2) is non-negative for $t'' \leq \theta_c'$. We have dropped the singular term in $l(t - t'')$ since it does not contribute for $t > \theta_c'$. Therefore

$$p_c(t) = \sum_{i=1}^{N} l_i q_i e^{-\alpha_i t}, \tag{8.3.12}$$

where

$$q_i = \int_0^{\theta_c'} dt'' e^{\alpha_i t''} q(t''), \quad i = 1, 2, \ldots, N, \tag{8.3.13}$$

are positive constants and the l_i are negative. Thus $p_c(t)$ is a sum of negative decaying exponentials. As t increase from θ_c', we therefore have $p_c(t) > p_o(t)$ in accordance with (8.2.3), at least if $\omega\theta_c' \leq 3\pi/2$ while $p_o(t)$, given by (8.3.1), is increasing in magnitude. If $\omega\theta_c' > 3\pi/2$ this condition does not obviously follow from general mathematical arguments. However, it must be true on physical grounds. Any solution for which it did not hold could not be accepted. One deduces that $p_c(t)$ becomes equal to $p_o(t)$

again before $t = 2\pi/\omega$ since $p_c(t)$ remains negative. Therefore the time of reopening can be written as $\theta'_s = (2\pi - \eta)/\omega$ where $\eta > 0$.

Let us now write down the explicit form of $p_c(t)$. It can be shown that $T_1(t,t')$, defined in (8.3.7) has the form (see (7.3.3))

$$T_1(t,t') = \sum_{i,j=1}^{N} \frac{l_i k_j}{\alpha_i - \beta_j} exp[-\alpha_i(t-\theta'_c) - \beta_j(\theta'_c - t')]. \tag{8.3.14}$$

Using this relation, one can show that

$$p_c(t) = p_m \sum_{i,j=1}^{N} \frac{l_i k_j}{\alpha_i - \beta_j} \frac{e^{-\alpha_i(t-\theta'_c)}}{\sqrt{\beta_j^2+\omega^2}} \left[\sin(\omega\theta'_c - \psi_j) + \sin\psi_j e^{-\beta_j\theta'_c}\right], \tag{8.3.15}$$

where $0 \leq \psi_j \leq \pi/2$ and

$$\tan\psi_j = \omega/\beta_j. \tag{8.3.16}$$

The time θ'_s of next opening is the next time at which $p_c(t) = p_o(t)$, $t > \theta'_c$. Thus the times θ'_c and θ'_s can be determined from (8.3.6), (8.3.10) and (8.3.15) by numerical solution of the non-linear equations involved.

8.4 The Steady-State Problem

It will now be assumed that a sufficiently long time has passed since the stress was first applied to allow steady-state conditions to be established. The deformation is presumed to have begun in the distant past. The sinusoidal behaviour specified by the first equation of (8.3.1) will be assumed. Consider first the case where the crack is closed. Let this be the first cycle of closure after $t = 0$, partly occupying the time interval $[\pi/\omega, 2\pi/\omega]$. We expect θ_1 will be somewhat larger than π/ω. Let us denote it by θ_c. Also, θ_2 will be in the vicinity of $t = 0$. We denote it by θ_s. Strictly, according to the usage in (8.2.8), this should be $\theta_s - \Delta$. However, for simplicity, we alter our notation. For steady-state conditions (recall (4.3.2))

$$\theta_{r+2} = \theta_r - \Delta, \quad r = 1, 2, \ldots; \quad \Delta = 2\pi/\omega. \tag{8.4.1}$$

Equation (8.2.6) can be written as

$$p_c(t) = p_m \int_{\theta_s}^{\theta_c} dt' \Pi_1(t,t') \sin(\omega t'), \tag{8.4.2}$$

where $\Pi_1(t,t')$ is given by (4.3.4).

Next, consider the time period just after $t = 0$ for which the crack is open, roughly $[0, \pi/\omega]$. If t is in this interval then $\theta_1 = \theta_s$ and $\theta_2 = \theta_c - \Delta$. Equation (8.2.9) can written as

$$q(t) = p_m \int_{\theta_s}^{t} dt' k(t-t') \sin(\omega t') + p_m \int_{\theta_s}^{\theta_c} dt' \Gamma_0(t,t') \sin(\omega t'), \tag{8.4.3}$$

where $\Gamma_0(t,t')$ is given by (4.3.6). As noted earlier, the infinite summations (4.3.4) and (4.3.6) can be explicitly evaluated without difficulty for the case of a standard linear solid - that is, for the case where $N = 1$. The results then obtained are in agreement with those of Graham and Sabin (1981), obtained by a different approach. For $N > 1$, one must solve the integral equations (4.3.18) for $\Pi_1(t,t')$ and (4.3.23) for $\Gamma_0(t,t')$, following the same approach as developed in section 7.3.

Again we have that $G(u,t')$ is given by (7.3.3). Also, instead of (7.3.4) we have

$$K(u,t') = \sum_{i,j=1}^{N} K'_{ij} exp\{-\alpha_i(u-\theta_1) - \beta_j(\theta_1 - t')\},$$

$$K'_{ij} = \frac{l_1 k_j}{\alpha_i - \beta_j} \frac{1}{1 - exp(-\beta_j \Delta)}. \tag{8.4.4}$$

To solve (4.3.18) we make the ansatz for $\Pi_1(t,t')$ of the form

$$\Pi_1(t,t') = \sum_{j=1}^{N} P_j(t) e^{\beta_j t'}, \tag{8.4.5}$$

which clearly obeys (4.3.17); on substitution into (4.3.18), this yields an algebraic equation which we express as follows

$$P = N + PAK_1, \tag{8.4.6}$$

where P is a row vector with components $P_j(t)$, N is a row vector with components

$$N_j(t) = \sum_{i=1}^{N} K'_{ij} exp\{-\alpha_i(t-\theta_1) - \beta_j\theta_1\}, \tag{8.4.7}$$

and A, K_1 are matrices with components

$$(A)_{mn} = (e^{\beta_m\theta_2 - \alpha_n(\Delta+\theta_2-\theta_1)} - e^{\beta_m\theta_3})/(\beta_m - \alpha_n),$$

$$(K_1)_{nj} = K'_{nj} e^{-\beta_j\theta_1}, \tag{8.4.8}$$

The formal solution of (8.4.6) is

$$P = N(I - AK_1)^{-1}. \tag{8.4.9}$$

Now, using (7.3.2), it is found that

$$(1 - e^{-\beta_j\Delta})(I - AK_1)_{ij}(t) = \delta_{ij} - k_j c_{ij} e^{\beta_i\theta_2 - \beta_j\theta_1}, \tag{8.4.10}$$

where

$$c_{ij} = \sum_{n=1}^{N} \frac{l_n e^{\alpha_n(\theta_1-\theta_2-\Delta)}}{(\beta_i - \alpha_n)(\alpha_n - \beta_j)}. \tag{8.4.11}$$

For N greater that 2, the evaluation of P is best carried out by resorting to numerical methods. For $N = 2$, manageable formulae may be written down. Using (7.3.2), it is found that the determinant of the right-hand side of (8.4.10) is given by

$$\Delta_P(t) = 1-\sum_{i=1}^{2} k_i c_{ii} e^{\beta_i(\theta_2-\theta_1)} + exp\{(\beta_1+\beta_2-\alpha_1-\alpha_2)(\theta_2-\theta_1+\Delta)-(\beta_1+\beta_2)\Delta\}. \quad (8.4.12)$$

It follows from (7.3.3) and (8.4.7)-(8.4.11) that

$$\Delta_P(t)e^{\beta_1\theta_1}P_1(t) = \lambda_1 k_1\{1 - k_2 c_{22} e^{\beta_2(\theta_2-\theta_1)}\} + \lambda_2 k_2 k_1 c_{12} e^{\beta_2(\theta_2-\theta_1)},$$
$$\Delta_P(t)e^{\beta_2\theta_1}P_2(t) = \lambda_1 k_1 k_2 c_{12} e^{\beta_1(\theta_2-\theta_1)} + \lambda_2 k_2\{1 - k_1 c_{11} e^{\beta_1(\theta_2-\theta_1)}\}, \quad (8.4.13)$$

where

$$\lambda_i = \sum_{n=1}^{N} \frac{l_n e^{-\alpha_n(t-\theta_1)}}{\alpha_n - \beta_i}. \quad (8.4.14)$$

In the case of a standard linear solid, $N = 1$ and the appropriate form for $P_1(t)$ is obtained by putting $\beta_2 \to \infty$. In particular

$$\Delta_P(t) = 1 - e^{(\beta_1-\alpha_1)(\theta_2-\theta_1)-\alpha_1\Delta} \quad (8.4.15)$$

in agreement with [14].

Proceeding analogously to the above case, we find that for the integral equation (4.3.23)

$$\sum_{n=1}^{\infty} k(t-t'+n\Delta) = \sum_{i=1}^{N} \frac{k_i e^{-\beta_i\Delta}}{1-e^{-\beta_i\Delta}} e^{-\beta_i(t-t')}, \qquad \theta_1(t) \le t' \le \theta_2(t)+\Delta, \quad (8.4.16)$$

and

$$L(u,t') = \sum_{i,j=1}^{N} L_{ij} exp\{-\alpha_i(u-\theta_2) - \beta_j(\theta_2 - t')\},$$

$$L_{ij} = \frac{l_i k_j}{\alpha_i - \beta_j} \frac{e^{-\beta_j\Delta}}{1-e^{-\beta_j\Delta}}. \quad (8.4.17)$$

Then

$$\Gamma_0(t,t') = \sum_{j=1}^{N} Q_j(t) e^{\beta_j t'}, \quad (8.4.18)$$

satisfies (4.3.24) and will satisfy (4.3.23) if

$$Q = M + QBL_1, \quad (8.4.19)$$

where Q is the row vector formed by $Q_j(t)$ and M is the row vector with components

$$M_j(t) = \sum_{i=1}^{N} \frac{k_i k_j}{1-e^{\beta_j\Delta}} D_{ij} exp\{\beta_i(\theta_1 - t) - \beta_j(\theta_2+\Delta)\},$$

$$D_{ij} = \sum_{n=1}^{N} \frac{l_n e^{\alpha_n(\theta_2-\theta_1)}}{(\beta_i - \alpha_n)(\alpha_n - \beta_j)}, \tag{8.4.20}$$

while L_1 and B are square matrices with components

$$(L_1)_{nj} = L_{nj} e^{-\beta_j \theta_2},$$

$$(B)_{mn} = \left\{ e^{(\beta_n-\alpha_n)\theta_1 + \alpha_n\theta_2} - e^{\beta_m\theta_2} \right\} / (\beta_m - \alpha_n). \tag{8.4.21}$$

The formal solution of (8.4.19) is

$$Q = M(1 - BL_1)^{-1}. \tag{8.4.22}$$

In this case, using (7.3.2), it is found that

$$(1 - e^{-\beta_j\Delta})(1 - BL_1)_{ij}(t) = \left\{ \delta_{ij} - k_j D_{ij} e^{\beta_i\theta_1 - \beta_j(\theta_2+\Delta)} \right\}. \tag{8.4.23}$$

For $N = 2$, the determinant of the right-hand side of (8.4.23) is

$$\Delta_Q(t) = 1 - \sum_{i=1}^{2} k_i D_{ii} e^{\beta_i(\theta_1-\theta_2-\Delta)} + exp\{(\beta_1+\beta_2-\alpha_1-\alpha_2)(\theta_1-\theta_2)-(\beta_1+\beta_2)\Delta\}. \tag{8.4.24}$$

It follows from (8.4.17) and (8.4.20)-(8.4.23) that

$$\Delta_Q(t) e^{\beta_1(\theta_2+\Delta)} Q_1(t) = k_1^2 e^{-\beta_1(t-\theta_1)} \{D_{11} - k_2 \Delta_D e^{\beta_2(\theta_1-\theta_2-\Delta)}\} + e^{-\beta_2(t-\theta_1)} D_{12} k_1 k_2,$$

$$\Delta_Q(t) e^{\beta_2(\theta_2+\Delta)} Q_2(t) = k_2^2 e^{-\beta_2(t-\theta_1)} \{D_{22} - k_1 \Delta_D e^{\beta_1(\theta_1-\theta_2-\Delta)}\} + e^{-\beta_1(t-\theta_1)} D_{12} k_1 k_2, \tag{8.4.25}$$

$$\Delta_D = D_{11} D_{22} - D_{12}^2. \tag{8.4.26}$$

For standard linear solid

$$\Delta_Q(t) = 1 - exp\{(\beta_1 - \alpha_1)(\theta_1 - \theta_2) - \beta_1\Delta\}, \tag{8.4.27}$$

and $Q_1(t)$ has the form corresponding to earlier results [14].

8.5 Numerical Results

We now present numerical solutions for the case $N = 2$. Fundamentally, the problem is: given values for J_1/J_0, J_2/J_0, β_1, β_2 in (2.1.38) determine θ_c, θ_s. However, in order to facilitate comparison with earlier work, we introduce the following equivalent set of variables: φ_1, φ_2, ψ_1, ψ_2 where φ_1 is the loss angle for $J_2 = 0$ and φ_2 is the loss angle where $J_1 = 0$. Explicitly

$$\tan(\varphi_i) = \frac{\omega k_i}{(\beta_i^2 + \omega^2)k_0 + \beta_i k_i}, \quad i = 1, 2. \tag{8.5.1}$$

Also ψ_1, ψ_2 are defined by (8.3.16). Note that $\varphi_i \leq \psi_i$, $i = 1, 2$. From (8.2.8) and (8.2.11) we see that the equations for θ_c and θ_s are

$$q(\theta_c) = 0; \quad p_c(\theta_s + \Delta) = p_o(\theta_s + \Delta). \tag{8.5.2}$$

From (8.2.6), (8.2.9), (8.4.5) and (8.4.18), we see these become

$$k_0 \sin(\omega\theta_c) + \sum_{i=1}^{2} k_i e^{-\beta_i\theta_c} I_i + \sum_{i=1}^{2} Q_i(\theta_s) I_i = 0,$$

$$\sin(\omega\theta_s) = \sum_{i=1}^{2} p_i(\theta_s + \Delta) I_i, \tag{8.5.3}$$

where

$$I_i = \int_{\theta_s}^{\theta_c} dt' e^{\beta_i t'} \sin(\omega t') = (\beta_i^2 + \omega^2)^{-1/2} e^{\beta_i t'} \sin(\omega t' - \psi_i)\Big|_{\theta_s}^{\theta_c}. \tag{8.5.4}$$

The quantites $P_i(t)$, $Q_i(t)$ are given explicitly by (8.4.13) and (8.4.25). Alternatively, they can be evaluted by numerical matrix inversion from (8.4.9) and (8.4.22). In fact both methods were used in the results presented here, for purposes of checking. The non-linear equations (8.5.3) were solved using the two variable Newton-Raphson technique with numerically calculated derivatives. Some results are presented on Figures 8.1-8.3 where, following Graham and Sabin (1981), we replace θ_c, θ_s by the angular delay Φ and the angular advance Θ defined by

$$\Phi = \omega\theta_c - \pi; \quad \Theta = -\theta_s. \tag{8.5.5}$$

In Figures 8.1(a), (b) results are presented for $\varphi_2 = 0$ which are in good agreement with those given in Graham and Sabin (1981). The remaining figures explore the variation of Φ, Θ with φ_2, ψ_2. Consideration of Figure 8.1(a),(b) would lead one intuitively to expect that Φ should increase with φ_2 and ψ_2 while Θ should pass through a maximum as φ_2 increases and increase as ψ_2 increases. Figures 8.1-8.3(a) show that Φ increases as φ_2 increases except at large values of φ_1 for which it remains virtually constant. Figures 8.1-8.3(b) indicate that, at low values of φ_1, Θ does indeed pass through a maximum as a function of φ_2. However, at larger values of φ_1, it seems to decline as φ_2 increases. Curves for two values of ψ_2 and roughly similar values of φ_2 are shown on Figures 8.3(a), (b). These suggest that Φ and Θ do indeed increase with ψ_2. However, note that in fact Θ as a function of ψ_1 changes over and actually decreases at large values of φ_2 and ψ_2.

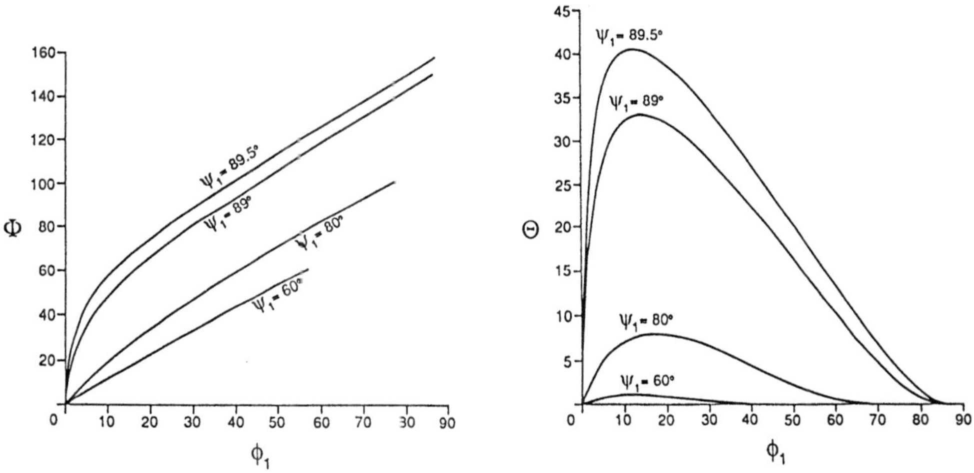

Figure 8.1: (a),(b). Plot of angular delay Φ and angular advance Θ as a function of φ_1 for various values of ψ_1 with $\varphi_2 = 0$.

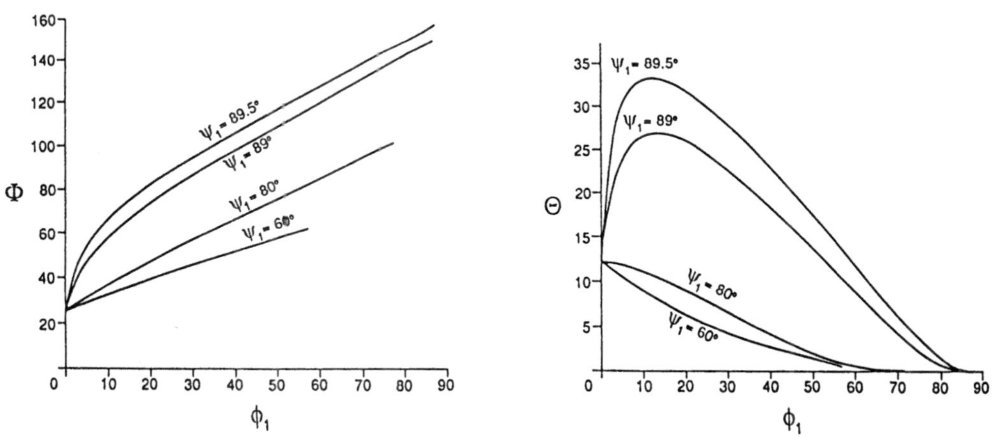

Figure 8.2: (a),(b). Plot of Φ and Θ as a function of φ_1 for various values of ψ_1 with $\varphi_2 = 10.2^o$ and $\psi_2 = 84.3^o$.

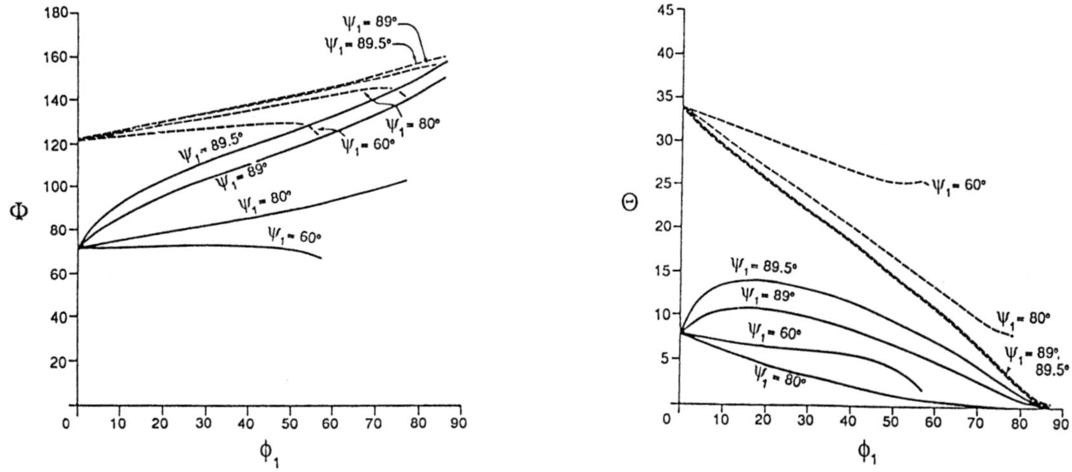

Figure 8.3: (a),(b). Plot of Φ and Θ as a function of φ_1 for various values of ψ_1 with $\varphi_2 = 42^o$ and $\psi_2 = 84.3^o$ (continuous lines); $\varphi_2 = 45^o$ and $\psi_2 = 89.9^o$ (broken lines).

Chapter 9

ASYMMETRIC STEADY-STATE SOLUTIONS FOR A CRACK IN A VISCOELASTIC FIELD IN PURE BENDING

9.1 Introduction

As a final example of an area-based approach, we consider a crack in a viscoelastic material subject to bending, such that partial closing occurs at one end only. In chapter 11, we consider closing at both ends, a problem which must be tackled by the point-based approach.

This type of problem was considered in the elastic case by Bowie and Freese (1976) and Comninou and Dundurs (1979). A central feature of interest was that the crack closed over part of the interval, including one end. The end which closed depended on the sign of the applied bending moment and switched instantly once the sign changed. For the case under consideration here, the closed interval was $[c/3,\ c]$ for positive bending moment, and the reflection of this in the y-axis for negative bending moment. The closures were instanteous over the entire extent of these regions, rather than gradually starting at the end and moving in.

The transient problem of a crack in a viscoelastic medium subject to a sinusoidal bending moment has been considered over the early history of loading [Graham (1982), Golden and Graham (1984)]. It was found that, in contrast to the elastic case, the crack can close gradually rather than suddenly. However, a frequency-dependent class of materials was isolated for which the crack closed suddenly at one end in two consecutive cycles, without closing at the other end. It was speculated [Golden and Graham (1984)] that a steady-state solution might exist, consisting entirely of asymmetric solutions of this kind, for certain limited types of materials. The conjecture is that such behaviour may exist under a sinusoidal loading with equal positive and negative maxima. If such were the case it would be quite surprising.

We will seek to check this conjecture. However, and more importantly, it is also possible to create one-sided partial closure in a wide class of materials, presumably any material, by applying a biased loading, in other words, one that is different in the positive and negative cycles. An example would be a loading that is sinusouidal but with a constant added. The broader aim of this chapter is to study behaviour under such an asymmetric loading. The developments described here are based on Golden and Graham (1991).

9.2 General Relations

Plane strain conditions are presumed to apply and we consider the particular cross-section of the medium lying in the xy plane. In this plane, let the crack which is assumed to be of fixed length, lie along the x-axis occupying $[-c, c]$ as in Figure 9.1. Let the material be subject to a bending moment such that the neutral axis is the y-axis. The ratio of this moment to the moment of inertia of the cross-section about the y-axis is termed $\eta(t)$. The sign convention is such that when $\eta(t)$ is positive, there will be compression across the positive x-axis, at least at points far from the origin. At this stage we simply assume that the behaviour of $\eta(t)$ is such as to cause closure only at the positive end. Later, we will settle on a specific behaviour.

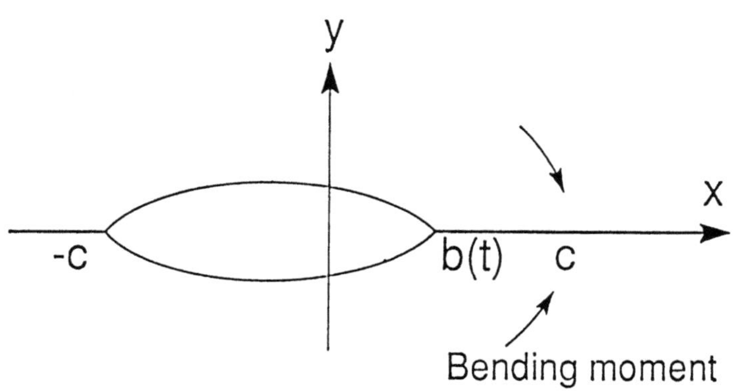

Figure 9.1: Partially closed crack along the x-axis.

Transferring the stresses to the crack face in the conventional manner [14], we have that on the open portion of the crack

$$p(x,t) = -\sigma_{22}(x,t) = -\eta(t)x. \tag{9.2.1}$$

The boundary conditions, neglecting friction between the faces, are (see (6.2.1), (6.2.2)) as follows. On the open portions of crack

$$u_2(x,0^+,t) - u_2(x,0^-,t) > 0,$$

$$\sigma_{22}(x,0^+,t) = \sigma_{22}(x,0^-,t) = \eta(t)x,$$

$$\sigma_{12}(x,0^+,t) = \sigma_{12}(x,0^-,t) = 0, \tag{9.2.2}$$

and on the closed portions

$$u_2(x,0^+,t) - u_2(x,0^-,t) = 0,$$

$$\sigma_{22}(x,0^+,t) = \sigma_{22}(x,0^-,t) < \eta(t)x,$$

$$\sigma_{12}(x,0^+,t) = \sigma_{12}(x,0^-,t) = 0, \tag{9.2.3}$$

in terms of the stresses tensor $\sigma_{ij}(x,y,t)$, $i,j = 1,2,3$ and the normal displacement $u_2(x,y,t)$. Because of the symmetry of the problem, $u_2(x,0^+,t) = -u_2(x,0^-,t)$. We denote $u_2(x,0^+,t)$ by $u(x,t)$ in the sequel and by $w(x,t)$ its derivative with respect to x. Using the approach presented in [Golden and Graham (1984)] rather than the more formal developments of section 6.2, we have

$$q(x,t) = -\frac{1}{\pi}\int_{-c}^{b(t)} dx' \frac{w(x',t)}{x'-x},$$

$$q(x,t) = \int_{-\infty}^{t} dt' k(t-t')p(x,t'), \tag{9.2.4}$$

where the open portion of crack at time t is $[-c, b(t)]$, $b(t) \leq c$ and $k(t)$ is, as before, the inverse of $l(t)$, given by (6.1.11) under the convolution product. Actually, (9.2.4) may be deduced by transferring the hereditary integral in (6.2.8) to the right-hand side and operating with $k(t-t')$ on (6.2.11). One then uses the Plemelj formulae. Consider a time t at which $b(t)$ is decreasing. We decompose $q(x,t)$ according to (4.1.9), with a somewhat adapted notation:

$$q(x,t) = \int_{W_q(t)} dt'\Gamma_q(t,t')q(x,t') + \int_{W_p(t)} dt'\Gamma_p(t,t')p(x,t'), \tag{9.2.5}$$

where $W_q(t)$ is the set of all times $t' \leq t$ such that $b(t') < b(t)$ and $W_p(t)$ is the set of $t' \leq t$ such the $b(t') \geq b(t)$. The transition times $\theta_l(t)$, $l = 1,2,\ldots$ are the earlier times, taken in order, at which $b(t') = b(t)$.

From (9.2.1) and the definition of $W_p(t)$ we have that the second term on the right-hand side of (9.2.5) is given by

$$\int_{W_p(t)} dt'\Gamma_p(t,t')p(x,t') = B(t)x, \quad -c \leq x \leq b(t),$$

$$B(t) = -\int_{W_p(t)} dt'\Gamma_p(t,t')\eta(t'). \tag{9.2.6}$$

Using (9.2.4) in (9.2.5) and remembering the definition of $W_q(t)$, gives

$$\frac{1}{\pi}\int_{-c}^{b(t)} dx' \frac{\{w(x',t) - v_c(x',t)\}}{x' - x} = -B(t)x, \quad -c \leq x \leq b(t),$$

$$v_c(x,t) = \int_{W_q(t)} dt'\Gamma_q(t,t')w(x,t'). \tag{9.2.7}$$

On inverting the Hilbert transform, we obtain the integral equation for $w(x,t)$

$$w(x,t) = \int_{W_q(t)} dt'\Gamma_q(t,t')w(x,t') + B(t)w_e(x,t), \quad -c \leq x \leq b(t), \tag{9.2.8}$$

where

$$w_e(x,t) = \frac{1}{m(x,t)}\left\{\frac{1}{2}\left(\frac{b(t)+c}{2}\right)^2 + x\left(\frac{b(t)-c}{2} - x\right)\right\},$$

$$m(x,t) = \{(b(t)-x)(x+c)\}^{1/2}. \tag{9.2.9}$$

The arbitrary constant that arises on inversion of finite Hilbert transforms can be put to zero because $u(x,t)$ vanishes at both ends of the open region and because, for $t' \in W_q(t)$, $b(t') < b(t)$.

Let $b(t) < c$, in other words let some partial closure be present at time t. Then for $t' \in W_q(t)$ partial closure will also be present. Thus, $w(x,t)$, $w(x,t')$ in (9.2.8) will have no singularities at the positive end. Therefore, $w_e(x,t)$ can have no singularities either, at $x = b(t)$. It follows that [14]

$$b(t) = c/3,$$

$$w_e(x,t) = w_0(x) = \frac{1}{3}\left(\frac{c/3 - x}{x+c}\right)^{1/2}(2c + 3x)d_1(x),$$

$$d_1(x) = \begin{cases} 1, & -c < x \leq c/3, \\ 0, & c/3 \leq x < c. \end{cases} \tag{9.2.10}$$

Thus, if partial closure occurs at the positive end only, $b(t)$ must be equal to the elastic value $c/3$. The only option allowed by (9.2.8) is a singularity at $x = c$, corresponding to a fully open crack.

If $b(t)$ is increasing at time t then we could construct a somewhat different equation for $w(x,t)$, based on decomposing $p(x,t)$ according to (4.1.3) rather than $q(x,t)$. The same argument goes through here also to give that the only possible behaviour is sudden reopening.

Therefore, we can only have instantaneous partial closing and reopening, of the kind that occurs in elastic theory, for a crack that is closing at only one end. The gradual closure found by [Golden and Graham (1984)] is a feature of cracks closing alternatively at both ends. The behaviour of a crack closing at only one end is schematically described by Figure 9.2.

The feature of instantaneous closing and reopening means that we are dealing with the degenerate case discussed after (5.2.9).

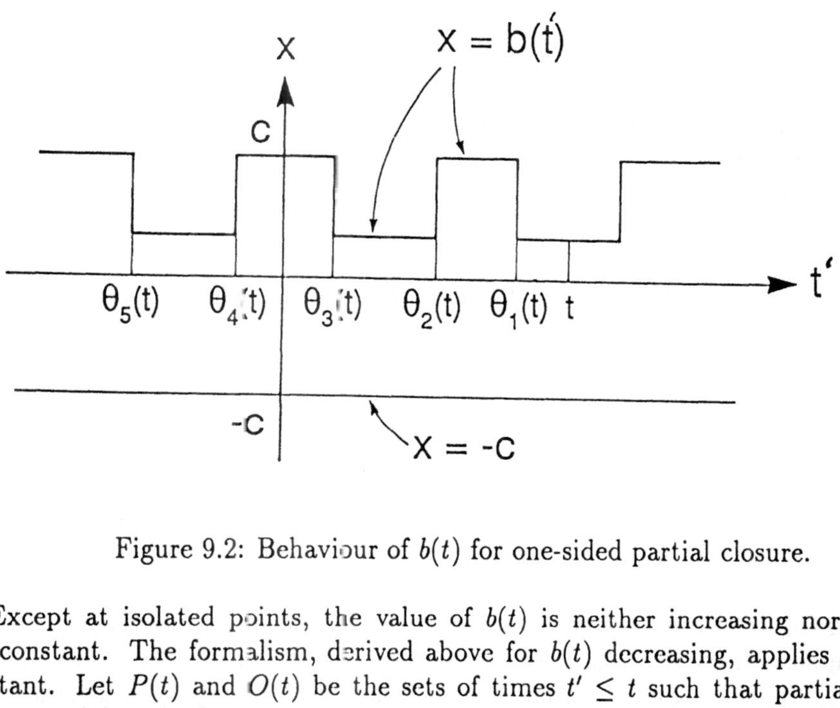

Figure 9.2: Behaviour of $b(t)$ for one-sided partial closure.

Except at isolated points, the value of $b(t)$ is neither increasing nor decreasing but constant. The formalism, derived above for $b(t)$ decreasing, applies also to $b(t)$ constant. Let $P(t)$ and $O(t)$ be the sets of times $t' \leq t$ such that partial closure is present and for which the crack is fully open, respectively. Let the crack be partially closed at time t. Then the times immediately preceding t must be regarded as belonging to $W_p(t)$. This suggests that we take $P(t) \subset W_p(t)$. Clearly also, $O(t) \subset W_p(t)$. Thus $W_p(t) = (-\infty, t]$ and $W_q(t)$ is empty. It follows from (9.2.8) and (9.2.10) that during times of partial closure

$$w(x,t) = -M(t)w_0(x),$$

$$M(t) = \int_{-\infty}^{t} dt' k(t-t')\eta(t'). \tag{9.2.11}$$

This result can in fact be derived more directly by observing that during partial closure

$$q(x,t) = -M(t)x, \quad -c \le x \le c/3. \tag{9.2.12}$$

Equation (9.2.11) follows by inverting the Hilbert transform in (9.2.4).

Applying similar reasoning to times when the crack is fully open, we see that it is natural to put $W_p(t) = O(t)$, $W_q(t) = P(t)$ so that from (9.2.8), (9.2.6)

$$w(x,t) = -A(t)w_0(x) + B(t)w_1(x),$$

$$w_1(x) = w_e(x,t)\,|_{b(t)=c} = \frac{1}{2}\frac{(c^2-2x^2)}{(c^2-x^2)^{1/2}}, \tag{9.2.13}$$

and

$$A(t) = \int_{P(t)} dt'\Gamma_q(t,t')M(t'),$$

$$B(t) = -\int_{O(t)} dt'\Gamma_p(t,t')\eta(t'), \quad t \in O(t), \tag{9.2.14}$$

where $\Gamma_p(t,t')$, $\Gamma_q(t,t')$ are defined as usual but where the $\theta_r(t)$ are now the times of opening and closing. They are independent of t except for discontinuous changes as t becomes equal to a new time of opening or closing. Decomposing $M(t), t \in O(t)$, analogously to (9.2.5), we obtain

$$M(t) = \int_{P(t)} dt'\Gamma_q(t,t')M(t') + \int_{O(t)} dt'\Gamma_p(t,t')\eta(t') = A(t) - B(t). \tag{9.2.15}$$

Let (9.2.13) hold at all times but where during times of partial closure we have, instead of (9.2.14)

$$A(t) = M(t); \quad B(t) = 0; \quad t \in P(t). \tag{9.2.16}$$

Note that it is always true that

$$A(t) - B(t) = M(t). \tag{9.2.17}$$

Once the fact of instantaneous opening and closing has been established, it is better to shift emphasis from $W_p(t)$ and $W_q(t)$ to $O(t)$ and $P(t)$ and regard the latter as fundamental. This is basically what is done when considering the degenerate case discussed after (5.2.9).

9.3 Equations for Displacement and Pressure

It follows from (9.2.13) that the displacement is given by

$$u(x,t) = \frac{1}{2}A(t)\left(\frac{c}{3}-x\right)^{3/2}(x+c)^{1/2}d_1(x) + \frac{1}{2}B(t)x(c^2-x^2)^{1/2}. \tag{9.3.1}$$

This quantity must be non-negative everywhere. We note that the coefficient of $B(t)$ is a solution of the elastic problem that is unacceptable because it can take negative values, which implies material overlap. However, it does have an important role in the viscoelastic problem. By considering special values of x in (9.3.1), it follows that $A(t)$ and $B(t)$ must be non-negative. Also, one deduces that, for $x < 0$

$$\frac{A(t)}{B(t)} \geq \frac{|x|(c+|x|)^{1/2}}{(c/3+|x|)^{3/2}}. \tag{9.3.2}$$

The expression on the right-hand side increases for increasing $|x|$ so that the inequality can be imposed at $x = -c$. It may be checked that the equality is the condition that closure begins at the negative end, which must be avoided. The condition that $u(x,t)$ remains non-negative and that no closure occurs at $x = -c$ is therefore that for all t

$$A(t) \geq 0, \quad B(t) \geq 0; \quad \frac{A(t)}{B(t)} \geq \frac{3}{4}\left(\frac{3}{2}\right)^{1/2}. \tag{9.3.3}$$

We have

$$v(x,t) = \int_{-\infty}^{t} dt' l(t-t') w(x,t') = -C(t) w_0(x) + D(t) w_1(x), \tag{9.3.4}$$

where

$$C(t) = \int_{-\infty}^{t} dt' l(t-t') A(t'); \quad D(t) = \int_{-\infty}^{t} dt' l(t-t') B(t'). \tag{9.3.5}$$

Extending the integral in (9.2.4) over the entire x-axis and transferring the hereditary integral from $q(x,t)$ to $w(x',t)$ we have that

$$\begin{aligned} p(x,t) &= -\frac{1}{\pi}\int_{-c}^{c} dx' \frac{v(x',t)}{x'-x} \\ &= C(t)[-x + g(x) d_2(x)] + D(t) x, \quad |x| < c, \end{aligned} \tag{9.3.6}$$

where

$$g(x) = \frac{(x+2c/3)\,(x-c/3)^{1/2}}{(x+c)^{1/2}},$$

$$d_2(x) = \begin{cases} 0, & x \in [-c, c/3], \\ 1, & x \in [c/3, c], \end{cases} \tag{9.3.7}$$

with the aid of Hilbert transforms give in [14]. Since (9.2.1) must hold on the open part of the crack, we have

$$C(t) - D(t) = \eta(t), \tag{9.3.8}$$

which follows anyway from (9.2.17) and (9.3.5). The boundary conditions (9.2.3) require that

$$C(t) > 0, \quad t \in P(t). \tag{9.3.9}$$

Note that

$$C(t) = 0, \quad t \in O(t). \tag{9.3.10}$$

Consider $D(t)$ as defined by (9.3.5). Decomposing this hereditary integral according to (4.1.3), again with slight notational alteration, we obtain

$$D(t) = \int_{P(t)} dt'\Pi_c(t,t')B(t') + \int_{O(t)} dt'\Pi_o(t,t')D(t'), \quad t \in P(t), \tag{9.3.11}$$

where

$$\Pi_o(t,t') = T_1(t,t')R(t';\theta_2,\theta_1) + T_3(t,t')R(t';\theta_4,\theta_3) + \cdots, \tag{9.3.12}$$

and θ_1, θ_2, $\cdots$ are the previous times of closing, opening etc. From (9.2.16), (9.3.8) and (9.3.10) we deduce from (9.3.11) that

$$\begin{aligned} D(t) &= -\int_{O(t)} dt'\Pi_o(t,t')\eta(t') \\ &= -\sum_{r=1,3,5,\ldots} \int_{\theta_{r+1}}^{\theta_r} dt' T_r(t,t')\eta(t'), \quad t \in P(t). \end{aligned} \tag{9.3.13}$$

We are now in a position to sketch a general strategy for solving the problem. Let the crack be open at time t and let the previous times of opening, closing ect., be θ_1, θ_2, ... These are presumed to be known. Then from (9.2.6), (9.2.15), (9.3.8) and (9.3.10)

$$\begin{aligned} B(t) &= -\sum_{r=0,2,4,\ldots} \int_{\theta_{r+1}}^{\theta_r} dt' N_r(t,t')\eta(t'), \\ A(t) &= M(t) + B(t), \\ C(t) &= 0, \\ D(t) &= -\eta(t), \end{aligned} \tag{9.3.14}$$

where $\theta_0 = t$. The time of next partial closing, t_c, is determined by

$$B(t_c) = 0. \tag{9.3.15}$$

During the closed phase

$$\begin{aligned} A(t) &= M(t), \\ B(t) &= 0, \\ C(t) &= \eta(t) + D(t), \end{aligned} \tag{9.3.16}$$

where $D(t)$ is given by (9.3.13). The time of next opening t_0 is determined by

$$C(t_0) = 0. \tag{9.3.17}$$

It is interesting to observe that, given the behaviour of $b(t)$ (see Figure 9.2), the Generalized Partial Correspondence Principle may be used to generate the third and fourth relations of (9.3.14) and first two of (9.3.16). There arises the question of continuity of the various coefficents at times of opening and closing. It is clear from (9.2.17) and (9.3.8) that if $B(t)$ and $C(t)$ are continuous then $A(t)$ and $D(t)$ are. Equation (9.3.15) and (9.3.17), together with (9.2.16) and (9.3.10), ensure continuity at times of closing for $B(t)$ and opening for $C(t)$ respectively. It remains to prove continuity for $B(t)$ at times of opening and $C(t)$ at times of closing. One can show, using results in [14] that

$$B(t_0^+) = -k_0 C(t_0) = 0; \quad C(t_c^+) = -l_0 B(t_c) = 0, \tag{9.3.18}$$

which demonstrates the required continuity.

The approach taken in section 9.2 is complicated by the fact that one wished to deduce the fact that closure was instantaneous, rather than assume it. If one were to start from (9.2.13) and (9.3.1), it is possible to derive the forms of $A(t)$ and $B(t)$ more directly. We leave this as a problem.

Problem 9.3.1 *Given that [(9.3.10), (9.2.16)]*

$$\int_{-\infty}^{t} dt' l(t-t')A(t') = 0, \quad t \in O(t),$$

$$A(t) = M(t), \quad t \in P(t), \tag{9.3.19}$$

derive the form of $A(t)$ in (9.2.14). Also, given that [(9.3.14)],

$$\int_{-\infty}^{t} dt' l(t-t')B(t') = -\eta(t), \quad t \in O(t), \tag{9.3.20}$$

$$B(t) = 0, \quad t \in P(t),$$

derive the form of $B(t)$ in (9.2.14). [Hint: decompose $A(t)$, $B(t)$.]

9.4 Steady-state Relations

Assume that $\eta(t)$ is periodic with period Δ and that sufficient time has passed for steady-state conditions to be established. Consider a given time period $[t_o, t_c]$ when the crack is completely open and the subsequent period $[t_c, t_o + \Delta]$ when it is partially closed. For $t \in (t_o, t_c]$, we have

$$\theta_1 = t_o; \quad \theta_2 = t_c - \Delta; \quad \theta_3 = t_o - \Delta; \quad \text{etc.}, \tag{9.4.1}$$

while for $t \in (t_c, t_o + \Delta]$

$$\theta_1 = t_c; \quad \theta_2 = t_o; \quad \theta_3 = t_c - \Delta; \quad \text{etc.} \tag{9.4.2}$$

The coefficients $A(t)$, $B(t)$, $C(t)$, $D(t)$ will all be periodic with period Δ. Recalling section 4.3, we see that $B(t)$, given by (9.3.14), becomes

$$B(t) = -\int_{t_o}^{t} dt' k(t-t')\eta(t') - \sum_{r=1}^{\infty}\int_{t_o}^{t_c} dt' N_{2r}(t, t'-r\Delta)\eta(t'), \quad t \in (t_o, t_c], \tag{9.4.3}$$

while (9.3.13) becomes

$$D(t) = -\sum_{r=0}^{\infty}\int_{t_o}^{t_c} dt' T_{2r+1}(t, t'-r\Delta)\eta(t'), \quad t \in (t_c, t_o+\Delta]. \tag{9.4.4}$$

The other coefficients are given by (9.3.14), (9.3.16).

The standard linear model

We explore these relationships for the case of a standard linear model. A unique Poission's ratio ν will be assumed. We put

$$l(t) = l_0\delta(t) + l_1 e^{-\alpha t}; \quad k(t) = k_0\delta(t) + k_1 e^{-\beta t} \tag{9.4.5}$$

where

$$k_0 l_0 = 1; \quad k_1 = -l_1/l_0^2; \quad \beta = \alpha - k_1/k_0. \tag{9.4.6}$$

Note that k_0, l_0, α and β are positive, while l_1 is negative. There is no need to resort to the integral equations derived in section 4.3 to carry out the summations in this case. The necessary relations can be shown directly without great difficulty. We leave this as a problem [14, 26].

Problem 9.4.1 *Show that under steady-state conditions*

$$N_{r+2}(t, t'-\Delta) = N_r(t, t')E, \quad t \neq t',$$

$$T_{r+2}(t, t'-\Delta) = T_r(t, t')E; \quad E = exp\{-(\alpha-\beta)(t_o - t_c) - \alpha\Delta\}. \tag{9.4.7}$$

Also for $t \in (t_c, t_o+\Delta]$ *show that*

$$T_1(t, t') = l_1 k_0 exp\{-\alpha(t-t_c) - \beta(t_c - t')\}. \tag{9.4.8}$$

Using these results in (9.4.3), (9.4.4), we obtain

$$B(t) = -k_0\eta(t) + k_1 e^{-\beta t} I(t, t_c) - \frac{k_1 e^{-\beta t} I(t_o, t_c)}{1-E}, \quad t \in (t_o, t_c],$$

$$D(t) = -\frac{l_1 k_0 exp\{-\alpha(t-t_c) - \beta t_c\} I(t_o, t_c)}{1-E}, \quad t \in (t_c, t_o+\Delta], \tag{9.4.9}$$

where

$$I(t_1, t_2) = \int_{t_1}^{t_2} dt' e^{\beta t'}\eta(t'). \tag{9.4.10}$$

The second term on the left-hand side of the first equation in (9.4.9) is a consequence of the fact that the non-singular part of $k(t-t') = N_0(t, t')$ in (9.4.3) is included in the infinite summation while the integration range of this term stops at t.

9.5 Asymmetrical Sinusoidal Moment

Let us put

$$\eta(t) = \eta_0(d + \sin(\omega t)), \quad 0 < d \leq 1, \qquad (9.5.1)$$
$$\Delta = 2\pi/\omega.$$

One can show that

$$\eta(t) \begin{cases} \geq 0, & 2n\pi - \gamma \leq \omega t \leq (2n+1)\pi + \gamma, \\ < 0, & (2n-1)\pi + \gamma \leq \omega t \leq 2n\pi - \gamma, \end{cases}$$

$$\gamma = \sin^{-1}(d), \qquad (9.5.2)$$

for all integer n. Partial closure can occur at the positive end only when $\eta(t)$ is positive, by virtue of (9.2.3). We consider a period of closure $[t_c, t_o + \Delta]$ in the time interval $[-\gamma/\omega, (\pi+\gamma)/\omega]$. The quantity $I(t_1, t_2)$ given by (9.4.10), has the form

$$I(t_1, t_2) = \eta_0 \left[\frac{d}{\beta}(e^{\beta t_2} - e^{\beta t_1}) + S(t_1, t_2) \right], \qquad (9.5.3)$$

$$S(t_1, t_2) = \int_{t_1}^{t_2} dt' e^{\beta t'} \sin(\omega t') = \left. \frac{e^{\beta t'} \sin(\omega t' - \psi)}{(\beta^2 + \omega^2)^{1/2}} \right|_{t_1}^{t_2},$$

where the phase ψ is defined by

$$\tan \psi = \omega/\beta. \qquad (9.5.4)$$

By virtue of (9.3.15), (9.3.17) and (9.4.9), the equations determining t_c, t_o are

$$k_0 \eta_0 (d + \sin \omega t_c) + \frac{k_1 e^{-\beta t_c} I(t_o, t_c)}{1 - E} = 0,$$

$$\eta_0(d + \sin \omega t_0) - \frac{l_1 k_0 exp\{-\alpha(t_o - t_c + \Delta) - \beta t_c\} I(t_o, t_c)}{1 - E} = 0, \qquad (9.5.5)$$

which must be solved numerically. The quantity $M(t)$, defined in (9.2.11), has the form

$$M(t) = \eta_0 \{ d\hat{k}(0) + \mathrm{Im}[\hat{k}(\omega) e^{i\omega t}] \}$$
$$= \eta_0 \{ d\hat{k}(0) + |\hat{k}(\omega)| \sin(\omega t - \varphi) \}, \qquad (9.5.6)$$

where $\hat{k}(\omega)$, a quantity proportional to the complex modulus for creep, has the form

$$\hat{k}(\omega) = k_0 + \frac{k_1}{\beta + i\omega}, \qquad (9.5.7)$$

and φ is the loss angle for the material defined by [see (2.1.29)]

$$\tan(\varphi) = \frac{k_1 \omega}{k_0(\beta^2 + \omega^2) + k_1 \beta}, \quad 0 \leq \varphi \leq \pi/2. \qquad (9.5.8)$$

Thus we have

$$M(t)=\eta_0\left\{d(k_0+k_1/\beta)+\left(k_0+\frac{k_1\beta}{\beta^2+\omega^2}\right)\sin(\omega t)-\frac{k_1\omega}{\beta^2+\omega^2}\cos(\omega t)\right\}$$

$$=\eta_0\left\{d(k_0+k_1/\beta)+\left[\left(k_0+\frac{k_1\beta}{\beta^2+\omega^2}\right)^2+\left(\frac{k_1\omega}{\beta^2+\omega^2}\right)^2\right]^{1/2}\sin(\omega t-\varphi)\right\}. \quad (9.5.9)$$

Condition (9.3.3) may be written as

$$\min_{t_0\le t\le t_c}\left[\frac{M(t)}{B(t)}\right]\ge\frac{3}{4}\left(\frac{3}{2}\right)^{1/2}-1=-0.08144, \quad (9.5.10)$$

together with the non-negativity requirements on $A(t)$, $B(t)$. The function $B(t)$ is given by (9.4.9) and (9.5.3).

To satisfy (9.3.9), (9.3.10), $C(t)$ must be non-negative at all times. This may be checked by using (9.3.16) and (9.4.9).

If we set

$$\Theta=\gamma-\pi-\omega t_0, \quad \Phi=\omega t_c+\gamma, \quad (9.5.11)$$

where

$$\Psi=\pi+2\gamma-\Theta-\Phi\ge 0, \quad (9.5.12)$$

then Θ is a measure of the 'angular advance' in the crack opening, before the effect of the applied moment on the positive x-axis changes from compression to tension, while Φ is 'angular delay' in the crack closing, once the effect of the applied moment has become the compressive on the positive x-axis (see [14], Chap. 4;6). The quantity Ψ measures the amount of the time that the crack is closed down on the right-hand side in each cycle. In terms of these quantities and the angles ψ, φ defined by (9.5.4) and (9.5.8), the equations (9.5.5) become

$$(1-E)(\sin(\gamma)+\sin(\Phi-\gamma))+f\sin(\gamma)(1-e^{(\Psi-2\pi)\cot(\psi)})$$
$$+f\cos(\psi)(\sin(\Phi-\gamma-\psi)+e^{(\Psi-2\pi)\cot(\psi)}\sin(\gamma-\psi-\Theta))=0, \quad (9.5.13)$$
$$(1-E)(\sin(\gamma)-\sin(\gamma-\Theta))+f\sin(\gamma)e^{-(1+f)\Psi/g}(1-e^{(\Psi-2\pi)\cot(\psi)})$$
$$+f\cos(\psi)e^{-(1+f)\Psi/g}(\sin(\Phi-\gamma-\psi)+e^{(\Psi-2\pi)\cot(\psi)}\sin(\gamma-\psi-\Theta))=0, \quad (9.5.14)$$

where E is given by (9.4.7) and the dimensionless quantities f,g are defined as

$$f=\frac{k_1}{k_0\beta}, \quad g=\frac{\omega}{\beta}=\tan(\psi). \quad (9.5.15)$$

We have

$$E=exp\left(\frac{-\Psi\sin(\varphi)}{\sin(\psi)\sin(\psi-\varphi)}-2\pi\cot(\psi)\right),$$

$$f = \frac{\sin(\varphi)}{\cos(\psi)\sin(\psi - \varphi)}; \quad \frac{1+f}{g} = \cot(\psi) + \frac{\sin(\varphi)}{\sin(\psi)\sin(\psi - \varphi)}. \tag{9.5.16}$$

Equations (9.5.13) and (9.5.14) may be combined to give

$$\sin(\gamma) + \sin(\Phi - \gamma) = e^{(1+f)\Psi/g}(\sin(\gamma) - \sin(\gamma - \Theta)). \tag{9.5.17}$$

Certain particular solutions of these equations are immediately available. For a Voigt material $\psi = \varphi$ and we deduce from (9.5.14) or (9.5.17) that $\Theta = 0$. Then (9.5.13) implies that Φ satisfies

$$(\cos(\varphi)\sin(\gamma - \varphi) - \sin(\gamma))e^{(2\gamma - \Phi - \pi)\cot(\psi)} + \cos(\varphi)\sin(\Phi - \gamma - \varphi) + \sin(\gamma) = 0. \tag{9.5.18}$$

Furthermore, by examining (9.5.10) with its right-hand side set equal to zero it is found that a sufficient condition for this solution to be valid is $\pi/2 - \gamma \le \varphi \le \pi/2$.

A Maxwell material is characterized by $\psi = \pi/2$. In this case equations (9.5.13), (9.5.17) reduce to

$$(1 - e^{-\Psi \tan\varphi})(\sin(\gamma) + \sin(\Phi - \gamma)) + (2\pi - \Psi)\tan(\varphi)\sin(\gamma)$$
$$- 2\tan(\varphi)\cos\left(\frac{\Theta - \Phi}{2}\right)\sin\left(\frac{\Psi}{2}\right) = 0, \tag{9.5.19}$$

$$\sin(\gamma) + \sin(\Phi - \gamma) = (\sin(\gamma) - \sin(\gamma - \Theta))e^{\Psi \tan\varphi}. \tag{9.5.20}$$

By examining (9.5.19), (9.5.20) it is found that $\Theta = \Phi$ when $\varphi = 0$ and $\Theta = 0$ when $\varphi = \pi/2$. Moreover, when $\gamma = 0$ an appropriate solution of (9.5.19), (9.5.20) satisfies $\Psi \equiv 0$. Also, the condition (9.5.10) for non-closure on the left is satisfied for all γ, $0 \le \gamma \le \pi/2$.

9.6 Numerical Results

Condition (9.5.10) guarantees that closure takes place only at one end. Equality in this condition implies a relation between the dimensionless parameters f, g for each value of d. Some examples are plotted on Figure 9.3(a). The same curves are plotted in Figure 9.3(b), using the variables φ, g where from (9.5.8) the loss angle φ is related to f, g by

$$\tan\varphi = \frac{fg}{1 + g^2 + f}, \quad 0 \le \varphi \le \pi/2. \tag{9.6.1}$$

Interestingly, these curves are somewhat reminiscent of the equivalent curves derived for the transient problem under a symmetric bending moment [Golden and Graham (1984)]. This is particularly true of those on Figure 9.3(b).

As d tends to zero, the curve moves off to infinity. This was indicated, though not of course proved, by the numerical results. To confirm such a statement, it is necessary

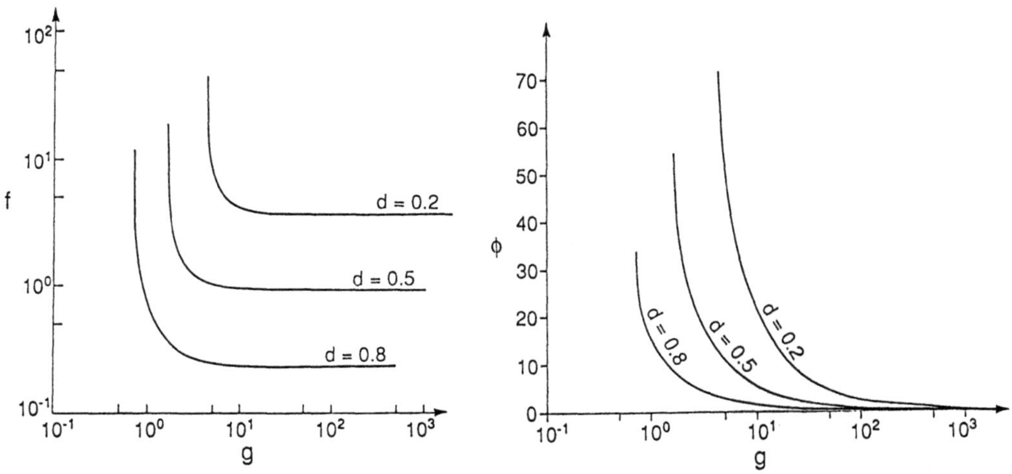

Figure 9.3: (a) Curves determined by condition (9.5.10) outside of which, partial closure on one side only, is possible. These are plotted as relations between f, g defined by (9.5.15). (b) Curves determined by condition (9.5.10) outside of which, partial closure on one side only, is possible. These are plotted as relations between g and φ, defined by (9.5.15) and (9.6.1) respectively.

to give an analytical proof. We will give a brief sketch of such an argument. What can be deduced from the numerical work is that if there are values of f, g for which (9.5.10) holds, then they must be very large. Thus we consider the ratio $M(t)/B(t)$ for $f \sim g \gg 1$. We see from (9.6.1) that $\tan\varphi \sim f/g$. The angle φ lies in the interval $[0, \pi/2]$. We can put $\psi = \pi/2$ since g is large. It will be clear later, on the basis of numerical evidence, that ωt_0 hovers just at or below $(-\pi)$. Also ωt_c lies above zero. Using these facts, it can be shown that

$$\frac{M(t)}{B(t)} \sim \frac{\sin(\omega t - \varphi)}{\sin(\omega t_c - \varphi) - \sin(\omega t - \varphi)}, \tag{9.6.2}$$

in the limit under consideration. By setting $\omega t = \omega t_c - \pi$ it is found that the constraint (9.5.10) is broken in the limit $d \sim 0$ for general viscoelastic materials. For the limiting case of a Maxwell material (see previous section) the constraint (9.5.10) is satisfied for each value of d. However $\Psi = 0$ at $d \sim 0$ so that the crack is always open.

Thus the question as to whether steady-state solutions, closing at one end, exist for a symmetric bending moment may be answered in the negative - a result in accord with intuition.

For $d \geq 1$ one-sided closure will of course always occur for both elastic and viscoelastic materials. In this case $B(t) = 0$ in the above formalism. One-sided closure under a bending moment that changes sign ($d < 1$) is possible only for viscoelasitc materials.

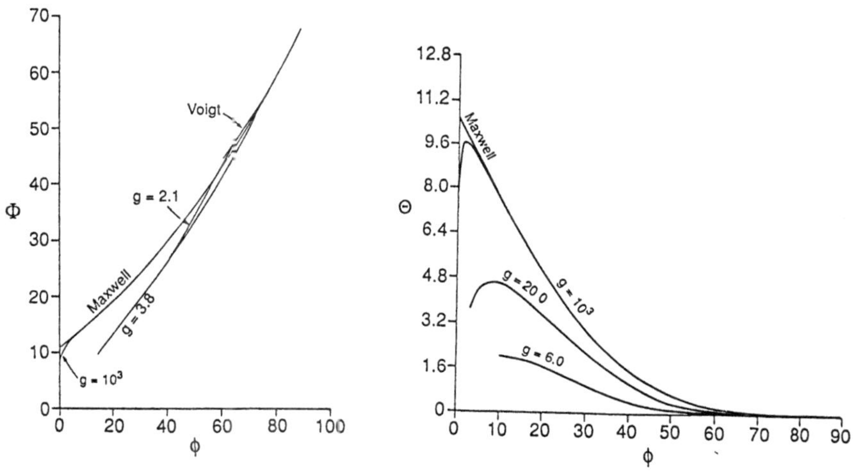

Figure 9.4: (a) The angular delay Φ defined by (9.5.11) as a function of the loss angle φ for various standard linear materials, with $d = 0.5$, $\gamma = \pi/6$. (b) The angular advance Θ defined by (9.5.11) as a function of the loss angle φ for various standard linear materials, with $d = 0.5$, $\gamma = \pi/6$.

The angular delay Φ and the angular advance Θ [see (9.5.11)], for a specific value of d and γ, may be determined by solving (9.5.5) or equivalently (9.5.13), (9.5.14). They are plotted on Figure 9.4(a) and (b) as functions of φ for various values of g and where condition (9.5.10) is observed. These curves show qualitatively similar behaviour to the corresponding quantities for a fixed length crack under a normal loading ([14], also chapter 8), though the range of variation of the two quantities is significantly less in the present case. In one curve on Figure 9.4(b), the characteristic 'hump' shape is truncated as a result of applying the constraint (9.5.10). Also, values of $\Theta(\varphi)$ corresponding to $g = 2.1, 3.8$ are negligible and have not been plotted on Figure 9.4(b).

It is in fact not surprising that the same qualitative behaviour occurs in these two quite different contexts, since this behaviour is a reflection of fundamental aspects of viscoelastic behaviour. The angular delay is a measure of the delay in closure once the stress has become compressive on the positive end of the crack while the angular advance is a measure of how early the crack opens before the stress has become tensile. Both of these phenomena are a result of creep effects from previous cycles tending to keep the crack open.

Golden and Graham (1991) discuss the required minmum magnitude of the bias term d for partial closure to occur on one side only for certain polymeric materials.

Chapter 10

THE PROBLEM OF A MOVING INDENTOR ON A VISCOELASTIC HALF-PLANE

We consider in this chapter the problem of a loaded indentor moving across a viscoelastic half-space, where plane strain conditions prevail. This problem was considered in general terms in section 6.1. We now present the detailed solution for one indentor on a lubricated half-plane.

This is the first of two examples of a point-based method, utilizing the integral equation (5.1.4), or in this case (6.1.37).

This problem has recently been considered in the inertial case, using a generalized version of the method described here [Golden and Graham (1994b)]. The inertial problem has also been considered by Walton (1984) for the case of power law viscoelastic behaviour.

10.1 The Integral Equation

Consider (6.1.37) for the case of a single moving load. Let the lowest point of the indentor have position $x_0(t)$ at time t. This specifies its overall motion. The displacement derivative in the contact region $C(t)$ will then have the form

$$u'(x,t) = f(x - x_0(t)), \tag{10.1.1}$$

where $f(x)$ is the slope of the indentor profile with the origin at the lowest point. The contact region $C(t)$ is in this case an interval $[a(t), b(t)]$. For a typical point $x \in C(t)$, the sets $W_u(x,t)$, $W_\sigma(x,t)$ are given by

$$W_u(x,t) = [t_1(x), t], \quad W_\sigma(x,t) = (-\infty, t_1(x)), \tag{10.1.2}$$

where $t_1(x)$ is defined as the (unique) solution of the equation

$$a(t_1(x)) = x, \quad x \in C(t). \tag{10.1.3}$$

In other words, it is the time at which the point x entered the contact interval. From (4.2.2) we have that the quantities Π_u, Π_σ, are given by

$$\Pi_u(t,t';x) = l(t-t')R(t';t_1(x),t),$$
$$\Pi_\sigma(t,t';x) = T_1(t,t';x)R(t';-\infty,t_1(x)), \tag{10.1.4}$$

where $T_1(t,t';x)$ is defined by (4.2.3) with x replacing $\underline{r}$. Then from (6.1.37) and (6.1.38) it follows that $v(x,t)$ obeys the equation

$$v(x,t) = \int_{-\infty}^{t_1(x)} dt' \int_{a(t')}^{b(t')} dx' K(x,x';t,t')v(x',t') + I(x,t), \tag{10.1.5}$$

where

$$K(x,x';t,t') = \frac{\sin(\pi\theta)n(x,t')T_1(t,t';x)}{\pi m(x',t')(x'-x)}, \tag{10.1.6}$$

by virtue of (6.1.24) and (6.1.25). This form emerges immediately if one uses (6.1.26) in deriving (6.1.37), instead of (6.1.19). Also

$$I(x,t) = \int_{t_1(x)}^{t} dt' l(t-t')u'(x,t'), \tag{10.1.7}$$

which is a known quantity, provided the contact interval, and consequently $t_1(x)$, is known. Two subsidiary conditions are required to determine $a(t)$, $b(t)$. These are (6.1.28), (6.1.31).

Equation (10.1.5) applies to general, transient conditions. It is not particularly amenable to analytic treatment. However, approximate solutions have been obtained, in the frictionless limit, applicable to times when transient effects have virtually decayed [Golden and Graham (1987a)]. The steady-state form of the equation is considerably simpler. It will be discussed later. In this context the term 'steady-state' implies conditions of uniform motion after transient effects have died away.

Let us consider briefly the case where there is more than one indentor. In each contact interval, we have an equation of the same form as (10.1.1), namely

$$u'(x,t) = f_i(x - x_{0i}(t)), \quad i - 1,2,\ldots,n, \quad x \in C_i(t), \tag{10.1.8}$$

where there are n indentors acting respectively on intervals $C_i(t) = [a_i(t), b_i(t)]$. The number of quantities $t_i(x)$ and their values depend on which contact interval x belongs to. If it lies in the leftmost interval $[a_1(t), b_1(t)]$, then (10.1.2) holds. If it lies in the next one, then there are three $t_i(x)$ and

$$W_u(x,t) = [t_3(x), t_2(x)] \bigcup [t_1(x), t],$$

$$W_\sigma(x,t) = (-\infty, t_3)\bigcup(t_2(x), t_1(x)), \tag{10.1.9}$$

and so on. Equation (6.1.37) now in fact becomes a set of coupled integral equations for the quantities $v^{(i)}(x,t)$, which are the function $v(x,t)$ evaluated in $C_i(t)$. The problem of several indentors has recently been considered by Fan, Golden and Graham (1994).

10.1.1 Steady-State Limit

We consider the single load case in the steady-state limit. Let the indentor have been moving in a negative x direction, at speed V, for a long time. If transient effects have died away then

$$a(t) = a_0 - Vt, \quad b(t) = b_0 - Vt, \tag{10.1.10}$$

so that, by virtue of (10.1.3)

$$t_1(x) = \frac{a_0 - x}{V}. \tag{10.1.11}$$

All quantities such as $u'(x,t)$, $v(x,t)$, $p(x,t)$ and so on, will be functions of $x + Vt$ rather than x, t separately. Also, the functions $n(x,t)$ and $m(x,t)$, given by (6.1.24) and (6.1.25), reduce to

$$m(x,t) = m(x + Vt), \quad n(x,t) = n(x + Vt), \tag{10.1.12}$$

where

$$m(u) = (u - a_0)^{1-\theta}(b_0 - u)^\theta, \quad n(u) = |a_0 - u|^{1-\theta}|b_0 - u|^\theta. \tag{10.1.13}$$

In (10.1.5), we change variables according to

$$x' \to y' = x' + Vt', \quad x \to y = x + Vt, \quad t' \to z = x + Vt', \tag{10.1.14}$$

giving

$$v(y) = \int_{a_0}^{b_0} dy' K(y,y')v(y') + I(y), \tag{10.1.15}$$

where

$$K(y,y') = \frac{\sin(\pi\theta)}{\pi}\int_{-\infty}^{a_0} \frac{dz}{V}\frac{T(y,z)n(z)}{(y'-z)m(y')},$$
$$T(y,z) = \int_z^{a_0} \frac{du}{V} l\left(\frac{y-u}{V}\right) k\left(\frac{u-z}{V}\right), \tag{10.1.16}$$

and

$$I(y) = \int_{a_0}^{y} \frac{dz}{V} l\left(\frac{y-z}{V}\right) u'(z), \tag{10.1.17}$$

where $u'(z)$ is $u'(x,t')$ expressed as function of $x + Vt'$ only. In the new coordinate system, the contact interval is stationary. Clearly, this is a coordinate system moving with the indentor.

It is convenient to choose length units and origin so that $[a_0, b_0] = [-1, 1]$ and time units such that $V = 1$. This means that any quantity of dimension length, which has value c in our initial units, becomes

$$c_1 = \frac{2c - (a_0 + b_0)}{b_0 - a_0}, \tag{10.1.18}$$

and any time in the theory of value τ in the initial units becomes

$$\tau_1 = \frac{\tau}{\lambda}, \quad \lambda = \frac{b_0 - a_0}{2V}, \tag{10.1.19}$$

in these units. We agree to use the same symbol for all quantities in terms of the new units. We also, for convenience, reintroduce the variables x, x', and y, now denoting space and time coordinates in the new moving, scaled reference system. In the light of (10.1.19), the functions $l(t)$, $k(t)$ in the initial units are replaced by

$$\lambda l(\lambda x), \quad \lambda k(\lambda x), \tag{10.1.20}$$

in the new units. These will be denoted by $l(x)$, $k(x)$, respectively. One may check that

$$\int_0^x dx' k(x - x') l(x') = \int_0^x dx' l(x - x') k(x') = \delta(x), \tag{10.1.21}$$

using a well-known property of the delta function. Note that, in terms of the physical relaxation and creep functions, under the proportionality hypothesis, one has, by virtue of (2.1.4), (2.1.6) and (6.1.14)

$$l(x) = \frac{1}{\varrho}\frac{d}{dx}[H(\lambda x)G(\lambda x)], \quad k(x) = \varrho\frac{d}{dx}[H(\lambda x)J(\lambda x)], \tag{10.1.22}$$

$$\varrho = \frac{\kappa - \alpha}{4}.$$

Equation (10.1.15) becomes

$$v(x) = \int_{-1}^{1} dx' K(x, x') v(x') + I(x), \tag{10.1.23}$$

where

$$K(x, x') = \frac{\sin(\pi\theta)}{\pi} \int_{-\infty}^{-1} dy \frac{T(x, y) n(y)}{(x' - y) m(x')}, \tag{10.1.24}$$

$$T(x, y) = \int_y^{-1} dy' l(x - y') k(y' - y), \quad |x| < 1, \quad y < -1$$

and

$$m(x') = (x' + 1)^{1-\theta}(1 - x')^{\theta}, \quad |x'| < 1,$$

$$n(y) = (-1 - y)^{1-\theta}(1 - y)^{\theta}, \quad y < -1. \tag{10.1.25}$$

Also,

$$I(x) = \int_{-1}^{x} dz l(x-z)u'(z). \tag{10.1.26}$$

The subsidiary conditions (6.1.28), (6.1.31) become, in the steady-state limit and for this coordinate system,

$$\int_{-1}^{1} dx \frac{v(x)}{m(x)} = 0,$$

$$q\int_{-1}^{1} dx \frac{xv(x)}{m(x)} = W_1,$$

$$W_1 = \frac{2W}{b_0 - a_0}. \tag{10.1.27}$$

The pressure, given by (6.1.27) becomes

$$p(x) = q\left[\frac{\sin(\pi\theta)m(x)}{\pi}\int_{-1}^{1} dx' \frac{v(x')}{(x'-x)m(x')} + v(x)\cos(\pi\theta)\right]. \tag{10.1.28}$$

In the frictionless limit, $\theta \to \frac{1}{2}$, so that

$$m(x) = (1-x^2)^{1/2}, \quad |x| < 1,$$

$$n(y) = (y^2-1)^{1/2}, \quad y < -1. \tag{10.1.29}$$

The kernel $K(x, x')$ reduces to

$$K(x,x') = \frac{1}{\pi}\int_{-\infty}^{-1} dy \frac{T(x,y)n(y)}{(x'-y)m(x')}, \tag{10.1.30}$$

while the load relation in (10.1.27) becomes

$$\int_{-1}^{1} dx \frac{v(x)x}{m(x)} = -W_1. \tag{10.1.31}$$

The pressure function, given by (10.1.28), reduces to

$$p(x) = -\frac{m(x)}{\pi}\int_{-1}^{1} dx' \frac{v(x')}{(x'-x)m(x')}. \tag{10.1.32}$$

It is interesting to observe that if the material is incompressible ($\kappa = 1$) or nearly so, then h, given by (6.1.14), is very large so that $\theta \to \frac{1}{2}$ and the solution has the same form as in the frictionless case. This is a significant simplification. Such an assumption is valid for a wide range of amorphous polymers, at temperatures well above their glass transition temperatures [Walton et al. (1978)].

10.2 Solution of the Integral Equation

The problem of a single indentor moving across a viscoelastic half-space is of considerable interest, largely because it offers a theoretical framework in which to investigate the phenomenon of hysteretic friction, which is discussed in more detail in section 10.3. If surface friction is neglected, the problem can be regarded as an indentor sliding across a lubricated half-plane, or a cylinder rolling over the plane. In the latter case, hysteretic friction is manifested as resistance to rolling, often termed rolling friction. This problem was first solved rigorously for a standard linear solid by Hunter (1961). He did so by transforming the equations into elastic form but where displacement and pressure are replaced these quantities acted upon by differential operators. This amounts to using the differential form of the constitutive relations. Morland (1962) considered the problem by formulating it as a set of dual integral equations. This is a generalization of the method of Sneddon (1951). It imposes no restriction the viscoelastic behaviour of the half-plane but there is no question of an analytic solution. Subsequently, Morland (1967, 1968), now using a methods closer to that of Hunter - the relationship between the methods is discussed by Golden (1977) - considered the problem of two viscoelastic cylinders rolling on each other, for the case where viscoelastic behaviour is described by a discrete spectrum model (Section 2.1).

Golden (1977) developed a method, essentially that described here, which is not restricted to particular types of materials. Later, Golden (1979, 1986), applied the method to the case where limiting friction is present.

For a power law material, an interesting method of solution has been developed by Walton et al. (1978). This has no apparent relationship with the method developed here. They derive, for steady-state conditions, an integral equation, which is a general Abel-type equation, and obtain explicit solutions. Walton (1984) generalized the method to apply to a material for which the shear relaxation function varies with depth also according to a power law.

This problem has also been considered in detail for an indentor moving over a layer of finite thickness, rather than a half-plane. We mention Alblas and Kuipers (1970), Margetson (1971, 1972), and Nachman and Walton (1978). Kalker (1975, 1977) reviews this topic in a systematic manner.

The discussion will be confined to the steady-state frictionless problem. In fact, however, the method works equally well for problems involving limiting friction [Golden (1979, 1986)]. Also, the transient problem was considered by Golden and Graham (1984).

The indentor is assumed to be smooth, with $u'(x)$ given by a polynomial

$$u'(x) = d(x) = \sum_{r=0}^{m} d_r x^r, \tag{10.2.1}$$

in the contact region. We are using the dimensionless moving coordinates introduced in section 10.1. The simplest case is that with a parabolic profile, given in dimensional moving coordinates by

$$u'(x) = -\frac{x}{R}, \quad a < x < b, \tag{10.2.2}$$

if the point of deepest indentation is at the origin. The contact interval is denoted by $[a, b]$, which is the interval $[a_0, b_0]$ of the previous section. The subscript is omitted for convenience. The quantity R is the radius of curvature at the origin. Returning to dimensionless coordinates, we have

$$u'(x) = d(x) = d_0 + d_1 x, \quad d_0 = -\frac{b+a}{2R}, \quad d_1 = -\frac{b-a}{2R}, \tag{10.2.3}$$

by virtue of (10.1.18). The quantity d_0 is proportional to the shift away from the origin of the indentor tip in this coordinate system. It is a measure of the asymmetry of the contact patch about the tip of the indentor. This form also applies approximately to the case where the indentor is cylindrical and the contact interval $(b-a)$ is small compared with the radius R of the cylinder. As we noted previously, the cylinder can be regarded as sliding or rolling.

The indentor profile given by (10.2.3) will be of most interest. However, for the moment, we keep the general form (10.2.1). Let

$$v(x) = \Delta(x) + q(x), \tag{10.2.4}$$

where

$$q(x) = \int_{-\infty}^{x} dx' l(x - x') d(x') = \int_{0}^{\infty} dy' l(y') d(x - y') = \sum_{r=0}^{m} q_r x^r, \tag{10.2.5}$$

and

$$\Delta(x) = \int_{-\infty}^{-1} dx' l(x - x')[u'(x') - d(x')]. \tag{10.2.6}$$

This is a decomposition into a polynominal part $q(x)$ and $\Delta(x)$, which in general has no polynomial part; in an exponential decay model, it decays exponentially with x, as we shall see. The polynomial $q(x)$ is known explicitly apart from the fact that it depends on the contact interval [see for example, (10.2.3)] which is not given a priori, but must be determined in the course of solving the problem.

We rewrite (10.1.23) in the form

$$\Delta(x) = \int_{-1}^{1} dx' K(x, x') \Delta(x') + N(x), \tag{10.2.7}$$

where

$$N(x) = \int_{-1}^{1} dx' K(x, x') q(x') - q(x) + I(x)$$

$$= \int_{-\infty}^{-1} dy T(x,y) Q(y) - \int_{-\infty}^{-1} dy l(x-y) d(y), \tag{10.2.8}$$

with the aid of (10.1.30), where

$$Q(y) = \frac{n(y)}{\pi} \int_{-1}^{1} dx' \frac{q(x')}{(x'-y)m(x')}. \tag{10.2.9}$$

The quantity $Q(y)$ can be conveniently evaluated in terms of the Chebyshev polynomials $T_l(x)$. Putting

$$q(x) = \sum_{l=0}^{m} a_l T_l(x), \tag{10.2.10}$$

we deduce from integrals given in [14] that for $y < -1$:

$$Q(y) = r(y) n(y) + q(y), \quad r(y) = \sum_{l=1}^{m} a_l U_{l-1}(y). \tag{10.2.11}$$

Now

$$\int_{-\infty}^{-1} dy T(x,y) q(y) = \int_{-\infty}^{-1} dy l(x-y) d(y), \tag{10.2.12}$$

by virtue of (10.1.21), (10.2.5) and definition of $T(x,y)$, given by (10.1.24). Therefore

$$N(x) = \int_{-\infty}^{-1} dy T(x,y) r(y) n(y). \tag{10.2.13}$$

We now specialize to particular viscoelastic models.

10.2.1 Discrete Spectrum Model

The problem will be solved for the case where the viscoelastic half-plane is characterized by a discrete spectrum model (Section 2.1). Therefore, from (10.1.19), (10.1.20) and (10.1.22), we have, for the frictionless case, relations similar to (7.3.1) and (7.3.2) derived from (2.1.39), (2.1.43), (2.1.44) and (6.1.11):

$$l(x) = l_0 \delta(x) + \sum_{i=1}^{N} l_i e^{-\alpha_i x}, \quad k(x) = k_0 \delta(x) + \sum_{i=1}^{N} k_i e^{-\beta_i x}, \tag{10.2.14}$$

where

$$l_0 = \frac{g_0}{1-\nu} = \frac{1}{1-\nu} \sum_{i=0}^{N} G_i, \quad k_0 = (1-\nu) h_0 = (1-\nu) J_0,$$

$$l_i = \frac{g_i}{1-\nu} = \frac{-\alpha_i G_i}{1-\nu}, \quad k_i = (1-\nu) h_i = (1-\nu) \beta_i J_i, \tag{10.2.15}$$

and

$$\alpha_i = \frac{b-a}{2V\tau_i}, \quad \beta_i = \frac{b-a}{2V\tau_i'}, \tag{10.2.16}$$

in terms of the decay constant τ_i, τ_i' for relaxation and creep. Also

$$l_0 = \frac{1}{k_0};$$

$$l_0 + \sum_{i=1}^{N} \frac{l_i}{\alpha_i - \beta_j} = 0, \quad j = 1,2,\ldots,N;$$

$$k_0 + \sum_{j=1}^{N} \frac{k_j}{\beta_j - \alpha_i} = 0, \quad i = 1,2,\ldots,N; \tag{10.2.17}$$

$$k_i = -\left(\sum_{j=1}^{N} \frac{l_j}{(\alpha_j - \beta_i)^2}\right)^{-1}, \quad i = 1,2,\ldots,N.$$

Note that from (10.2.15), the quantities l_i, $i = 1,2,\ldots,N$ are negative while the k_i, $i = 1,2,\ldots,N$ are positive.

The quantity $T(x,y)$ has the form

$$T(x,y) = \sum_{i,j=1}^{N} A_{ij} exp(-\alpha_i x + \beta_j y),$$

$$A_{ij} = \frac{l_i k_j}{\alpha_i - \beta_j} exp(-\alpha_i + \beta_j), \quad |x| < 1, \quad y < -1, \tag{10.2.18}$$

which may be shown with the aid of (10.2.17). The kernel $K(x,x')$, given by (10.1.30), therefore has the form

$$K(x,x') = \frac{-1}{\pi m(x')} \sum_{i,j=1}^{N} A_{ij} e^{-\alpha_i x} J(\beta_j, x'), \tag{10.2.19}$$

$$J(\beta,x) = \int_{-\infty}^{-1} dy \frac{n(y)e^{\beta y}}{y - x} = -m(x)\beta e^{\beta x} \int_x^1 dx' \frac{e^{-\beta x'}}{m(x')} E(\beta,x') + E(\beta,x),$$
$$E(\beta,x) = xK_0(\beta) - K_1(\beta), \quad |x| < 1,$$

in terms of modified Bessel functions with imaginary arguments $K_n(\beta)$, where the evaluation of the integral is carried out in [14]. Clearly, there is no singularity in the kernel at $x = x'$, in agreement with the more general conclusion of section 6.1. However, as noted in that context, the factor $m(x')$ in the denominator of $K(x,x')$ is not cancelled by $J(\beta_j,x')$ so that the kernel has a singularity and is not in fact square-integrable as it stands, which is the requirement that would ensure the applicability of Fredholm theory. Fortunately, the transformation $x = \cos\theta$ gives an integral equation in the variable θ which is non-singular, and Fredohlm theory applies. We will not, however, have occasion to call upon the full machinery of this theory. It is clear from (10.2.19) that the kernel is separable (or degenerate or of the Pincherle-Goursat type), so that the equation may be reduced without difficulty to a set of algebraic equations, which we shall now proceed to do. From (10.2.6), we have that

$$\Delta(x) = \sum_{i=1}^{N} C_i e^{-\alpha_i x},$$

$$C_i = l_i \int_{-\infty}^{-1} dx' e^{\alpha_i x'} [u'(x') - d(x')]. \tag{10.2.20}$$

Therefore

$$\int_{-1}^{1} dx' K(x,x')\Delta(x') = \sum_{i,j,k=1}^{N} A_{ij}\Gamma_{jk}C_k e^{-\alpha_i x}, \tag{10.2.21}$$

with

$$\Gamma_{jk} = \frac{1}{\pi}\int_{-\infty}^{-1} dy \int_{-1}^{1} dx' \frac{n(y) exp[\beta_j y - \alpha_i x']}{(x'-y)m(x')} = \Gamma(\beta_j, \alpha_k), \tag{10.2.22}$$

where $\Gamma(\beta_j, \alpha_k)$ is shown in [14] to be given by

$$\Gamma(\beta_j, \alpha_k) = -\frac{\alpha_k \chi(\beta_j, \alpha_k) + e^{\alpha_k - \beta_j}}{\alpha_k - \beta_j},$$

$$\chi(\beta_j, \alpha_k) = -[I_0(\alpha_k)K_1(\beta_j) + I_1(\alpha_k)K_0(\beta_j)], \tag{10.2.23}$$

in terms of the modified Bessel function $I_n(\alpha)$, $K_n(\beta)$.

Problem 10.2.1

Show that matrix $B_{ij} = exp(-\beta_i + \alpha_j)/(\beta_i - \alpha_j)$ is the inverse of A_{ij}.

Hint: split denominator of the product into partial fractions and use (10.2.17).

It follows from Problem (10.2.1) that the exponential term in $\Gamma(\beta_j, \alpha_k)$ contributes simply $\Delta(x)$ in (10.2.21), which cancels the left-hand side of (10.2.7). Remembering (10.2.13) and (10.2.18), we see that (10.2.7) reduces to

$$\sum_{i,j=1}^{N} A_{ij} E_j e^{-\alpha_i x} = 0,$$

$$E_j = \sum_{k=1}^{N} \frac{C_k \alpha_k \chi(\beta_j, \alpha_k)}{\alpha_k - \beta_j} - \int_{-\infty}^{-1} dy e^{\beta_j y} r(y) n(y). \tag{10.2.24}$$

This must be true for all $|x| < 1$. Therefore, we must have

$$\sum_{j=1}^{N} A_{ij} E_j = 0. \tag{10.2.25}$$

Since A_{ij} is invertible, it follows that the E_j are zero, giving

$$\sum_{k=1}^{N} \frac{C_k \alpha_k \chi(\beta_j, \alpha_k)}{\alpha_k - \beta_j} = \int_{-\infty}^{-1} dy e^{\beta_j y} r(y) n(y). \tag{10.2.26}$$

So, for a discrete spectrum model, (10.2.7) degenerates to the relatively simple form (10.2.26). This is a set of linear equations for the C_k in terms of the parameters α_i, β_i, $i = 1, 2, \ldots, N$. Since from (10.2.17) it follows that $\alpha_i \neq \beta_j$, $i, j = 1, 2, \ldots, N$, there is no reason to expect difficulties in obtaining unique solutions, except perhaps for very special materials.

The conditions (10.1.27) for $\theta = \frac{1}{2}$ and (10.1.31) become, in terms of the modified Bessel functions with imaginary argument $I_n(\alpha)$,

$$\sum_{n=1}^{N} C_n I_0(\alpha_n) = -\frac{1}{\pi} \int_{-1}^{1} dx \frac{q(x)}{m(x)},$$

$$\sum_{n=1}^{N} C_n I_1(\alpha_n) - \frac{1}{\pi} \int_{-1}^{1} dx \frac{q(x) x}{m(x)} = \frac{W_1}{\pi} \tag{10.2.27}$$

by virtue of (10.2.4), (10.2.20) and standard integrals [14]. Equation (10.2.5) gives, for a cylindrical indentor:

$$q(x) = q_0 + q_1 x,$$

$$q_0 = d_0 \left(l_0 + \sum_{i=1}^{N} \frac{l_i}{\alpha_i} \right) - d_1 \sum_{i=1}^{N} \frac{l_i}{\alpha_i^2},$$

$$q_1 = d_1 \left(l_0 + \sum_{i=1}^{N} \frac{l_i}{\alpha_i} \right). \tag{10.2.28}$$

In this case, from (10.2.10), (10.2.11) we have

$$r(x) = q_1, \tag{10.2.29}$$

so that (10.2.26) becomes

$$\sum_{k=1}^{N} \frac{C_k \alpha_k \chi(\beta_j, \alpha_k)}{\alpha_k - \beta_j} = \frac{K_1(\beta_j) q_1}{\beta_j}. \tag{10.2.30}$$

These relations were first given by Morland (1968). The constraints (10.2.27) become

$$q_0 + \sum_{n=1}^{N} C_n I_0(\alpha_n) = 0, \quad -\frac{1}{2} q_1 + \sum_{n=1}^{N} C_n I_1(\alpha_n) = \frac{W_1}{\pi}. \tag{10.2.31}$$

Equations (10.2.30), (10.2.31) together determine the constants C_k and the parameters d_0, d_1 which give the contact region.

An expression for the contact pressure may also be given. From (10.2.4), (10.2.20), and (10.2.28), we have

$$v(x) = \sum_{i=1}^{N} C_i e^{-\alpha_i x} + q_0 + q_1 x. \tag{10.2.32}$$

The polynomial contribution to (10.1.32) gives, with the aid of standard integrals [14]:

$$p_1(x) = \frac{m(x)}{\pi} \int_{-1}^{1} dx' \frac{q_0 + q_1 x'}{(x - x')m(x')} = -q_1 m(x), \quad |x| < 1, \tag{10.2.33}$$

while, with the aid of an integral evaluated in that reference the contribution to (10.1.32) of the exponential terms may be evaluated, giving the combined expression

$$p(x) = -q_1 m(x) - \sum_{i=1}^{N} C_i \alpha_i e^{-\alpha_i x} \int_{-1}^{x} e^{\alpha_i y} [y I_0(\alpha_i) - I_1(\alpha_i)] \frac{dy}{m(y)}. \tag{10.2.34}$$

It is easy to check that this expression in bounded, in fact zero, at $x = \pm 1$. It must also be confirmed that pressure is positive everywhere in the interval $[-1, 1]$. It would seem that in general, this can only be done numerically, for specified configurations.

Let us write the above equations for the standard linear model. The viscoelastic functions are given by (2.1.24) and (2.1.31). The quantities l_0, l_1, k_0, k_1 are related to the physical coefficients by (10.2.15). Equation (10.2.30) gives that

$$C = \frac{K_1(\beta) q_1 (1 - \beta/\alpha)}{\beta \chi(\beta, \alpha)}, \tag{10.2.35}$$

where subscripts have been dropped and α, β are the inversé dimensionless decay constants given by

$$\alpha = \frac{b - a}{2V\tau} = -\frac{d_1 R}{V\tau}, \quad \beta = \frac{b - a}{2V\tau'} = -\frac{d_1 R}{V\tau'}, \tag{10.2.36}$$

by virtue of (10.2.3) and (10.2.16). Equations (2.1.31) and (2.1.32) give that

$$1 - \frac{\beta}{\alpha} = \frac{G_1}{G_0 + G_1}. \tag{10.2.37}$$

Also, (10.2.28) and (10.2.15) yield

$$q_0 = \frac{1}{1 - \nu} \left(d_0 G_0 + \frac{d_1 G_1}{\alpha} \right) = \frac{1}{1 - \nu} \left(d_0 G_0 - \frac{V\tau G_1}{R} \right),$$

$$q_1 = \frac{d_1 G_0}{1 - \nu}. \tag{10.2.38}$$

Equation (10.2.31) gives

$$q_0 = -\frac{K_1(\beta) I_0(\alpha) q_1}{\beta \chi(\beta, \alpha)} \frac{G_1}{G_0 + G_1},$$

$$1 + \frac{2W_1}{\pi q_1} = 2\frac{K_1(\beta)I_1(\alpha)}{\beta\chi(\beta,\alpha)}\frac{G_1}{G_0+G_1}. \qquad (10.2.39)$$

The second relations of (10.2.38) and (10.2.39) give an implicit equation for d_1 or $(b-a)$. Once this is evaluated, d_0 or $(b+a)$ and C can be calculated without difficulty from (10.2.38), the first relation of (10.2.39) and (10.2.35). The quantities $v(x)$ and $p(x)$ are then obtainable from (10.2.32) and (10.2.34). Minor restructuring of (10.2.39) gives relations identical to those of Hunter (1961), allowing for the different convention for velocity direction.

Some numerical solutions are presented in that reference and, for the case of two decay times, by Morland (1968). For a discussion of the small viscoelasticity approximation see [14].

10.3 Hysteretic Friction

The phenomenon of hysteretic friction arises when for example a loaded indentor is moved across a viscoelastic half-space. As a result of the time-dependent deformation of the viscoelastic material, the indentor experiences a resisting force. The reason for this can be viewed in either of two equivalent ways: (a) the time-dependent deformation results in a dissipation of mechanical energy into heat; or (b) the recovery after deformation is not instantaneous so that the resisting pressure on the indentor is greater than that in the direction of motion. It may be shown [14] that for an indentor on a lubricated surface, under a load W, moving along the negative x axis, the hysteretic friction coefficient is given by

$$f_H = \frac{1}{W}\int_C ds\, p(\underline{r},t)\frac{\partial}{\partial x}u(\underline{r},t), \qquad (10.3.1)$$

in terms of the pressure $p(\underline{r},t)$ and the normal displacement $u(\underline{r},t)$. Adapting this formula to steady-state, plane strain condition in terms of dimensionless coordinates, we obtain

$$f_H = \frac{1}{W_1}\int_{-1}^{1} dx\, p(x)u'(x),$$

$$W_1 = \int_{-1}^{1} dx\, p(x) = \frac{2W}{b-a} = \frac{-W}{d_1 R}. \qquad (10.3.2)$$

The quantity W is now the load per unit length. For a cylindrical indentor, characterized by (10.2.3), f_H has the form

$$f_H = d_0 + \frac{2\delta}{\pi}\int_{-1}^{1} dx\, p(x)x, \qquad (10.3.3)$$

where

$$\delta = \frac{\pi d_1}{2W_1} = -\frac{\pi d_1^2 R}{2W} = -\frac{1}{l_0}\left(\frac{d_1}{d_{i1}}\right)^2, \qquad (10.3.4)$$

$$d_{i1} = -\left(\frac{2W}{\pi l_0 R}\right)^{1/2},$$

where d_{i1} is the value of d_1 in the instantaneous elastic limit. Let us now express $p(x)$ in terms of $v(x)$, using (10.1.32). On interchanging the order of integration, we obtain

$$f_H = d_0 + \frac{2\delta}{\pi}\int_{-1}^{1}\frac{dx}{m(x)}v(x)J(x), \quad J(x) = \frac{1}{\pi}\int_{-1}^{1}dx'\frac{m(x')x'}{x'-x}, \tag{10.3.5}$$

where $m(x)$ is given by (10.1.29). Evaluating $J(x)$ from integrals in [14] gives

$$J(x) = \frac{1}{2} - x^2 = -\frac{1}{2}T_2(x). \tag{10.3.6}$$

in terms of the Chebyshev polynomials. However, by virtue of (10.1.27), we can write f_H as

$$f_H = d_0 + \frac{2\delta}{\pi}\int_{-1}^{1}\frac{dx}{m(x)}v(x)(e - x^2), \tag{10.3.7}$$

where the value of e is arbitrary. Let us return to the choice $e = \frac{1}{2}$. Then, (10.3.6) and the orthogonality relations for the Chebyshev polynomials give that the integral in (10.3.7) will annihilate $q(x)$, the polynominal part of $v(x)$, as given by (10.2.4) and (10.2.28). It follows that

$$f_H = d_0 + \frac{2\delta}{\pi}\int_{-1}^{1}\frac{dx}{m(x)}\Delta(x)\left(\frac{1}{2} - x^2\right). \tag{10.3.8}$$

For a discrete spectrum model, $\Delta(x)$ is given by (10.2.20). Substituting this form gives,

$$f_H = d_0 - \delta\sum_{i=1}^{N} C_i I_2(\alpha_i), \tag{10.3.9}$$

which can be evaluated if (10.2.30), (10.2.31) are solved. Numerical results have been presented by Hunter (1961) and Morland (1968). Characteristically, this quantity turns out to be hump shaped, as a function of velocity, in qualitative agreement with experiment. The results of Hunter (1961) are presented in Figure 10.1.

Discussion of this phenomenon in the presence of surface friction and for small viscoelasticity may be found in [14].

The methodology presented here may be generalized to the corresponding inertial problem for the steady-state case [Golden and Graham (1994b)].

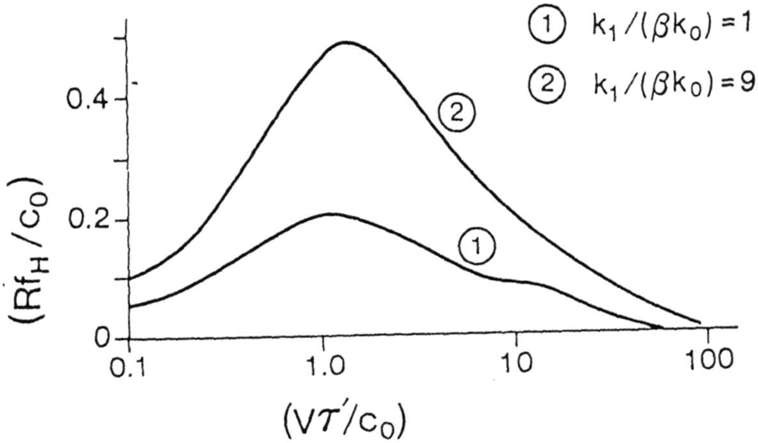

Figure 10.1: Variation of coefficient of friction f_H with velocity: c_0 is the semi-contact width for a stationary cylinder [after Hunter (1961)].

Chapter 11

STEADY-STATE SOLUTIONS FOR A VISCOELASTIC CRACK UNDER AN ALTERNATING BENDING MOMENT

11.1 Introduction

In this chapter, we consider a second example of a problem where a point-based approach must be adopted. The transient problem of a crack in a viscoelastic medium subject to a sinusoidal bending moment has been considered over the early history of loading [14], [Graham (1982), Golden and Graham (1984)]. In the present chapter, we consider the steady-state limit of this problem, after transient effects have died away. Given the complexity of the behaviour, even over the first cycle or two, it may be anticipated that analytical solutions are not likely to emerge.

It has been shown [14], [Graham (1982), Golden and Graham (1984)] that sudden partial closure can occur as in elastic theory and also gradual partial closure, which is a specifically viscoelastic effect. Here, we will look for solutions which close gradually at both ends, in an alternating fashion. Solutions involving closure at only one end were obtained in chapter 9. They can be obtained by area-based methods.

The developments in this chapter are based on [Golden, Graham and Trummer (1994)].

11.2 The General Integral Equation

Plane strain condition are presumed to apply and we consider the particular cross-section of the medium lying in the xy plane. In this plane, let the crack, which is assumed to be of fixed length, lie along the x-axis occupying $[-c, c]$ as in Figure 11.1. Let the material be subject to a bending moment such that the neutral axis is the y-axis. The ratio of this moment to the moment of inertia of the cross-section about the y-axis [Sokolnikolf (1956)] is termed $\eta(t)$. The sign convention is such that when $\eta(t)$ is positive, there will be compression across the positive x-axis, at least at points far from the origin. At this stage, we assume that $\eta(t)$ is fluctuating from positive to negative values. Later, we will settle on simple sinusoidal behaviour.

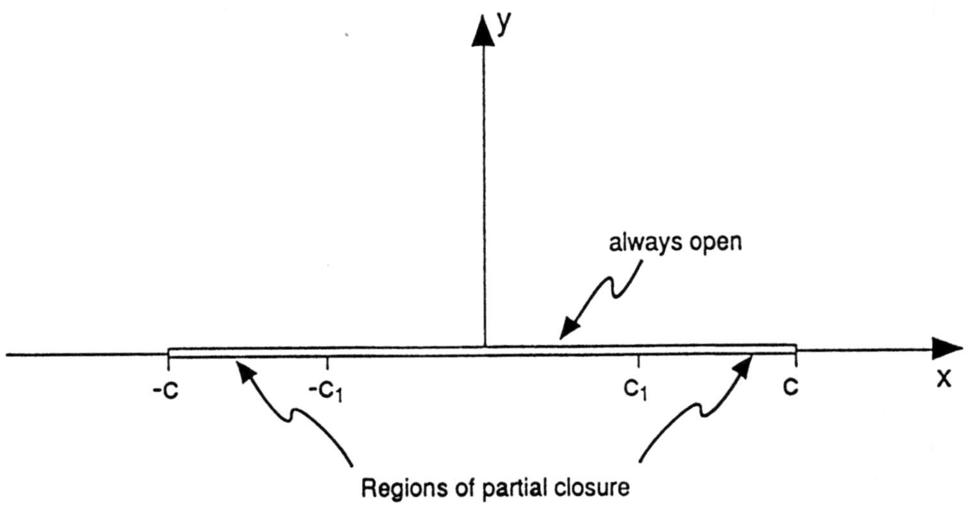

Figure 11.1: Crack profile along x-axis.

The transfer of the stresses to the crack face giving on the open portion of the crack (9.2.1) and the boundary conditions (9.2.2), (9.2.3) apply unchanged to the present problem. The quantities $u(x,t)$, $p(x,t)$ and $q(x,t)$ also carry the same meaning.

It should be emphasized that the boundary conditions apply to the problem with the external loading transferred to the crack face. In particular, the normal stress in the original physical problem is $\sigma_{22}(x,y,t) - \eta(t)x$ which must be negative or zero on the crack face. We start with (see (9.2.4))

$$q(x,t) = -\frac{1}{\pi}\int_{-a(t)}^{b(t)} dx' \frac{u'(x',t)}{x'-x},$$

$$q(x,t) = \int_{-\infty}^{t} dt' k(t-t') p(x,t'), \tag{11.2.1}$$

where the open portion of crack at time t is $[-a(t), b(t)]$, $a(t) \leq c$, $b(t) \leq c$. Inverting the finite Hilbert transform in (11.2.1) gives

$$u'(x,t) = \frac{1}{\pi m(x,t)} \int_{-a(t)}^{b(t)} dx' \frac{m(x',t)q(x',t)}{x'-x}, \tag{11.2.2}$$

where

$$m(x,t) = \begin{cases} (c^2 - x^2)^{1/2}, & a(t) = b(t) = c, \\ \left[\frac{c-x}{a(t)+x}\right]^{1/2}, & a(t) < c, \quad b(t) = c, \\ \left[\frac{c+x}{b(t)-x}\right]^{1/2}, & a(t) = c, \quad b(t) < c, \\ [(b(t)-x)(x+a(t))]^{-1/2}, & a(t) < c, \quad b(t) < c, \end{cases} \tag{11.2.3}$$

which builds in the correct singular or otherwise behaviour at open and partially closed ends. These various options are deduced from the basic inverse Hilbert transform formula ($m(x,t) = [(b(t)-x)(x+a(t)]^{1/2}$) by using the subsidiary conditions

$$\int_{-a(t)}^{c} dx \left[\frac{c-x}{a(t)+x}\right]^{-1/2} q(x,t) = 0, \quad a(t) < c, \quad b(t) = c,$$

$$\int_{-c}^{b(t)} dx \left[\frac{c+x}{b(t)-x}\right]^{1/2} q(x,t) = 0, \quad a(t) = c, \quad b(t) < c,$$

$$\int_{-a(t)}^{b(t)} dx \frac{q(x,t)}{[(a(t)+x)(b(t)-x)]^{1/2}} = 0, \quad a(t) < c, \quad b(t) < c, \tag{11.2.4}$$

which state essentially that no singularity occurs in $u'(x,t)$ at partially closed ends; this follows from (11.2.2), putting $m(x,t) = [(b(t)-x)(x+a(t))]^{1/2}$.

Thus we can see from (11.2.2) that if $q(x,t)$ is known over the open portion of the crack, then the displacement derivative is determined on the crack face. Also, using (11.2.1), (11.2.2), an integral given in [14] and (11.2.4), we obtain $q(x,t)$ off the open portion, in terms of $q(x,t)$ on the open portion, in the form

$$q(x,t) = \frac{\epsilon(x,t)}{\pi n(x,t} \int_{-a(t)}^{b(t)} dx' \frac{m(x',t)q(x',t)}{x'-x}, \quad x \notin [-a(t), b(t)], \tag{11.2.5}$$

where

$$\epsilon(x,t) = \begin{cases} 1, & x > b(t), \\ -1, & x < -a(t), \end{cases} \tag{11.2.6}$$

and $m(x,t)$ is given by (11.2.3) while

$$n(x,t) = \begin{cases} (x^2 - c^2)^{1/2}, & a(t) = b(t) = c, \\ \left[\frac{x-c}{x+a(t)}\right]^{1/2} & a(t) < c, \quad b(t) = c, \\ -\left[\frac{c+x}{x-b(t)}\right]^{1/2}, & a(t) = c, \quad b(t) < c, \\ -[(x-b(t))(x+a(t)]^{-1/2}, & a(t) < c, \quad b(t) < c. \end{cases} \tag{11.2.7}$$

If $q(x,t)$ is known for all times then $p(x,t)$ may be deduced from the second equation of (11.2.1) to be

$$p(x,t) = \int_{-\infty}^{t} dt' l(t-t')q(x,t'). \tag{11.2.8}$$

Thus all quantities are known everywhere on the x-axis if $q(x,t)$ is given on the open portion of the crack at all times. it is then not difficult, in principle, to determine stresses and displacements everywhere [14]. Therefore, if $q(x,t)$ is known on the open crack face at all times, the problem is solved. We now set about deriving an integral equation satisfied by $q(x,t), x \in [-a(t), b(t)]$. Decomposing $q(x,t)$ according to the point-based prescription outlined in section 4.2 we obtain

$$q(x,t) = \int_{W_q(x,t)} dt' \Gamma_q(t,t';x)q(x,t') + \int_{W_p(x,t)} dt' \Gamma_p(t,t';x)p(x,t'),$$

$$x \in [-a(t), b(t)], \tag{11.2.9}$$

where $W_q(x,t)$ is the set of all times $t' \leq t$ such that $x \notin (-a(t'), b(t'))$ and $W_p(x,t)$ is the set of all times $t' \leq t$ such that $x \in (-a(t'), b(t'))$. Also, $\Gamma_q(t,t';x)$ and $\Gamma_p(t,t';x)$ are given by $\Gamma_u(t,t')$ and $\Gamma_v(t,t')$ defined by (4.1.10) where the transition times $t_r(x)$, $r = 1, 2, \ldots$, are now dependent on x, as outlined in section 4.2. The time $t_1(x)$ is the last time prior to t that x enters the open portion, $t_2(x)$ the immediately previous time that it leaves the open portion and so on.

Substituting (11.2.5) into the first term on the right-hand side of (11.2.9) gives the following integral equation for $q(x,t), x \in [-a(t), b(t)]$:

$$q(x,t) = \int_{W_q(x,t)} dt' \int_{-a(t')}^{b(t')} dx' K(x,x';t,t')q(x',t') + I(x,t), \tag{11.2.10}$$

where

$$K(x,x';t,t') = \frac{\epsilon(x,t')m(x',t')\Gamma_q(t,t';x)}{\pi n(x,t')(x'-x)},$$

$$I(x,t) = \int_{W_p(x,t)} dt' \Gamma_p(t,t';x)p(x,t'). \tag{11.2.11}$$

Note that $I(x,t)$ is known apart from the fact that the transition times $t_r(x)$ have yet to be determined. Because of the definition of $W_q(x,t)$, x' and x do not become equal in the first term of (11.2.10), except at transition times. These occurrences produce logarithmic singularities which are integrable.

Apart from (11.2.10), there are the subsidiary conditions (11.2.4) which must be satisfied by $q(x,t)$. These in fact are the extra constraints which allow determination of the progress of partial closure.

11.3 Steady-State Conditions under a Sinusoidal Loading

Let us now assume that a bending moment such that

$$\eta(t) = \eta_0 \sin(\omega t), \tag{11.3.1}$$

has been acting for a long time, sufficient to allow transient effects to die away, so that steady-state conditions prevail. The crack is symmetrical about the y-axis. We assume that partial closure takes place as far as c_1 (Figure 11.1) on the positive side and $-c_1$ on the negative side, in an alternating fashion. Thus, there are three subdivisions of $[-c, c]$, which must be considered separately, namely $[c_1, c]$, $[-c, -c_1]$ in both of which closure takes place, and $[-c_1, c_1]$ which is never closed. In this latter interval

$$q(x,t) = -\eta_0 x \int_{-\infty}^{t} dt' k(t-t') \sin(\omega t') = -\eta_0 x Q(t), \quad -c_1 \leq x \leq c_1, \tag{11.3.2}$$

$$Q(t) = \mathrm{Im}[e^{i\omega t}\hat{k}(\omega)],$$

where $\hat{k}(\omega)$ is given by (8.3.3). For a material with constant Poisson's ratio the quantity $\hat{k}(\omega)$ is proportional to the complex modulus for creep, which is the inverse of the complex modulus for relaxation (see (8.3.4)).

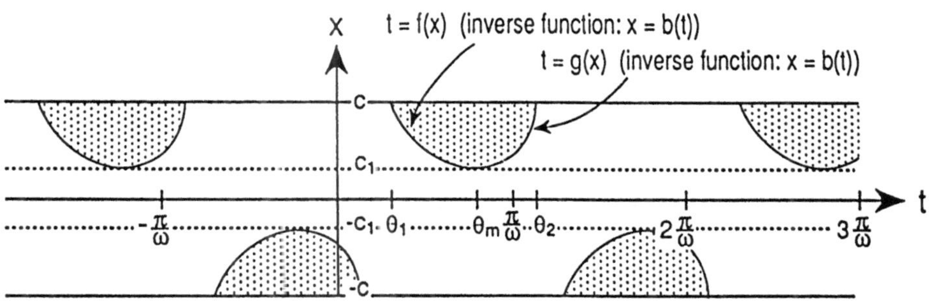

Figure 11.2: A priori possible behaviour of the quantity $b(t)$ schematically portrayed with related quantities indicated.

We now consider regions in which closure occurs. The expected patterns of partial closure are sketched schematically in Figure 11.2, based on the findings of Graham

(1982), Golden and Graham (1984). Actually, it emerges that the actual partial closure patterns are much more square-shaped than suggested by Figure 11.2. However, there are advantages in basing the discussions of analytical developments on the schematic shapes presented on that figure. We focus on the time interval $[0, 2\pi/\omega]$. Consider the closure at the positive end, finishing just after $t = \pi/\omega$. We agree that the time at which this period of closure begins is denoted by θ_1, and where it ends by θ_2. It reaches its maximum at θ_m. All physical quantities, and in particular $q(x,t)$, are periodic in

$$\Delta = 2\pi/\omega. \tag{11.3.3}$$

Also, because the crack is symmetrical with respect to the y-axis and the bending moment varys symmetrically in time we have

$$q(-x,t) = q(x, t \pm \pi/\omega). \tag{11.3.4}$$

This extra symmetry is not in fact essential to the method developed here but it results in considerable simplification of the mathematics and so we build it in without further ado. It means essentially that we need only consider the positive interval $[c_1, c]$.

On Figure 11.2, we have assumed, guided by earlier work, that there is a period for which the crack is fully open, between partial closures. It is not clear, a priori whether this is the only possibility or whether the opposite or either can occur. This will emerge from the solution. The progress of closing and reopening may be characterized by two functions, indicated on Figure 11.2:

$$\begin{aligned} t &= f(x), \quad t = g(x), \\ \theta_1 &= f(c), \quad \theta_m = f(c_1), \\ \theta_m &= g(c_1), \quad \theta_2 = g(c). \end{aligned} \tag{11.3.5}$$

These are inverses of the function $x = b(t)$ so that

$$b(f(x)) = x, \quad b(g(x)) = x, \quad c_1 \leq x \leq c. \tag{11.3.6}$$

We now wish to write down the integral equation (11.2.10) under the steady-state assumption for $x \in [c_1, b(t)]$. Consider first the case where $t \in [\theta_m, 2\pi/\omega]$. Note that $a(t') = c$ for $t' \in W_q(x,t)$. Using the periodicity of $q(x,t)$, we obtain

$$q(x,t) = \int_{f(x)}^{g(x)} dt' \int_{-c}^{b(t')} dx' C(x,x';t')\Sigma_1(t,t';x)q(x',t') + I_1(x,t), \tag{11.3.7}$$

where from (11.2.3) and (11.2.7)

$$C(x,x';t') = \frac{m(x',t')}{\pi n(x,t')(x'-x)} = -\frac{1}{\pi}\left[\frac{x-b(t')}{x+c}\right]^{1/2}\left[\frac{x'+c}{b(t')-x'}\right]^{1/2}\frac{1}{(x'-x)}, \tag{11.3.8}$$

and

$$\Sigma_1(t,t';x) = \sum_{k=0}^{\infty} N_{2k+1}(t,t'-k\Delta;x),$$

$$I_1(x,t) = \int_{g(x)}^{f(x)+\Delta} dt' K_1(t,t';x)p(x,t'), \tag{11.3.9}$$

$$K_1(t,t';x) = \sum_{k=0}^{\infty} N_{2k}(t,t'-k\Delta;x).$$

Note that in the first term of $K_1(t,t';x)$, the integral stops at t, since this term is equal to $k(t-t')$, which vanishes for $t' > t$. This will be relevant in the next section when we come to sum such series.

Next consider the case where $t \in [0,\theta_m]$. We obtain

$$q(x,t) = \int_{f(x)}^{g(x)} dt' \int_{-c}^{b(t')} dx' C(x,x';t')\Sigma_2(t,t';x)q(x',t') + I_2(x,t), \tag{11.3.10}$$

where $C(x,x';t')$ is given by (11.3.8) and

$$\Sigma_2(t,t';x) = \sum_{k=1}^{\infty} N_{2k-1}(t,t'-k\Delta;x),$$

$$I_2(x,t) = \int_{g(x)-\Delta}^{f(x)} dt' K_2(t,t';x)p(x,t'), \tag{11.3.11}$$

$$K_2(t,t';x) = \sum_{k=0}^{\infty} N_{2k}(t,t'-k\Delta;x).$$

Note that K_2 has the same form as K_1, though (x,t) is in a different region.

We now wish to transform (11.3.7) and (11.3.10) into integral equations over $[c_1,c]$, using (11.3.2) and (11.3.4). Equation (11.3.7) becomes

$$q(x,t) = \int_{f(x)}^{g(x)} dt' \int_{c_1}^{c} dx' \Sigma_1(t,t';x)\{C(x,x';t')q(x',t')H(b(t')-x')$$

$$+ C(x,-x';t')q(x',t' \pm \pi/\omega)\} + L_1(x,t) + I_1(x,t), \tag{11.3.12}$$

where

$$L_1(x,t) = \int_{f(x)}^{g(x)} dt' \Sigma_1(t,t';x) \int_{-c_1}^{c_1} dx' C(x,x';t')q(x',t'). \tag{11.3.13}$$

In (11.3.13), $q(x',t')$ is given by (11.3.2). Also, the sign in $q(x',t' \pm \pi/\omega)$ in (11.3.12) is chosen so that $0 \le t' \pm \pi/\omega \le 2\pi/\omega$, since we are considering only this interval. The function $H(x)$ is the Heaviside step-function. Similarly (11.3.10) becomes

$$q(x,t) = \int_{f(x)}^{g(x)} dt' \int_{c_1}^{c} dx' \Sigma_2(t,t';x)\{C(x,x';t')q(x',t')H(b(t')-x')$$

$$+ C(x, -x'; t')q(x', t' \pm \pi/\omega)\} + L_2(x,t) + I_2(x,t), \tag{11.3.14}$$

where

$$L_2(x,t) = \int_{f(x)}^{g(x)} dt'\Sigma_2(t,t';x) \int_{-c_1}^{c_1} dx' C(x,x';t')q(x',t'). \tag{11.3.15}$$

The space integral in the second term of (11.3.13) and (11.3.15) can be carried out with the aid of the formula

$$N(x,t') = \frac{1}{\pi}\left[\frac{x-b(t')}{x+c}\right]^{1/2} \int_{-c_1}^{c_1} dx' \left[\frac{x'+c}{b(t')-x'}\right]^{1/2} \frac{x'}{x'-x}$$

$$= \frac{1}{\pi}\left[\frac{x-b(t')}{x+c}\right]^{1/2}\left[x + \frac{b(t')+c}{2}\right]\left\{\cos^{-1}\left[\frac{c-b(t')-2c_1}{b(t')+c}\right] - \cos^{-1}\left[\frac{c-b(t')+2c_1}{b(t')+c}\right]\right\}$$

$$+\frac{1}{\pi}\left[\frac{x-b(t')}{x+c}\right]^{1/2}\left[\sqrt{(c-c_1)(b(t')+c_1)} - \sqrt{(c+c_1)(b(t')-c_1)}\right]$$

$$-\frac{2x}{\pi}\left\{\tan^{-1}\left[\sqrt{\frac{x+c}{x-b(t')}}\sqrt{\frac{b(t')+c_1}{c-c_1}}\right] - \tan^{-1}\left[\sqrt{\frac{x+c}{x-b(t')}}\sqrt{\frac{b(t')-c_1}{c+c_1}}\right]\right\}, \tag{11.3.16}$$

giving

$$L_i(x,t) = \eta_0 \int_{f(x)}^{g(x)} dt'\Sigma_i(t,t';x)Q(t')N(x,t'), i = 1,2. \tag{11.3.17}$$

The subsidiary condition appropriate when closure is taking place at the positive end has the form

$$\int_{-c}^{b(t)} dx \left[\frac{c+x}{b(t)-x}\right]^{1/2} q(x,t)$$

$$= \int_{c_1}^{c} dx \left\{\left[\frac{c+x}{b(t)-x}\right]^{1/2} q(x,t)H(b(t)-x) + \left[\frac{c-x}{b(t)+x}\right]^{1/2} q(x,t\pm\pi/\omega)\right\}$$

$$-\eta_0 Q(t)\int_{-c_1}^{c_1} dx \left[\frac{c+x}{b(t)-x}\right]^{1/2} x = 0 \tag{11.3.18}$$

11.4 The Standard Linear Model

We now specialize to the case where the medium is a standard linear solid with a unique Poisson's ratio. In particular, this allows us to easily carry out the infinite summations in (11.3.9), (11.3.11). The functions $l(t)$, $k(t)$ have the form given by (9.4.5) and (9.4.6). Adapting the results of Problem 9.4.1, we obtain

$$\Sigma_1(t,t';x) = \frac{k_1 l_0}{1-E(x)} exp\{-\beta(t-g(x)) - \alpha(g(x)-t')\},$$

$$E(x) = exp\{(\alpha - \beta)(f(x) - g(x)) - \beta\Delta\}. \tag{11.4.1}$$

Also, it can be shown by similar techniques that

$$K_1(t, t'; x) = k_0\delta(t - t') + \frac{k_1 e^{-\beta(t-t')}}{1 - E(x)}, \tag{11.4.2}$$

from which $I_1(x, t)$ can be deduced to be

$$I_1(x,t) = k_0 p(x,t) + \frac{k_1}{1 - E(x)} \int_{g(x)}^{f(x)+\Delta} dt' e^{-\beta(t-t')} p(x,t') - k_1 \int_t^{f(x)+\Delta} dt' e^{-\beta(t-t')} p(x,t'), \tag{11.4.3}$$

on recalling the remark after (11.3.9). Since $p(x,t) = -\eta(t)x$ in the region of integration of (11.4.3), we can, on using (11.3.1), give a more explicit expression for $I_1(x,t)$. Let

$$M(t) = \int dt' e^{\beta t'} \sin(\omega t') = \frac{e^{\beta t} \sin(\omega t - \psi)}{(\beta^2 + \omega^2)^{1/2}}, \tag{11.4.4}$$

where ψ is defined by the relations

$$\tan\psi = \frac{\omega}{\beta}, \quad 0 \leq \psi \leq \pi/2. \tag{11.4.5}$$

Then

$$I_1(x,t) = k_0 p(x,t) - \frac{\eta_0 x k_1 e^{-\beta t}}{1 - E(x)} \{E(x)M(f(x) + \Delta) - M(g(x)) + (1 - E(x))M(t)\}. \tag{11.4.6}$$

The quantity $L_1(x,t)$ can be evaluated by numerical quadrature, using (11.3.17) and (11.4.1) once the function $b(t)$ and its inverses are known.

Furthermore, one can show that

$$\Sigma_2(t, t'; x) = \frac{k_1 l_0}{1 - E(x)} exp\{-\beta(t - g(x) + \Delta) - \alpha(g(x) - t')\}$$

$$= e^{-\beta\Delta}\Sigma_1(t, t'; x), \tag{11.4.7}$$

and $I_2(x,t)$ has the same form as (11.4.6) but with t replaced by $(t + \Delta)$. Similarly,

$$L_2(x,t) = e^{-\beta\Delta} L_1(x,t), \tag{11.4.8}$$

which follows from (11.4.7).

The quantity $Q(t)$ defined by (11.3.2) has the form

$$Q(t) = \left[k_0 + \frac{k_1\beta}{\beta^2 + \omega^2}\right] \sin(\omega t) - \frac{k_1\alpha}{\beta^2 + \omega^2} \cos(\omega t), \tag{11.4.9}$$

which determines $q(x,t)$ for $-c_1 < x < c_1$.

11.5 Conclusion

The numerical solution of (11.3.12) and (11.3.14) is carried through by Golden, Graham and Trummer (1994) and plots of the pressure function and $b(t)$ are presented . While smooth rather than sudden closing always occurs, the closing action is extremely rapid for medium and large values of β.

A simple approximate method of solution is also proposed which has a surprising degree of universality.

Acknowledgement: Work on these lectures was carried out at the Dublin Institute for Advanced Studies. The lectures were typed by Dr. H.Z. Fan. The research was supported by funds provided by the Natural Sciences and Engineering Research Council of Canada.

Bibliography

[1] Alblas, J.B., Kuipers, M.: The contact problem of a rigid cylinder rolling on a thin viscoelastic layer, Int. J. Eng. Sci., **8** (1970), 363-380.

[2] Bowie, O.L., Freese, C.E.: On the 'overlapping' problem in crack analysis, Eng. Fracture Mech., **8** (1976), 373-379.

[3] Comninou, M., Dundurs, J.: On the frictional contact in crack analysis. Eng. Fracture Mech., **12** (1979), 117-123.

[4] Fan, H.Z., Golden, J.M., Graham, G.A.C.: The problem of several indentors moving on a viscoelastic half-plane, Submitted for publication, 1994.

[5] Ferry, J.D.: Viscoelastic Properties of Polymers, 2nd ed., Wiley, New York 1970.

[6] Galin, L.A.: Contact Problems in the Theory of Elasticity (Ed. I.N. Sneddon) [English translation by H. Moss], Department of Mathematics, North Carolina State University, Releigh 1961

[7] Gladwell, G.M.L: Contact Problem in the Classical Theory of Elasticity, Sijthoff and Noordhoff, Alphen aan den Rijn 1980.

[8] Golden, J.M.: Hysteretic friction of a plane punch on a half-plane with arbitrary viscoelastic behaviour, Q.J. Mech. Appl. Math. **30** (1977), 23-49.

[9] Golden, J.M.: The problem of a moving rigid punch on an unlubricated viscoelastic half-plane, Q. J. Mech. Appl. Math. **32** (1979), 25-52.

[10] Golden, J.M.: Frictional viscoelastic contact problems, Q.J. Mech. Appl. Math. **39** (1986), 125-137.

[11] Golden, J.M., Graham, G.A.C.: Crack in a viscoelastic field of pure bending, Int. J. Eng. Sci. **22** (1984), 801-811.

[12] Golden, J.M., Graham, G.A.C.: The transient quasi-static plane viscoelastic moving load problem, Int. J. Eng. Sci. **25** (1987a), 65-84.

[13] Golden, J.M., Graham, G.A.C.: The steady-state plane normal viscoelastic contact problem, Int. J. Eng. Sci. **25** (1987b), 277-291.

[14] Golden, J.M., Graham, G.A.C.: Boundary Value Problem in Linear Viscoelasticity, Springer-Verlag, Berlin, Heidelberg 1988.

[15] Golden, J.M., Graham, G.A.C.: A fixed length crack in a sinusoidally loaded general viscoelastic medium, in: Continuum Mechanics and Its Applications (Eds. G.A.C. Graham and S.K. Malik), Hemisphere Publishing Corporation, Washington, D.C. 1989, 171-188.

[16] Golden, J.M., Graham, G.A.C.: Stress, strain and area-controlled modes for the steady-state normal viscoelastic contact problem, in: Proceedings of the 11th International Annual Conference of the Canadian Applied Mathematices Society (Ed. M. Rahman), Computational Mechanics Publications, Southampton 1990, 739-753.

[17] Golden, J.M., Graham, G.A.C.: Asymmetric steady-state solutions for a crack in a viscoelastic field of pure bending, Int. J. Eng. Sci. **29** (1991), 223-235.

[18] Golden, J.M., Graham, G.A.C.: Energy loss in steady-state solutions to viscoelastic crack problems, Canadian Applied Mathematics Quarterly **2** (1994a), 1-26.

[19] Golden, J.M., Graham, G.A.C.: The viscoelastic moving contact problem with inertial effects included, Submitted for publication, (1994b).

[20] Golden, J.M., Graham, G.A.C., Lan, Q.: Three-dimensional steady-state indentation problem for a general viscoelastic material, Q. Appl. Math., to appear, 1994.

[21] Golden, J.M., Graham. G.A.C., M. Trummer: Steady-state solutions for a viscoelastic crack under an alternating bending moment, Int. J. Eng. Sci. **32** (1994), 899-924.

[22] Graham, G.A.C.: The contact problem in the linear theory of viscoelastcity, Int. J. Eng. Sci. **3** (1965), 27-46.

[23] Graham, G.A.C.: The contact problem in the linear theory of viscoelastcity where the time dependent contact area has any number of maxima and minima, Int. J. Eng. Sci. **5** (1967), 495-514.

[24] Graham, G.A.C.: The correspondence principle of linear viscoelasticity theory for mixed boundary value problems involving time-dependent boundary regions, Q. Appl. Math. **26** (1968), 167-174.

[25] Graham, G.A.C.: Stresses and displacements in cracked linear viscoelastic bodies that are acted upon by alternating tensile and compressive loads, Int. J. Eng. Sci. **14** (1976), 1135-1142.

[26] Graham, G.A.C.: Viscoelastic crack in a field of pure bending, Mech. Res. Commun. **9** (1982), 219-226.

[27] Graham, G.A.C., Golden, J.M.: The three-dimensional steady-state viscoelastic indentation problem, Int. J. Eng. Sci. **26** (1988a), 121-126.

[28] Graham, G.A.C., Golden, J.M.: The generalized partial correspondence principle in linear viscoelasticity, Q. Appl. Math.**46** (1988b), 527-538; **49** (1991), 397.

[29] Graham, G.A.C., Sabin, G.C.W.: The correspondence principle of linear viscoelasticity for problems that involve time-dependent regions, Int. J. Eng. Sci. **11** (1973), 123-140.

[30] Graham, G.A.C., Sabin, G.C.W.: Steady state solutions for a cracked standard linear viscoelastic body, Mech. Res. Commun. **8** (1981), 361-368.

[31] Green, A.E., Zerna, W.: Theoretical Elasticity, Oxford University Press, 1968.

[32] Gross, G.: Mathematical Structure of the Theories of Viscoelasticity, Hermann & Cie, Paris 1953.

[33] Gurtin, M.E., Sternberg, E.: On the linear theory of viscoelasticity, Arch. Rat. Mech. Anal. **2** (1962), 291-356.

[34] Harvey, R.B.: On the deformation of a viscoelastic cylinder rolling without slipping, Q. J. Mech. Appl. Math. **28** (1975), 1-24.

[35] Hunter, S.C.: The Hertz problem for a rigid spherical indentor and a viscoelastic half-space, J. Mech. Phys. Solids **8** (1960), 219-234.

[36] Hunter, S.C.: The rolling contact of a rigid cylinder with a viscoelastic half-space, J. Appl. Mech. **28** (1961), 611-617.

[37] Kalker, J.J.: Aspects of contact mechanics, in: The Mechanics of the Contact Between Deformable Bodies (Eds. D. de Pater, J.J. Kalker), Delft University Press 1975, 1-25.

[38] Kalker, J.J.: A survey of the mechanics of contact between solid bodies, J. Appl. Math. Phys. (ZAMP) **57** (1977), 13-17.

[39] Kikuchi, N., Oden, J.T.: Contact Problems in Elasticity: A Study of Variational Inequalities and Finite Element Methods, SIAM, Philadelphia 1988.

[40] Margetson, J.: Rolling contact of a smooth viscoelastic strip between rotating rigid cylinders, Int. J. Mech. Sci. **13** (1971), 207-215.

[41] Margetson, J.: Rolling contact of a rigid cylinder over a smooth elastic or viscoelastic layer, Acta Mech. **13** (1972), 1-9.

[42] Morland, L.W.: A plane problem of rolling contact in linear viscoelasticity theory, J. Appl. Mech. **29** (1962), 345-352.

[43] Morland, L.W.: Exact solutions for rolling contact between viscoelastic cylinders, Q. J. Mech. Appl. Math. **20** (1967), 73-106.

[44] Morland, L.W.: Rolling contact between dissimilar viscoelastic cylinders, Q. Appl. Math. **25** (1968), 363-376.

[45] Muskhelishvili, N.I.: Some Basic Problems of the Mathematical Theory of Elasticity, 4th ed. [English translation by J.R.M. Radok], Noordhoff, Groningen 1963.

[46] Nachman, A., Walton, J.R.: The sliding of a rigid indentor over a power law viscoelastic layer, J. Appl. Mech. **45** (1978), 111-113.

[47] Oden, J. T., Lin, T.L.: On the general rolling contact problem for finit deformations of a viscoelastic cylinder, Computer Methods in Applied Mechanics and Engineering **57** (1986), 297-367.

[48] Sabin, G.C.W., Graham, G.A.C.: The normal aging viscoelastic contact problem, Int. J. Eng. Sci. **18** (1980), 751-757.

[49] Sih, G.C.(ed): Methods of Analysis and Solution of Crack Problems, Noorhoff, Leyden 1973.

[50] Sih, G.C., Liebowitz, H.: Mathematical theories of brittle fracture, in: Fracture: An Advanced Treaties, Vol. II (Ed. H. Liebowitz), Academic, New York 1968, 67-190.

[51] Sneddon, I.N.: Fourier Transforms, McGraw-Hill, New York 1951.

[52] Sokolnikoff, I.S.: Mathematical Theory of Elasticity, 2nd ed., McGraw-Hill, New York 1956.

[53] Ting, T.C.T.: Contact problems in the linear theory of viscoelasticity, J. Appl. Mech. **35** (1968), 248-254.

[54] Ting, T.C.T.: A mixed boundary value problem in viscoelasticity with time-dependent boundary regions, in: Proseeding of the Eleventh Midwestern Mechanics Conference (Eds. H.J. Weiss, D.F. Young, W.F. Riley, T.R. Rogge), Iowa University Press 1969, 591-598.

[55] Walton, J.R.: The sliding with Coulomb friction of a rigid indentor over a power-law inhomogeneous linearly viscoelastic half-plane, J. Appl. Mech. **51** (1984), 289-293.

[56] Walton, J.R., Nachman, A. Schapery, R.A.: The sliding of a rigid indentor over a power-law viscoelastic half-space, Q. J. Mech. Appl. Math. **31** (1978), 296-321.

WAVE SOLUTIONS IN LINEAR VISCOELASTIC MATERIALS

A. Morro
DIBE, University of Genoa, Genoa, Italy

Abstract

Dissipative materials are modelled as linear viscoelastic and the thermodynamic restrictions are investigated in detail. For definiteness, attention is mainly confined to solids. Some concepts about thermodynamics and dissipativity in materials with memory are re-visited. Next wave propagation is considered and emphasis is given to the consequences of thermodynamic restrictions on the amplitude evolution. Waves are described as surface discontinuities, time-harmonic waves, rays. The known properties of surface discontinuities are recalled to emphasize the exponential attenuation induced by thermodynamics. Particular attention is then addressed to the description of time-harmonic inhomogeneous waves and asymptotic rays in isotropic and anisotropic solids. Recent results are exhibited along with some work in progress.

1. Introduction

A joint research on linear viscoelasticity (cf. [1]) has produced a simple and exhaustive scheme of thermodynamic restrictions. The scheme generalizes previous works and unifies seemingly unrelated relations in a set of properties which are consequences of a single inequality on the relaxation function. This modelling is certainly of interest in connection with the thermodynamic analysis of constitutive equations. Further, the model provides a general framework for the investigation of any problem involving dissipative materials. In particular, important developments are made more systematic within the study of existence-uniqueness properties and the derivation of wave solutions (cf. [2]).

In this paper we consider wave solutions in viscoelastic materials (solids and fluids); for definiteness, attention is mainly confined to solids. For the purpose of a self-contained presentation, we start with the essentials of modelling and thermodynamic analysis. In this connection we show also that some views about thermodynamics, memory effects and dissipativity are fallacious. Next attention is addressed to wave solutions. Emphasis is given to the connection between thermodynamic restrictions and properties of wave solutions. First discontinuity waves are examined thus revisiting a subject which is well known in the literature. Next time-harmonic waves are dealt with in different cases. Detailed results are established about waves in uniform, isotropic solids; the solutions are in fact inhomogeneous waves characterized by complex-valued wave vectors such that the real part and the imaginary one are not parallel. Qualitative consequences are obtained also for anisotropic solids. Finally, anisotropic, non-uniform, pre-stressed solids are considered and solutions are examined as asymptotic rays.

Notation. The reader is supposed to be acquainted with the essentials of continuum mechanics. Direct tensor notation is used throughout unless, for particular reasons, the component notation looks convenient. The space under consideration is the three-dimensional Euclidean point space $\mathcal{E}$. Vectors are elements of the associated translation space V and are denoted by boldface minuscules. Second-order tensors are meant as linear transformations of V into V. Sym is the set of symmetric (second-order) tensors while sym denotes the symmetric part of a tensor. Lin stands for the set of all tensors; Lin(Sym) is the set of all linear transformations of Sym into Sym; Lin^+ is the subset of Lin whose elements have a positive determinant. Boldface majuscules usually denote elements of Lin or Lin(Sym); $\mathbf{1}$ is the identity of Lin. The symbols tr and det stand for the trace and the determinant of tensors while the superscript T denotes transpose. Letting $\mathbf{u}, \mathbf{v} \in V$, we write $\mathbf{u} \cdot \mathbf{v}$ for the standard inner product. If $\mathbf{L}, \mathbf{M} \in \text{Lin}$, then $\mathbf{L} \cdot \mathbf{M}$ stands for $\text{tr}(\mathbf{L}\mathbf{M}^T)$. If $\mathbf{M}$ is an element of Lin or Lin(Sym), then the writing $\mathbf{M} > 0$ (< 0) means that $\mathbf{M}$ is positive (negative) definite in V or in Sym. For any two vectors $\mathbf{u}, \mathbf{v}$, $\mathbf{u} \otimes \mathbf{v}$ denotes the tensor product. A superposed dot denotes the material time derivative while ∇ and $\nabla \cdot$ stand for the gradient and the divergence. Moreover, $\mathbb{R}, \mathbb{R}^+, \mathbb{R}^{++}, \mathbb{R}^-$ are the set of reals, positive reals, strictly positive reals, and negative reals, respectively.

2. Modelling and thermodynamics of viscoelastic solids

Consider a solid body occupying a three-dimensional region Ω in a stress-free placement (or configuration). A point of the body is identified with the position vector $\mathbf{x}$ in the reference placement Ω.

Linear (infinitesimal) viscoelasticity and especially the model of linear viscoelastic solid can be traced back to Boltzmann [3] who, roughly speaking, considered an elastic material with memory. He elaborated the model of (linear) viscoelastic solid on the basis of the following assumptions.

At any point $\mathbf{x}$ of the body, the stress at any time t depends upon the strain at all preceding times. If the strain at all preceding times is in the same direction, then the effect is to reduce the corresponding stress. The influence of a previous strain on the stress depends on the time elapsed since that strain occurred and is weaker for those strains that occurred long ago. Such properties make the model of solid elaborated by Boltzmann a material with fading memory. In addition, Boltzmann made the assumption that a superposition of the influence of previous strains holds, which means that the stress-strain relation is linear.

A formal derivation of the stress-strain relation, within the framework of materials with memory, can be given (cf., e.g., [1], ch. 3). The result is that the Cauchy stress $\mathbf{T}$, at time t, is given by

$$\mathbf{T}(t) = \mathbf{G}_0\mathbf{E}(t) + \int_0^\infty \mathbf{G}'(u)\mathbf{E}(t-u)du$$

where $\mathbf{E} \in \mathrm{Sym}$ is the infinitesimal strain tensor and $\mathbf{G}_0, \mathbf{G}(u) \in \mathrm{Lin}(\mathrm{Sym})$. Also, $\mathbf{G}_0$ is called the instantaneous elastic modulus and is taken to be bounded. The assumption $\mathbf{G}' \in L^1(\mathbb{R}^+)$ allows the definition of the relaxation function

$$\mathbf{G}(u) = \mathbf{G}_0 + \int_0^u \mathbf{G}'(\xi)d\xi$$

and the limit value $\mathbf{G}_\infty = \mathbf{G}(\infty)$, called equilibrium elastic modulus. The solid character of the body is modelled by the condition

$$\mathbf{G}_\infty > 0$$

in Sym. Since $\mathbf{G}' \in L^1(\mathbb{R}^+)$ we have $\mathbf{G}'(\infty) = 0$. In all relations, the possible dependence of $\mathbf{E}$ and the constitutive quantities $\mathbf{G}_0, \mathbf{G}'$ on the point $\mathbf{x}$ is understood and not written.

Though the second law of thermodynamics can be stated for approximate cycles (cf. [1], ch. 3), here it is enough to consider the statement for cycles. For simplicity we let cycles occur in the interval $[0, d)$. Denote by $\mathcal{T}(\mathbf{E}^t)$ the constitutive functional. In the linear approximation the stress power is $\mathbf{T}\cdot\dot{\mathbf{E}}$ and hence the second law requires that the constitutive functional $\mathcal{T}(\mathbf{E}^t)$ satisfy

$$\int_0^d \mathcal{T}(\mathbf{E}^t)\cdot\dot{\mathbf{E}}(t)dt > 0 \tag{1}$$

for every non-trivial cycle. To exploit the inequality (1) we consider a strain function

$$\tilde{\mathbf{E}}(t) = \mathbf{E}_1 \cos\omega t + \mathbf{E}_2 \sin\omega t$$

where $\omega \in \mathbb{R}^{++}$, $\mathbf{E}_1, \mathbf{E}_2 \in \mathrm{Sym}$. The period d is any integer times $2\pi/\omega$. Substitution into the inequality (1) and integration with respect to t yields

$$\mathbf{E}_1 \cdot [\mathbf{G}_0^T - \mathbf{G}_0]\mathbf{E}_2 - \int_0^\infty [\mathbf{E}_1 \cdot \mathbf{G}'(u)\mathbf{E}_1 + \mathbf{E}_2 \cdot \mathbf{G}'(u)\mathbf{E}_2] \sin\omega u \, du$$

$$-\int_0^\infty \mathbf{E}_1 \cdot [\mathbf{G}'(u) - \mathbf{G}'^T(u)]\mathbf{E}_2 \cos\omega u\, du > 0$$

for every $\omega \in \mathbb{R}^{++}$ and every $\mathbf{E}_1, \mathbf{E}_2 \in \mathrm{Sym}$.

By the Riemann-Lebesgue lemma, the limit $\omega \to \infty$ makes the integrals vanish. Hence the arbitrariness of $\mathbf{E}_1, \mathbf{E}_2 \in \mathrm{Sym}$ implies that

$$\mathbf{G}_0 = \mathbf{G}_0^T. \tag{2}$$

Substitution and the limit $\omega \to 0$ yields

$$\mathbf{E}_1 \cdot [\mathbf{G}_\infty^T - \mathbf{G}_\infty]\mathbf{E}_2 \geq 0.$$

The arbitrariness of $\mathbf{E}_1, \mathbf{E}_2$ yields

$$\mathbf{G}_\infty = \mathbf{G}_\infty^T.$$

Hence we are left with the inequality

$$\int_0^\infty [\mathbf{E}_1 \cdot \mathbf{G}'(u)\mathbf{E}_1 + \mathbf{E}_2 \cdot \mathbf{G}'(u)\mathbf{E}_2] \sin\omega u\, du + \int_0^\infty \mathbf{E}_1 \cdot [\mathbf{G}'(u) - \mathbf{G}'^T(u)]\mathbf{E}_2 \cos\omega u\, du < 0. \tag{3}$$

If, in particular, we let $\mathbf{E}_1 = \mathbf{E}_2$, which we can, we obtain that, in Sym,

$$\mathbf{G}'_s(\omega) < 0, \qquad \omega \in \mathbb{R}^{++}, \tag{4}$$

where

$$\mathbf{G}'_s(\omega) = \int_0^\infty \mathbf{G}'(u) \sin\omega u\, du$$

is the half-range Fourier sine transform of $\mathbf{G}'$. Obviously, if $\mathbf{G}'(u) = \mathbf{G}'^T(u)$, $u \in \mathbb{R}^+$, then the inequality (3) is equivalent to the negative definiteness of $\mathbf{G}'_s$. The symmetry of $\mathbf{G}'(u)$ for $u \neq 0, \infty$ need not be true but is usually taken to hold.

The physical meaning of (4) is made clear by the observation that, if $\mathbf{E}(t) = \mathbf{E}\sin\omega t$ and $d = 2\pi/\omega$,

$$\int_0^d \mathcal{T}(\mathbf{E}^t) \cdot \dot{\mathbf{E}}(t)dt = -\pi\mathbf{E} \cdot \mathbf{G}'_s(\omega)\mathbf{E},$$

whence $-\pi\mathbf{E} \cdot \mathbf{G}'_s(\omega)\mathbf{E}$ is the energy dissipated in one period. That is why $-\mathbf{G}'_s$ is also called the loss modulus.

The inequality (4) implies a set of important relations. For later convenience we examine only the consequence on $\mathbf{G}'_0 := \mathbf{G}'(0)$. To this end observe that $\mathbf{G}'$ is defined on $\mathbb{R}^+$. Fourier inversion formula in terms of $\mathbf{G}'_s$ only can be established by carrying $\mathbf{G}'$ over to $\mathbb{R}^-$ by letting

$$\mathbf{G}'(-u) = -\mathbf{G}'(u), \qquad u \in \mathbb{R}^{++}$$

and defining $\mathbf{G}'_0$ as the right limit of $\mathbf{G}'$. Fourier inversion formula thus provides

$$\mathbf{G}'(u) = \frac{2}{\pi}\int_0^\infty \sin\omega u\, \mathbf{G}'_s(\omega)d\omega, \qquad u \in \mathbb{R}^{++}.$$

Integration with respect to u yields

$$\mathbf{G}(u) - \mathbf{G}_0 = \frac{2}{\pi}\int_0^\infty \frac{1-\cos\omega u}{\omega}\mathbf{G}'_s(\omega)d\omega.$$

The negative definiteness of $\mathbf{G}'_s$ implies that

$$\mathbf{G}(u) - \mathbf{G}_0 < 0, \qquad u \in \mathbb{R}^{++}.$$

The limit as $u \to \infty$ yields

$$\mathbf{G}_\infty - \mathbf{G}_0 < 0 \tag{5}$$

while the limit of $[\mathbf{G}(u) - \mathbf{G}_0]/u$ as $u \to 0$ yields

$$\mathbf{G}'_0 \leq 0, \tag{6}$$

in Sym.

Let $\mathbf{G}''(u) = d\mathbf{G}'(u)/du$ and assume $\mathbf{G}'' \in L^1(\mathbb{R}^+)$. An integration by parts and the condition $\mathbf{G}'(\infty) = 0$ yield

$$\mathbf{G}'_s(\omega) = \frac{1}{\omega}\mathbf{G}'_0 + \frac{1}{\omega}\int_0^\infty \mathbf{G}''(u)\cos\omega u\, du.$$

As $\omega \to \infty$, by the Riemann-Lebesgue lemma the integral vanishes. Hence, by (4), a large enough ω allows us to conclude that

$$\mathbf{G}'_0 < 0. \tag{7}$$

Accordingly, the additional hypothesis $\mathbf{G}'' \in L^1(\mathbb{R}^+)$ makes the inequality (6) into (7).

The inequality (3), along with the symmetry property (2), is also sufficient for the (second law) inequality (1) to hold. The detailed proof is given in [1], §3.3.

If the solid is isotropic the tensors $\mathbf{G}_0$ and $\mathbf{G}'(u)$ are represented by two scalar values μ_0, λ_0 and $\mu'(u), \lambda'(u)$ as

$$\mathbf{T}(t) = 2\mu_0\mathbf{E}(t) + 2\int_0^\infty \mu'(u)\mathbf{E}(t-u)du + \left[\lambda_0 \mathrm{tr}\,\mathbf{E}(t) + \int_0^\infty \lambda'(u)\mathrm{tr}\,\mathbf{E}(t-u)du\right]\mathbf{1}. \tag{8}$$

In such a case (2) is obviously true while (3) becomes

$$\mu'_s(\omega) < 0, \qquad 2\mu'_s(\omega) + 3\lambda'_s(\omega) < 0, \qquad \forall\omega \in \mathbb{R}^{++}. \tag{9}$$

The analogue of (6) is

$$\mu'_0 \leq 0, \qquad 2\mu'_0 + 3\lambda'_0 \leq 0. \tag{10}$$

Models of dissipative solids are often based on combinations of springs and dashpots. A rather general model is the standard linear solid (also named after Zener). In one-dimensional form, it is represented by

$$\dot{T} + \frac{c_2}{\nu}T = (c_1 + c_2)\dot{E} + \frac{c_1 c_2}{\nu}E,$$

where c_1, c_2, ν are suitable parameters. If $c_1 = 0$ we obtain the Maxwell (fluid) model, if $c_2 \to \infty$ we obtain the Kelvin-Voigt model. Hence we consider the three-dimensional version in the form

$$\mathbf{T} + \tau\dot{\mathbf{T}} = \mathbf{C}\mathbf{E} + \mathbf{D}\dot{\mathbf{E}}, \tag{11}$$

where $\tau > 0$ may be viewed as a relaxation time and $\mathbf{C}, \mathbf{D} \in \text{Lin(Sym)}$. A relation of the form (11) is considered, e.g., in [4]. Also with a view to applications, it is of interest to investigate the thermodynamic restrictions on the constitutive tensors $\mathbf{C}, \mathbf{D}$. Integration of (11) on $(-\infty, t]$ yields

$$\mathbf{T}(t) = \frac{1}{\tau}\mathbf{C}\int_{-\infty}^{t} \exp[-(t-\xi)/\tau]\mathbf{E}(\xi)d\xi + \frac{1}{\tau}\mathbf{D}\int_{-\infty}^{t} \exp[-(t-\xi)/\tau]\dot{\mathbf{E}}(\xi)d\xi.$$

An integration by parts and a change of variable yields

$$\mathbf{T}(t) = \frac{1}{\tau}\mathbf{D}\mathbf{E}(t) + \frac{1}{\tau^2}(\tau\mathbf{C} - \mathbf{D})\int_{0}^{\infty} \exp(-u/\tau)\mathbf{E}(t-u)du. \tag{12}$$

This shows that the standard linear solid is in fact a particular case of viscoelastic solid where

$$\mathbf{G}_0 = \frac{1}{\tau}\mathbf{D}, \qquad \mathbf{G}'(u) = \frac{1}{\tau^2}(\tau\mathbf{C} - \mathbf{D})\exp(-u/\tau).$$

Hence we have

$$\mathbf{G}_\infty = \mathbf{C}$$

and $\mathbf{G}'_s$ turns out to be

$$\mathbf{G}'_s(\omega) = (\tau\mathbf{C} - \mathbf{D})\frac{\omega}{1+\omega^2\tau^2}.$$

Accordingly,

$$\mathbf{G}_0 > \mathbf{G}_\infty > 0 \qquad \Longleftrightarrow \qquad \mathbf{D} > \tau\mathbf{C} > 0$$

while

$$\mathbf{G}'_s(\omega) < 0 \qquad \Longleftrightarrow \qquad \tau\mathbf{C} - \mathbf{D} < 0.$$

In conclusion, compatibility with thermodynamics is summarized by

$$\mathbf{D} > \tau\mathbf{C}, \qquad \mathbf{C} = \mathbf{C}^T, \qquad \mathbf{D} = \mathbf{D}^T$$

while the solid character is given by

$$\mathbf{C} > 0.$$

3. Clausius-Planck inequality and dissipation

Denote by θ the absolute temperature, η the entropy, and ϵ the internal energy. Also, let $w = \mathbf{T}\cdot\mathbf{L}$ be the stress power. One way of stating the second law of thermodynamics is through the Clausius-Planck inequality which reads

$$\delta := \theta\dot{\eta} - \big(\dot{\epsilon} - \frac{w}{\rho}\big) \geq 0$$

δ being called *internal dissipation* (cf. [5], Lect. 2). Let

$$\boldsymbol{\tau} = \big(\frac{1}{\rho}\mathbf{T}\mathbf{F}^{-T}, -\eta\big), \qquad \boldsymbol{\lambda} = (\mathbf{F}, \theta).$$

In terms of the free energy $\psi = \epsilon - \theta\eta$, we have

$$\delta = \boldsymbol{\tau} \cdot \dot{\boldsymbol{\lambda}} - \dot{\psi}.$$

Consider the constitutive equations for ψ and $\boldsymbol{\tau}$ in the form

$$\psi(t) = \mathcal{F}(\boldsymbol{\lambda}^t), \qquad \boldsymbol{\tau}(t) = \boldsymbol{\mathcal{T}}(\boldsymbol{\lambda}^t).$$

Differentiability of the functional $\mathcal{F}$ yields

$$\delta = (\boldsymbol{\tau} - D_{\boldsymbol{\lambda}}\mathcal{F}) \cdot \dot{\boldsymbol{\lambda}} + \mathcal{D}$$

where $D_{\boldsymbol{\lambda}}$ is the derivative with respect to the present value of the history $\boldsymbol{\lambda}^t$ and $-\mathcal{D} = d\mathcal{F}(\boldsymbol{\lambda}^t|\,{}_r\dot{\boldsymbol{\lambda}}^t)$ is the Fréchet derivative at the history $\boldsymbol{\lambda}^t$ in the direction of the restricted history ${}_r\dot{\boldsymbol{\lambda}}^t$. Along with the continuity of $\boldsymbol{\mathcal{T}}$ this implies that (cf. [5], Lect. 3) $\mathcal{F}$ and $\boldsymbol{\mathcal{T}}$ satisfy the inequality $\delta \geq 0$ for every time evolution if and only if

$$\boldsymbol{\mathcal{T}} = D_{\boldsymbol{\lambda}}\mathcal{F},$$

$$\mathcal{D} \geq 0.$$

We can write

$$\boldsymbol{\tau} \cdot \dot{\boldsymbol{\lambda}} = \dot{\psi} + \mathcal{D}$$

and say that the generalized power $\boldsymbol{\tau} \cdot \dot{\boldsymbol{\lambda}}$ consists of two contributions. The first one is the recoverable power in that it results in the time derivative of the free energy. The second one is then non-recoverable and is non-negative for every process. One might then be tempted to say that $\mathcal{D}$ is the effective measure of dissipation and that $\mathcal{D} = 0$ for every process characterizes nondissipative bodies. In particular, since viscoelasticity models dissipative bodies we expect that $\mathcal{D} > 0$ for viscoelastic solids. Otherwise, $\mathcal{D} = 0$ is expected to correspond to elastic solids.

In materials with memory these conclusions are inconsistent. We show that such is the case for linear viscoelasticity where θ is a parameter and

$$\boldsymbol{\tau} \to \mathbf{T}, \qquad \boldsymbol{\lambda} \to \mathbf{E}.$$

Let Φ_r be the space of restricted histories ${}_r\mathbf{E}^t$. A functional $\mathcal{F}$ on $\Phi = \mathrm{Sym} \times \Phi_r$ is a free energy if

a) $\mathcal{F}$ is continuous on Φ, differentiable with respect to the present value of $\mathbf{E}^t$ and

$$\boldsymbol{\mathcal{T}} = D_{\mathbf{E}}\mathcal{F}(\mathbf{E}, {}_r\mathbf{E}^t);$$

b) for each value of ξ such that $\dot{\mathbf{E}}(t+\xi)$ is continuous, $\mathcal{F}$ satisfies the inequality

$$\dot{\mathcal{F}}(\mathbf{E}^{t+\xi}) \leq \boldsymbol{\mathcal{T}}(\mathbf{E}^{t+\xi}) \cdot \dot{\mathbf{E}}(t+\xi);$$

c) the functional $\mathcal{F}$ is minimal in correspondence with constant histories in that, if $\mathbf{E}^t$ is any history ending at the value $\mathbf{E}$,

$$\mathcal{F}(\mathbf{E}^t) \geq \mathcal{F}(\mathbf{E}^\dagger).$$

Let $\mathcal{T}$ be the standard functional of linear viscoelasticity, viz.

$$\mathbf{T}(t) = \mathbf{G}_0 + \int_0^\infty \mathbf{G}'(s)\mathbf{E}(t-s)ds,$$

where $\mathbf{G}' \in L^1(\mathbb{R}^+) \cap L^2(\mathbb{R}^+)$. Property (a) of $\mathcal{F}$ and the form of $\mathcal{T}$ yield

$$\mathcal{F}(t) = \tfrac{1}{2}\mathbf{E}(t)\mathbf{G}_0\mathbf{E}(t) + \mathbf{E}(t)\cdot\int_0^\infty \mathbf{G}'(s)\mathbf{E}(t-s)ds + \hat{\mathcal{F}}({}_r\mathbf{E}^t)$$

where $\hat{\mathcal{F}}$ is an arbitrary functional on Φ_r. Non-uniqueness holds because (b) and (c) are not sufficient to determine the functional $\hat{\mathcal{F}}$. Indeed, cf. [6, 1], the set of admissible free energies is a convex set with properly defined minimal and maximal free energies. Moreover, the maximal free energy $\mathcal{F}_M$ is given by

$$\mathcal{F}_M(\mathbf{E}^t) = \mathcal{F}(\mathbf{E}(0)^\dagger) + \int_0^t \mathcal{T}(\mathbf{E}^\xi)\cdot\dot{\mathbf{E}}(\xi)d\xi$$

whence

$$\dot{\mathcal{F}}_M(t) = \mathcal{T}(\mathbf{E}^t)\cdot\dot{\mathbf{E}}(t).$$

Incidentally non-uniqueness of thermodynamic potentials in materials with memory was first shown by Day [7] in connection with heat conductors with memory.

Consider the functional Ψ defined by

$$\Psi(\mathbf{E}^t) = \tfrac{1}{2}\mathbf{E}(t)\cdot\mathbf{G}_\infty\mathbf{E}(t) + \tfrac{1}{2}\int_0^\infty\int_0^\infty [\mathbf{E}^t(s_1)-\mathbf{E}(t)]\cdot\mathbf{G}_{12}(|s_1-s_2|)[\mathbf{E}^t(s_2)-\mathbf{E}(t)]ds_1ds_2$$

where the subscripts 1, 2 of $\mathbf{G}$ denote differentiation with respect to s_1, s_2, viz.

$$\mathbf{G}_{12} = \frac{\partial^2\mathbf{G}}{\partial s_1\partial s_2}.$$

Moreover we let $\mathbf{G}$ be symmetric. We can write

$$\begin{aligned}\Psi(\mathbf{E}^t) &= \tfrac{1}{2}\mathbf{E}(t)\cdot\mathbf{G}_\infty\mathbf{E}(t) + \tfrac{1}{2}\int_0^\infty\int_0^\infty \mathbf{E}^t(s_1)\cdot\mathbf{G}_{12}(|s_1-s_2|)\mathbf{E}^t(s_2)ds_1ds_2\\ &\quad -\mathbf{E}(t)\cdot\int_0^\infty\int_0^\infty \mathbf{G}_{12}(|s_1-s_2|)\mathbf{E}^t(s_2)ds_1ds_2\\ &\quad +\tfrac{1}{2}\mathbf{E}(t)\cdot\Big[\int_0^\infty\int_0^\infty \mathbf{G}_{12}(|s_1-s_2|)ds_1ds_2\Big]\mathbf{E}(t).\end{aligned}$$

Because

$$\int_0^\infty\int_0^\infty \frac{\partial^2\mathbf{G}}{\partial s_1\partial s_2}(|s_2-s_1|)ds_1ds_2 = -\int_0^\infty \frac{\partial\mathbf{G}}{\partial s_1}(s_1)ds_1 = \mathbf{G}_0-\mathbf{G}_\infty,$$

substitution yields

$$\begin{aligned}\Psi(\mathbf{E}^t) &= \tfrac{1}{2}\mathbf{E}(t)\cdot\mathbf{G}_0\mathbf{E}(t) + \mathbf{E}(t)\cdot\int_0^\infty \mathbf{G}'(s_1)\mathbf{E}^t(s_1)ds_1\\ &\quad +\tfrac{1}{2}\int_0^\infty\int_0^\infty \mathbf{E}^t(s_1)\cdot\mathbf{G}_{12}(|s_2-s_1|)\mathbf{E}^t(s_2)ds_1ds_2\end{aligned}$$

whence we have $\partial\Psi/\partial\mathbf{E} = \mathbf{T}$. Now we evaluate the time derivative,

$$\dot{\Psi}(\mathbf{E}^t) = \mathbf{T}\cdot\dot{\mathbf{E}} + \mathbf{E}(t)\cdot\int_0^\infty \mathbf{G}'(s)\dot{\mathbf{E}}^t(s)ds + \int_0^\infty\int_0^\infty \dot{\mathbf{E}}^t(s_1)\mathbf{G}_{12}(|s_2 - s_1|)\mathbf{E}^t(s_2)ds_1 ds_2.$$

Consider the last integral. Integration by parts gives

$$\begin{aligned}\int_0^\infty\int_0^\infty & \dot{\mathbf{E}}^t(s_1)\cdot\mathbf{G}_{12}(|s_2 - s_1|)\mathbf{E}^t(s_2)ds_2\,ds_1 \\ = & \int_0^\infty ds_1\dot{\mathbf{E}}^t(s_1)\cdot\Big[-\mathbf{G}'(s_1)\mathbf{E}(t) + \int_0^\infty \mathbf{G}_1(|s_2 - s_1|)\dot{\mathbf{E}}^t(s_2)ds_2\Big]\end{aligned}$$

The first integral in the right side cancels. The second integral vanishes in that

$$\frac{\partial\mathbf{G}}{\partial s_1} = \operatorname{sgn}(s_1 - s_2)\,\mathbf{G}'(|s_1 - s_2|).$$

Hence $\dot{\Psi}(\mathbf{E}^t) = \mathbf{T}\cdot\dot{\mathbf{E}}$.

The last step is to show that the functional Ψ is minimal at constant histories. In this regard we need the following result.

Proposition. Let $\mathbf{W} \in L^2(\mathbb{R}^+)$ and $\mathbf{K} \in L^1(\mathbb{R}^+)$; $\mathbf{W}$ has values in Lin and $\mathbf{K}$ maps Lin into Lin. The integral

$$\int_0^\infty \mathbf{W}(t)\cdot\int_0^t \mathbf{K}(t-\tau)\mathbf{W}(\tau)d\tau\,dt$$

is positive definite if and only if the cosine transform $\mathbf{K}_c$ is positive definite in Lin.

Now observe that

$$\mathbf{G}_{12} = -2\delta(s_1 - s_2)\mathbf{G}'(|s_1 - s_2|) - \mathbf{G}''(|s_1 - s_2|)$$

and hence

$$\begin{aligned}\int_0^\infty\int_0^\infty & \mathbf{W}(s_1)\cdot\mathbf{G}_{12}(|s_1 - s_2|)\mathbf{W}(s_2)ds_2\,ds_1 \\ = & -2\int_0^\infty \mathbf{W}(t)\cdot\int_0^t [\mathbf{G}''(t-\tau) + \mathbf{G}_0'\delta(t-\tau)]\mathbf{W}(\tau)d\tau\,dt.\end{aligned}$$

Accordingly, $\mathbf{G}_{12}$ is a positive definite kernel if and only if the cosine transform of $\mathbf{G}''(t-\tau) + \mathbf{G}_0'\delta(t-\tau)$ is negative definite. Now, an integration by parts gives

$$\mathbf{G}_c''(\omega) = \int_0^\infty \mathbf{G}''(\xi)\cos\omega\xi\,d\xi = \mathbf{G}'(\xi)\cos\omega\xi\big|_0^\infty + \omega\int_0^\infty \mathbf{G}'(\xi)\sin\omega\xi\,d\xi$$

whence

$$\mathbf{G}_c''(\omega) + \mathbf{G}_0' = \omega\mathbf{G}_s'(\omega) < 0.$$

Hence, because

$$\int_0^\infty [\mathbf{G}''(\xi) + \mathbf{G}'(\xi)\delta(\xi)]\cos\omega\xi\,d\xi = \mathbf{G}_c''(\omega) + \mathbf{G}_0',$$

it follows that

$$\int_0^\infty \int_0^\infty \mathbf{W}(s_1) \cdot \mathbf{G}_{12}(|s_1 - s_2|)\mathbf{W}(s_2)ds_2ds_1$$

is positive definite and hence vanishes only at $\mathbf{W} = 0$. Since $\mathbf{W}(s) = \mathbf{E}^t(s) - \mathbf{E}(t)$, the minimality of the functional at constant histories follows. This proves that $\Psi(\mathbf{E}^t)$ is a free energy functional, indeed the maximal one.

Letting $\psi(t) = \Psi(\mathbf{E}^t)$ we have

$$\delta = \mathcal{D} = 0,$$

namely the internal dissipation vanishes. Meanwhile, for the *same* viscoelastic body we have an infinite set of free energies for each of which $\delta \neq 0$, and δ depends on the free energy functional. This conclusion makes the interpretation of δ, as a measure of the dissipation of the body, inconsistent. Also, $\delta = 0$ does not mean that the solid is elastic; δ vanishes for every relaxation function $\mathbf{G}$, compatible with thermodynamics and such that its derivatives up to second order are integrable. What is crucial for the vanishing of δ is the choice of the (maximal) free energy. Also, this shows that thermodynamic restrictions based on the Clausius-Planck inequality may be affected by non-uniqueness of the free energy functional.

4. Discontinuity waves

Lately, the connection between thermodynamic restrictions and wave propagation properties has been investigated in isotropic, viscoelastic bodies [8]. The dissipativity of the material results in a decay of the amplitude in the direction of propagation. This aspect is qualitatively common to plane discontinuities, inhomogeneous waves, Green's functions, and rays.

This section shows how the memory property of the viscoelastic model results in the attenuation of discontinuity waves. Useful references to the subject are the article [9], as far as acceleration waves are concerned, and the paper [10] about shock waves. The body is supposed to be homogeneous, namely the relaxation function $\mathbf{G}$ is taken to be independent of $\mathbf{x}$.

We consider waves of order $m \geq 2$ characterized by singular surfaces $\mathcal{S}$ on $\Omega \times \mathbb{R}$ as follows.

1) The displacement $\mathbf{u}$ on $\Omega \times \mathbb{R}$ is $m-1$ times continuously differentiable while the m-th derivatives may possess jump discontinuities across $\mathcal{S}$.

2) The body force $\mathbf{b}$ is $m-2$ continuously differentiable.

The waves of order $m = 2$ are called acceleration waves. The density ρ is continuous across $\mathcal{S}$ and hence constant on $\Omega \times \mathbb{R}$.

The direction of propagation is given by the unit normal $\mathbf{n}$ to $\mathcal{S}$ and U denotes the (local) speed of propagation. For any function f on $\Omega \times \mathbb{R}$ the symbol $[\![f]\!]$ denotes the jump of f across $\mathcal{S}$, namely the difference $f^- - f^+$ between the limit values f^- and f^+ behind and ahead of $\mathcal{S}$. Parameterize the points of $\mathcal{S}$ by a pair of coordinates $\Gamma = (\Gamma_1, \Gamma_2)$ and describe the motion of the points of $\mathcal{S}$ through $\mathbf{x} = \mathbf{x}(\Gamma, t)$. Consider $f^-(\mathbf{x}(\Gamma, t), t)$ and let d/dt be the time derivative at constant Γ, i. e. the displacement (Thomas) derivative. We have

$$\frac{df^-}{dt} = \frac{\partial f^-}{\partial t} + \nabla f^- \cdot \frac{d\mathbf{x}}{dt},$$

and similarly for f^+. Let $d\mathbf{x}/dt = U\mathbf{n}$ and $\dot{f} = \partial f/\partial t$. Subtraction yields

$$\frac{d}{dt}[\![f]\!] = [\![\dot{f}]\!] + U\mathbf{n}\cdot[\![\nabla f]\!].$$

Meanwhile, evaluation of the derivative along a line on $\mathcal{S}$ gives

$$\frac{df^-}{dl} = \nabla f^- \cdot \frac{d\mathbf{x}}{dl}, \qquad \frac{df^+}{dl} = \nabla f^+ \cdot \frac{d\mathbf{x}}{dl}.$$

Subtraction yields

$$\frac{d[\![f]\!]}{dl} = [\![\nabla f]\!]\cdot\frac{d\mathbf{x}}{dl}.$$

If $[\![f]\!] = 0$ we obtain the kinematic condition of compatibility

$$[\![\dot{f}]\!] + U\mathbf{n}\cdot[\![\nabla f]\!] = 0$$

and the geometric condition of compatibility

$$[\![\nabla f]\!] = [\![df/dn]\!]\mathbf{n}$$

where $[\![df/dn]\!]$ is the jump of the normal derivative. Comparison gives

$$[\![\nabla f]\!] = [\![df/dn]\!]\mathbf{n} = -\frac{1}{U}\mathbf{n}[\![\dot{f}]\!].$$

The analysis of shock waves ($m = 1$) can be performed by a similar procedure, relative to a reference configuration (cf. [10]). For shock waves, though, a nonlinear model might be more appropriate and that is why we restrict attention to $m \geq 2$.

About the (local) propagation speed we have the following result (cf. [11]).

Theorem. The speed of propagation U of a wave is independent of the order m of the wave and satisfies the eigenvalue problem (Fresnel-Hadamard propagation condition)

$$(\mathbf{Q}(\mathbf{n}) - \rho U^2\mathbf{1})\mathbf{q} = 0, \qquad \mathbf{q} \in V,$$

where $\mathbf{Q}(\mathbf{n})$ is the acoustic tensor defined as

$$\mathbf{Q}(\mathbf{n})\mathbf{w} = [\mathbf{G}_0(\mathbf{w}\otimes\mathbf{n})]\mathbf{n}, \qquad \mathbf{w} \in V.$$

Proof. Let $m = 2$. By the equation of motion we have

$$\rho[\![\ddot{\mathbf{u}}]\!] = [\![\nabla\cdot\mathbf{T}]\!].$$

Hence, since $\mathbf{G}$ is independent of $\mathbf{x}$ and $[\![\dot{\mathbf{u}}]\!] = 0$, we can write

$$\mathbf{n}\mathbf{G}_0[\![\dot{\mathbf{E}}]\!] + \mathbf{n}[\![\int_0^\infty \mathbf{G}'(s)\,\dot{\mathbf{E}}(t-s)\,ds]\!] + \rho U[\![\ddot{\mathbf{u}}]\!] = 0.$$

By letting $f = \dot{\mathbf{u}}$, we find that

$$[\![\dot{\mathbf{E}}]\!] = -\frac{1}{U}\mathrm{sym}(\mathbf{n}\otimes\mathbf{q})$$

where $\mathbf{q} = [\![\ddot{\mathbf{u}}]\!]$ is the polarization of the wave. Substitution yields the desired result.

Now let $m > 2$. Consider the $(m-2)$-th time derivative of the equation of motion. Observe that $\mathbf{u}^{(m-1)} := d^{m-1}\mathbf{u}/dt^{m-1}$ is continuous across $\mathcal{S}$ whence, for example,

$$[\![\mathbf{E}^{(m-1)}]\!] = -\frac{1}{U}\mathrm{sym}(\mathbf{n}\otimes\mathbf{q})$$

where $\mathbf{q} = [\![\mathbf{u}^{(m)}]\!]$. Substitution completes the proof. □

By thermodynamics, $\mathbf{G}_0$ is symmetric. Hence $\mathbf{Q}$ is symmetric and there exist three, not necessarily distinct, propagation speeds. Moreover, $\mathbf{G}_0$ is also positive definite in Sym and then

$$\mathbf{v}\cdot[\mathbf{G}_0(\mathbf{v}\otimes\mathbf{w})]\mathbf{w} > 0$$

for any non-zero vectors $\mathbf{v}, \mathbf{w}$. The propagation speeds are real and non-zero.

The eigenvalues of $\mathbf{Q}(\mathbf{n})$ depend on $\mathbf{n}$ only. Hence the propagation speed U is constant if $\mathbf{n}$ is constant. Consider plane waves of the form

$$\mathbf{u}(\mathbf{x},t) = \bar{\mathbf{u}}(x,t)$$

where $x = \mathbf{x}\cdot\mathbf{n}$. In connection with the wave of order m we define the induced jump γ as

$$\gamma = -[\![\frac{\partial^{m+1}\bar{\mathbf{u}}}{\partial x^{m+1}}]\!](-U)^{m+1}$$

and let

$$\mathbf{q} = [\![\bar{\mathbf{u}}^{(m)}]\!] = [\![\frac{\partial^{m}\bar{\mathbf{u}}}{\partial x^{m}}]\!](-U)^{m}$$

be the polarization. Let $\mathbf{G}_0'$ be symmetric and define the *attenuation tensor* $\mathbf{Q}'(\mathbf{n})$ by

$$\mathbf{Q}'(\mathbf{n})\mathbf{w} = [\mathbf{G}_0'(\mathbf{w}\otimes\mathbf{n})]\mathbf{n}, \qquad \mathbf{w}\in V.$$

The symmetry of $\mathbf{G}_0'$ in Sym implies that of $\mathbf{Q}'$ in V. The following result is essential to the investigation of the attenuation of the polarization in time.

Theorem. The polarization $\mathbf{q}$ of a plane wave of order m which propagates in a homogeneous medium with speed U is governed by the differential equation

$$\mathbf{Q}'(\mathbf{n})\mathbf{q} - 2\rho U^2\frac{d\mathbf{q}}{dt} = \mathbf{Q}(\mathbf{n})\gamma - \rho U^2\gamma.$$

Proof. Let $m = 2$. Differentiate the equation of motion with respect to x. By the definition of $\mathbf{Q}$ we can write

$$\rho[\![\frac{\partial\ddot{\bar{\mathbf{u}}}}{\partial x}]\!] = \mathbf{Q}(\mathbf{n})[\![\frac{\partial^3\bar{\mathbf{u}}}{\partial x^3}]\!] + [\![\frac{\partial}{\partial x}\int_0^\infty \nabla\cdot(\mathbf{G}'(s)\frac{\partial\mathbf{u}}{\partial\mathbf{x}}(t-s))\,ds]\!].$$

Examine the last term. Because $\mathbf{G}'$ is independent of $\mathbf{x}$ and the integral is continuous across the wave we have

$$[\![\frac{\partial}{\partial x}\int_0^\infty \nabla\cdot[\mathbf{G}'(s)\frac{\partial\mathbf{u}}{\partial\mathbf{x}}(t-s)]\,ds]\!] = -\frac{1}{U}[\![\frac{\partial}{\partial t}\int_0^\infty \mathbf{G}'(s)\frac{\partial^2\mathbf{u}}{\partial\mathbf{x}\partial\mathbf{x}}(t-s)\,ds]\!].$$

Since $\partial^2\mathbf{u}/\partial\mathbf{x}\partial\mathbf{x}$ suffers a jump discontinuity at the wave, to evaluate the jump we apply a procedure elaborated in [12]. We obtain

$$[\![\frac{\partial}{\partial t}\int_0^\infty \mathbf{G}'(s)\frac{\partial^2\mathbf{u}}{\partial\mathbf{x}\partial\mathbf{x}}(t-s)\,ds]\!] = \lim_{\lambda\to 0_+}\frac{1}{\lambda}\int_0^\lambda \{\mathbf{G}_0'[\![\frac{\partial^2\mathbf{u}}{\partial\mathbf{x}\partial\mathbf{x}}]\!] + O(\lambda)\}\,ds = \mathbf{G}_0'\,[\![\frac{\partial^2\mathbf{u}}{\partial\mathbf{x}\partial\mathbf{x}}]\!].$$

Hence we have

$$\rho U^3[\![\frac{\partial\ddot{\mathbf{u}}}{\partial x}]\!] = \mathbf{Q}\gamma - \mathbf{Q}'\mathbf{q}.$$

On applying twice geometric and kinematic compatibility conditions we arrive at

$$[\![\frac{\partial\ddot{\mathbf{u}}}{\partial x}]\!] = -\frac{2}{U}\frac{d\mathbf{q}}{dt} + \frac{1}{U}\gamma.$$

Substitution provides the desired result.

If $m > 2$ we differentiate the equation of motion $m-1$ times with respect to x. Then we can parallel step by step the procedure for $m = 2$ to obtain the desired result. □

Concerning the evolution of $\mathbf{q}$, it is apparent that things are very simple when γ is such that

$$\mathbf{Q}\gamma = \rho\, U^2\gamma.$$

When this is the case we say that the wave is *axially similar*. For example, in one-dimensional bodies $\mathbf{Q}$, γ, and $\mathbf{q}$ reduce to real numbers and then all waves are axially similar. Really, such waves are not confined to the one-dimensional case. As we show in a moment, waves in isotropic solids are axially similar.

Theorem. At a plane, axially similar wave of order m, propagating in the direction $\mathbf{n}$ with speed U, the polarization $\mathbf{q}$ is given by

$$\mathbf{q}(t) = \exp(\frac{t}{2\,\rho\,U^2}\mathbf{Q}')\,\mathbf{q}(0).$$

Proof. The result follows from the evolution equation for $\mathbf{q}$ and the hypothesis $\mathbf{Q}\gamma - \rho U^2\gamma = 0$. Also, the symmetry of $\mathbf{Q}'$ allows $\mathbf{q}$ to be written as

$$\mathbf{q}(t) = \sum_{i=1}^{3}\alpha_i\mathbf{a}_i\exp(\mu_i t)$$

where $2\mu_i\rho U^2$ and $\mathbf{a}_i$ are the i-th eigenvalue and eigenvector of $\mathbf{Q}'$ while the coefficients α_i are determined by the initial conditions. □

This result emphasizes the central role of $\mathbf{Q}'$ in the evolution of $\mathbf{q}$. By thermodynamics, $\mathbf{G}_0'$ is negative semidefinite in Sym. Hence it follows that $\mathbf{Q}'$ is negative semidefinite in V, i. e.

$$\mathbf{w}\cdot\mathbf{Q}'\,\mathbf{w} \le 0$$

for any non-zero $\mathbf{w}\in V$. The eigenvalues μ_i are then negative and $\mathbf{q}$ decays in time.

To complete the scheme consider the case $\mathbf{Q}(\mathbf{n})\gamma \neq \rho U^2\gamma$; the corresponding waves are called *axially dissimilar*. For such waves the time evolution of $\mathbf{q}$ is given by a theorem which is due to Fisher [13].

Theorem. At a plane, axially dissimilar wave of order m, propagating in the direction $\mathbf{n}$ with speed U, the time dependence of the polarization $\mathbf{q}$ is given as follows.
i. If ρU^2 is a distinct eigenvalue of $\mathbf{Q}$, with eigenvector $\mathbf{e}$, then

$$\mathbf{q}(t) = \alpha\,\mathbf{e}\,\exp(\zeta t), \qquad \zeta = \frac{\mathbf{e}\cdot\mathbf{Q}'\mathbf{e}}{2\rho U^2}.$$

ii. If ρU^2 is a double eigenvalue of $\mathbf{Q}$ then

$$\mathbf{q}(t) = \sum_{\nu=1}^{2} \alpha_\nu \mathbf{a}_\nu \exp(\zeta_\nu t)$$

where $2\zeta_\nu \rho U^2$ is the eigenvalue corresponding to eigenvector $\mathbf{a}_\nu$ of the reduced attenuation matrix $Q_{\alpha\beta} = \mathbf{e}_\alpha \cdot \mathbf{Q}'\mathbf{e}_\beta$, $\alpha, \beta = 1, 2$, with $\mathbf{e}_1, \mathbf{e}_2$ two orthonormal basis vectors in the principal plane of $\mathbf{Q}$.
Proof. If ρU^2 is a distinct eigenvalue of $\mathbf{Q}$ then there exists only one unit vector $\mathbf{e}$ such that $\mathbf{Q}\mathbf{e} = \rho\, U^2 \mathbf{e}$. Then

$$\mathbf{e}\cdot(\mathbf{Q}\gamma - \rho\, U^2\gamma) = 0$$

and

$$(\mathbf{Q}'\mathbf{q} - 2\,\rho\, U^2 \frac{d\mathbf{q}}{dt})\cdot\mathbf{e} = 0.$$

If the polarization $\mathbf{q}$ is parallel to $\mathbf{e}$, $\mathbf{q}(t) = \beta(t)\mathbf{e}$, then

$$(\mathbf{e}\cdot\mathbf{Q}'\mathbf{e})\beta - 2\,\rho\, U^2 \frac{d\beta}{dt} = 0$$

and the desired result follows.

If $\rho\, U^2$ is a double eigenvalue of $\mathbf{Q}$ then there are two orthogonal unit vectors $\mathbf{e}_1, \mathbf{e}_2$ such that each eigenvector $\mathbf{b}$ of $\mathbf{Q}$, with eigenvalue ρU^2, can be written as $\mathbf{b} = b_\gamma \mathbf{e}_\gamma$. Again we have

$$(\mathbf{Q}'\mathbf{e}_\alpha b_\alpha - 2\,\rho\, U^2 \mathbf{e}_\alpha \frac{db_\alpha}{dt})\cdot\mathbf{e}_\gamma = 0.$$

Therefore letting $Q'_{\alpha\beta} = \mathbf{e}_\beta \cdot \mathbf{Q}'\mathbf{e}_\alpha$, we obtain

$$Q'_{\alpha\beta} b_\alpha - 2\,\rho\, U^2 \frac{db_\beta}{dt} = 0$$

whence the result follows.

Of course, if $\rho\, U^2$ is a triple eigenvalue of $\mathbf{Q}$ then any vector γ is an eigenvector of $\mathbf{Q}$ and the corresponding waves are axially similar. □

For isotropic solids,

$$\mathbf{Q}(\mathbf{n})\,\mathbf{w} = \mu_0 \mathbf{w} + (\mu_0 + \lambda_0)(\mathbf{w}\cdot\mathbf{n})\mathbf{n}$$

and then $\mathbf{n}$ is an eigenvector of $\mathbf{Q}$, with eigenvalue $2\mu_0 + \lambda_0$, while any vector orthogonal to $\mathbf{n}$ is an eigenvector of $\mathbf{Q}$ with (double) eigenvalue μ_0. Accordingly, it follows from the Fresnel-Hadamard condition that *longitudinal waves,* for which

$$\mathbf{n} \times \mathbf{q} = 0,$$

can exist with propagation speed

$$U_L = \sqrt{\frac{2\mu_0 + \lambda_0}{\rho}}$$

and *transverse waves,* for which

$$\mathbf{n} \cdot \mathbf{q} = 0,$$

can exist with propagation speed

$$U_T = \sqrt{\frac{\mu_0}{\rho}}.$$

Incidentally, both $\mu_0 > 0$ and $2\mu_0 + \lambda_0 > 0$ are guaranteed by the positive definiteness of $\mathbf{G}_0$. As regards $\mathbf{Q}'$ we have

$$\mathbf{Q}'(\mathbf{n})\mathbf{w} = \mu_0' \, \mathbf{w} + (\mu_0' + \lambda_0')(\mathbf{w} \cdot \mathbf{n})\mathbf{n}.$$

It is evident that $\mathbf{Q}$ and $\mathbf{Q}'$ commute and have the same characteristic manifolds; the eigenvectors are the normal $\mathbf{n}$ (with distinct eigenvalue) and any pair of vectors orthogonal to $\mathbf{n}$ (and each other). This implies that, for isotropic bodies, all waves are axially similar. Hence we have the following result.

Corollary. Plane waves of order m propagating in an isotropic viscoelastic body are either longitudinal or transverse.
i. At longitudinal waves $\boldsymbol{\gamma}$ and $\mathbf{q}$ are both parallel to $\mathbf{n}$ and

$$\mathbf{q}(t) = q_0 \mathbf{n} \exp\left(\frac{\lambda_0' + 2\mu_0'}{2(\lambda_0 + 2\mu_0)} t\right).$$

ii. At transverse waves $\mathbf{q}$ and $\boldsymbol{\gamma}$ are both orthogonal to $\mathbf{n}$ and

$$\mathbf{q}(t) = q_0 \boldsymbol{\tau} \exp\left(\frac{\mu_0'}{2\mu_0} t\right), \qquad \boldsymbol{\tau} \cdot \mathbf{n} = 0.$$

The thermodynamic requirements $\mu_0' \leq 0$ and $\lambda_0' + 2\mu_0' \leq 0$ - cf. (10) - imply the decay of $\mathbf{q}$ in time.

5. Time-harmonic waves in isotropic bodies

Wave propagation in linear models of continuous bodies is naturally investigated via time-harmonic waves. Physically, this corresponds to wave propagation induced by a source which is a given function of time t. Time-harmonic waves can then be viewed as the Fourier components of the real wave.

In uniform, possibly dissipative, bodies the linearity of the model allows the space dependence of the waves to be taken in the typical form of plane waves. In this context the general solution is in the form of inhomogeneous waves, which means that the wave vector is complex valued; in essence, the real part describes the phase, the imaginary part describes the amplitude. Among the literature on inhomogeneous waves some papers need to be mentioned. Seemingly, the first analysis of properties of inhomogeneous waves is given in a paper of Lockett [14]. Buchen [15] investigates physical properties, such as energy flux and density, reflection and refraction amplitudes, of waves in linear viscoelasticity. In two companion papers [16, 17], Currie et al. examine the existence of Rayleigh waves, in viscoelastic solids, that are a superposition of inhomogeneous waves. Waves in fluids are considered, e.g., in [18, 19] in connection with the Navier-Stokes model of viscous fluid. The reference [2] provides a list of works on the subject.

Sometimes the identification between results of different authors about time-harmonic waves lead to seeming contradictions because of differences in sign. This is due to the fact that the time dependence is factorized in the form $\exp(-i\omega t)$ by someones and $\exp(i\omega t)$ by others. Since the time dependence is represented by a common factor, one might be tempted to say that the remaining part is independent of the use of $\exp(-i\omega t)$ or $\exp(i\omega t)$. Yet, it is not so. Consider the constitutive equation of linear viscoelasticity and let

$$\mathbf{u} = \mathbf{u}_\omega \exp(-i\omega t), \qquad \mathbf{T} = \mathbf{T}_\omega \exp(-i\omega t).$$

Accordingly, $\mathbf{E} = \mathbf{E}_\omega \exp(-i\omega t)$. Substitution gives

$$\mathbf{T}_\omega = [\mathbf{G}_0 + \mathbf{G}'_c(\omega) + i\mathbf{G}'_s(\omega)]\mathbf{E}_\omega.$$

Hence we have

$$\mathbf{T}_\omega = \mathbf{G}_\omega \mathbf{E}_\omega, \qquad \operatorname{Im} \mathbf{G}_\omega < 0,$$

for all $\omega \in \mathbb{R}^{++}$. Now, instead, we represent the time dependence through $\exp(i\omega t)$ and let

$$\mathbf{u} = \tilde{\mathbf{u}}_\omega \exp(i\omega t), \qquad \mathbf{T} = \tilde{\mathbf{T}}_\omega \exp(i\omega t),$$

and $\mathbf{E} = \tilde{\mathbf{E}}_\omega \exp(i\omega t)$. Substitution yields

$$\tilde{\mathbf{T}}_\omega = [\mathbf{G}_0 + \mathbf{G}'_c(\omega) - i\mathbf{G}'_s(\omega)]\tilde{\mathbf{E}}_\omega.$$

Hence,

$$\tilde{\mathbf{T}}_\omega = \tilde{\mathbf{G}}_\omega \tilde{\mathbf{E}}_\omega, \qquad \operatorname{Im} \tilde{\mathbf{G}}_\omega > 0,$$

for all $\omega \in \mathbb{R}^{++}$.

It is common practice to write the vector amplitudes merely as $\mathbf{u}, \mathbf{T}, \mathbf{E}$ without any mention of the time factor to which they refer. Hence it is to emphasize that, while $\mathbf{G}'_s(\omega) < 0$, irrespective of the time factor, the sign of the imaginary part of the tensor relating $\mathbf{E}$ and $\mathbf{T}$ depends on the time factor chosen.

We now examine time-harmonic waves in isotropic solids. We write the constitutive relation in the form (8) and, by thermdynamics, μ', λ' satisfy (9), (10). Accordingly, letting

$$\mu(\omega) = \mu_0 + \int_0^\infty \mu'(\xi) \exp(i\omega\xi) d\xi, \qquad \lambda(\omega) = \lambda_0 + \int_0^\infty \lambda'(\xi) \exp(i\omega\xi) d\xi.$$

we can write (9) as

$$\operatorname{Im}\mu(\omega) < 0, \qquad \operatorname{Im}(2\mu + \lambda)(\omega) < 0, \qquad \omega \in \mathbb{R}^{++}. \tag{13}$$

Disregard the effect of body forces and consider the equation of motion

$$\rho\ddot{\mathbf{u}} = \nabla \cdot \mathbf{T}.$$

Consider time-harmonic displacements in the form

$$\mathbf{u}(\mathbf{x}, t) = \boldsymbol{\mathcal{U}}(\mathbf{x}) \exp(-i\omega t),$$

ω being the angular frequency. Substitution yields

$$\mathbf{T}(t) = [\mu(\nabla\boldsymbol{\mathcal{U}} + \nabla\boldsymbol{\mathcal{U}}^T) + \lambda(\nabla \cdot \boldsymbol{\mathcal{U}})\mathbf{1}] \exp(-i\omega t).$$

Let μ, λ, ρ be independent of the position $\mathbf{x}$. The equation of motion becomes

$$-\rho\omega^2\boldsymbol{\mathcal{U}} = \mu\Delta\boldsymbol{\mathcal{U}} + (\mu + \lambda)\nabla(\nabla \cdot \boldsymbol{\mathcal{U}}).$$

Now take $\boldsymbol{\mathcal{U}}$ in the form of plane waves, viz.

$$\boldsymbol{\mathcal{U}}(\mathbf{x}) = \tilde{\mathbf{u}} \exp(i\mathbf{k} \cdot \mathbf{x})$$

where $\mathbf{k}$ is a complex-valued vector (we call it the wave vector). Hence the equation of motion becomes

$$(-\rho\omega^2 + \mu\mathbf{k} \cdot \mathbf{k})\tilde{\mathbf{u}} + (\mu + \lambda)(\mathbf{k} \cdot \tilde{\mathbf{u}})\mathbf{k} = 0.$$

Two possibilities occur. First, $\tilde{\mathbf{u}}$ and $\mathbf{k}$ are linearly dependent, namely $\tilde{\mathbf{u}} = \beta\mathbf{k}$, $\beta \in \mathbb{C}$. Irrespective of β, this solution holds with

$$\mathbf{k} \cdot \mathbf{k} = \frac{\rho\omega^2}{2\mu + \lambda}.$$

Second, $\tilde{\mathbf{u}}$ and $\mathbf{k}$ are linearly independent. The coefficients of $\tilde{\mathbf{u}}$ and $\mathbf{k}$ must vanish, namely $\mathbf{k} \cdot \tilde{\mathbf{u}} = 0$ and

$$\mathbf{k} \cdot \mathbf{k} = \frac{\rho\omega^2}{\mu}.$$

Quite naturally, the first solution is viewed as longitudinal wave, the second one as transverse. Since $\mathbf{k}$ is complex, let $\mathbf{k} = \mathbf{l} + i\mathbf{m}$, with $\mathbf{l}$ and $\mathbf{m}$ real. We have

$$\mathbf{l}^2 - \mathbf{m}^2 = \frac{\rho\omega^2}{|2\mu + \lambda|^2}\operatorname{Re}(2\mu + \lambda), \qquad 2\mathbf{l} \cdot \mathbf{m} = -\frac{\rho\omega^2}{|2\mu + \lambda|^2}\operatorname{Im}(2\mu + \lambda),$$

for longitudinal waves and

$$\mathbf{l}^2 - \mathbf{m}^2 = \frac{\rho\omega^2}{|\mu|^2}\operatorname{Re}\mu, \qquad 2\mathbf{l} \cdot \mathbf{m} = -\frac{\rho\omega^2}{|\mu|^2}\operatorname{Im}\mu,$$

for transverse waves. In either case, it follows from thermodynamics that

$$\mathbf{l} \cdot \mathbf{m} > 0.$$

From

$$\mathbf{u}(\mathbf{x},t) = \tilde{\mathbf{u}} \exp[i(\mathbf{l}\cdot\mathbf{x} - \omega t)] \exp(-\mathbf{m}\cdot\mathbf{x}),$$

letting $\mathbf{x} = x\mathbf{l}/|\mathbf{l}|$ we see that $|\mathbf{u}|$ decays with x, namely along the direction of the phase propagation.

The requirements on $\mathbf{l}, \mathbf{m}$ are in fact two equations in the three unknowns $\mathbf{l}^2, \mathbf{m}^2$ and the angle ϑ that $\mathbf{m}$ subtends from $\mathbf{l}$; thermodynamics implies that ϑ is acute. Since only two equations hold for three unknowns, further information is needed to specify a plane wave. This need of information is obvious on the physical side in that the way of generation of the wave has to be specified. It is of interest to examine the determination of the angle ϑ at a material discontinuity surface.

Consider a plane surface Σ, with unit normal $\mathbf{n}$, that separates two homogeneous half-spaces in welded contact. Because of the discontinuity surface, four inhomogeneous waves originate at (any point of) Σ: a longitudinal and a transverse wave for each medium. A reflection-refraction problem consists in the determination of the reflected and transmitted (or refracted) waves in terms of the properties of the two media and the incident wave. The boundary conditions governing the process of reflection and refraction are the continuity of displacement and traction across Σ. Denote by a superposed $\check{}$ any quantity pertaining to transmitted waves. The continuity of the displacement and of the traction implies that the phase $\mathbf{k}\cdot\mathbf{x}$ at any point of Σ takes a common value for all waves. Letting $\mathbf{x}$ be the position vector of a current point of Σ relative to an origin in Σ, we have

$$\mathbf{k}^i\cdot\mathbf{x} = \mathbf{k}_L^r\cdot\mathbf{x} = \mathbf{k}_T^r\cdot\mathbf{x} = \check{\mathbf{k}}_L\cdot\mathbf{x} = \check{\mathbf{k}}_T\cdot\mathbf{x}, \qquad \forall\mathbf{x} : \mathbf{n}\cdot\mathbf{x} = 0,$$

where the superscript r labels quantities pertaining to the reflected waves. Represent any vector $\mathbf{x}$, such that $\mathbf{n}\cdot\mathbf{x} = 0$, in the form $\mathbf{x} = \mathbf{n}\times\mathbf{y}$. The set of vectors $\mathbf{x}$ is spanned by letting $\mathbf{y}$ be any vector in $\mathbb{R}^3$. The arbitrariness of $\mathbf{y}$ yields

$$\mathbf{k}^i\times\mathbf{n} = \mathbf{k}_L^r\times\mathbf{n} = \mathbf{k}_T^r\times\mathbf{n} = \check{\mathbf{k}}_L\times\mathbf{n} = \check{\mathbf{k}}_T\times\mathbf{n}. \tag{14}$$

We can regard (14) as the general form of Snell's law. It follows from the separate contributions of the real and imaginary parts of (14) that the planes $(\mathbf{l}, \mathbf{n})$ and $(\mathbf{m}, \mathbf{n})$ are invariant, namely all pairs $(\mathbf{l}, \mathbf{n})$ determine a common plane and so do the pairs $(\mathbf{m}, \mathbf{n})$. Moreover, (14) tells us that the tangential component of the real part of the wave vectors is common to all waves. So is the tangential component of the imaginary part of the wave vectors.

Time-harmonic plane waves with $\mathbf{l}\times\mathbf{m} \neq 0$ are called genuinely inhomogeneous [2]. That waves are usually inhomogeneous follows at once from the following consideration. Assume that the half-space supporting the incident wave is elastic and that $\mathbf{m}^i = 0$. If the other half-space is viscoelastic, solutions are allowed with $\mathbf{m} \neq 0$. Now by Snell's law $\mathbf{l}\times\mathbf{n}$ is required to equal $\mathbf{k}^i\times\mathbf{n}$. Meanwhile Snell's law requires that $\mathbf{m}^r, \check{\mathbf{m}}$ are parallel to $\mathbf{n}$. Hence the condition on $\mathbf{k}\cdot\mathbf{k}$ determines the angle between $\mathbf{l}, \mathbf{m}$ and the value of $|\mathbf{m}|$, whence the value of $|\mathbf{l}|$.

Dissipative fluids may be modelled through viscoelasticity; in the linear approximation we can write

$$\mathbf{T}(t) = -p(\rho(t))\mathbf{1} + 2\int_0^\infty \eta(s)\mathbf{D}(t-s)ds + \big[\int_0^\infty \nu(s)\mathrm{tr}\,\mathbf{D}(t-s)ds\big]\mathbf{1}$$

where $\mathbf{D} = \text{sym}(\nabla\dot{\mathbf{u}})$. We let $\eta, \nu \in L^1(\mathbb{R}^+)$ and require that they approach zero as $s \to \infty$ [1]. This model is compatible with the second law of thermodynamics if and only if

$$\eta_c(\omega) > 0, \qquad 2\eta_c(\omega) + 3\nu_c(\omega) > 0, \qquad \forall\omega \in \mathbb{R}^{++}.$$

If the fluid is incompressible then only the restriction $\eta_c > 0$ holds.

Look for time-harmonic, inhomogeneous solutions. Two wave solutions exist which still may be viewed as longitudinal and transverse (cf. [2], §3.3). Let p_ρ denote the derivative of p with respect to ρ and assume that $p_\rho > 0$. Straightforward calculations yield

$$\mathbf{l}^2 - \mathbf{m}^2 = \rho\omega^2 \frac{\rho p_\rho + \omega(2\eta_s + \nu_s)}{[\rho p_\rho + \omega(2\eta_s + \nu_s)]^2 + [\omega(2\eta_c + \nu_c)]^2}, \qquad \frac{\rho\omega\eta_s}{\eta_c^2 + \eta_s^2},$$

$$2\mathbf{l}\cdot\mathbf{m} = \rho\omega^3 \frac{2\eta_c + \nu_c}{[\rho p_\rho + \omega(2\eta_s + \nu_s)]^2 + [\omega(2\eta_c + \nu_c)]^2}, \qquad \frac{\rho\omega\eta_c}{\eta_c^2 + \eta_s^2},$$

depending on whether longitudinal or transverse waves are considered. For definiteness consider the longitudinal waves and, for simplicity, let $\mathbf{l} \times \mathbf{m} = 0$.

In seismology, the quality factor Q is widely used to characterize the dissipative property of the medium. The wave speed c and the attenuation rate α are defined such that

$$k = \frac{\omega}{c} + i\alpha.$$

Hence it follows that

$$\alpha = \frac{\omega}{2cQ}.$$

Letting the fluid be weakly dissipative, i. e. $\alpha^2 \ll (\omega/c)^2$, we have

$$c^2 = p_\rho, \qquad \alpha = \frac{2\eta_c + \nu_c}{2\rho p_\rho}\omega^2$$

whence

$$Q = \frac{\rho\sqrt{p_\rho}}{(2\eta_c + \nu_c)\omega}.$$

Experimental data (cf. [20]) show that $Q \simeq \omega^{-1}$ for a variety of fluids. We examine the properties of a specific model of fluid (cf. [21]).

Let η, ν be taken in the form

$$\eta(s) = \eta_0 \exp(-bs), \qquad \nu(s) = \nu_0 \exp(-bs), \qquad \eta_0, \nu_0, b > 0.$$

A simple calculation yields

$$\eta_c(\omega) = \frac{\eta_0 b}{b^2 + \omega^2}$$

and similarly for ν. Substitution gives

$$Q = \frac{1}{\omega}\frac{\rho\sqrt{p_\rho}}{2\eta_0 + \nu_0}\frac{b^2 + \omega^2}{b}.$$

It is obvious that $Q \simeq \omega^{-1}$ inasmuch as $\omega \ll b$. Since b physically represents the inverse of a relaxation time, this result signifies that $Q \simeq \omega^{-1}$ as long as the period of the wave is much bigger than the relaxation time, which is likely to be the case in the seismological framework. To the author's knowledge this is the first explanation for the ω^{-1} dependence of Q.

6. Time-harmonic waves in anisotropic solids

The wide literature on waves in anisotropic bodies deals essentially with elastic solids (cf. [22-27] and refs. therein). Account for dissipative effects in solids is given by Hayes and Rivlin [28, 29] in the form of inhomogeneous waves. To obtain definite results, though, they have recourse to the approximation that the imaginary parts of the complex moduli are sufficiently small compared with the real parts. Various approaches and applications of wave propagation in viscoelastic, possibly anisotropic, media are shown in a book edited by Mainardi [30]. In particular, in [30] Hayes generalizes a result by Buchen and proves that, also in anisotropic solids, the projection of the mean energy flux in the direction of the imaginary part of the wave vector equals the mean dissipation (over twice the frequency). This section is based on a joint work with Caviglia [31] and provides general aspects of inhomogeneous wave propagation in anisotropic, viscoelastic materials.

As a consequence of the second law of thermodynamics, the imaginary and real parts of $\mathbf{G}_\omega$ are suitably related in the space of symmetric tensors. Accordingly we first prove that the divergence of a commonly used energy flux is negative. This in turn is shown to imply that the projection of the energy flux along the imaginary part of the wave vector is positive (a result which traces back to Hayes in [30]) and, what is more, that the wave amplitude decays in the direction of the energy flux. Further, by the assumption that the viscoelastic tensor is symmetric and the real part is positive definite, also the projection of the energy flux along the real part of the wave vector is shown to be positive. Analogous results are proved for viscoelastic fluids. While the real and imaginary parts of the wave vector subtend an acute angle in isotropic bodies, proof is given that such need not be the case in anisotropic bodies thus providing the striking result that the amplitude may increase in the direction of phase propagation.

We consider time-harmonic displacement fields in linear viscoelastic solids. If the time dependence is factorized as $\exp(-i\omega t)$, the stress-strain relation becomes

$$\mathbf{T}(\mathbf{x},t) = \mathbf{G}_\omega \mathbf{E}(\mathbf{x},t).$$

The equation of motion takes the form

$$-\rho\omega^2 \mathbf{u} = \nabla \cdot (\mathbf{G}_\omega \mathbf{E}). \tag{15}$$

Let $\mathbf{H} = \mathbf{G}_0 + \mathbf{G}'_c(\omega)$ and $\mathbf{K} = \mathbf{G}'_s(\omega)$ and hence $\mathbf{G}_\omega = \mathbf{H} + i\mathbf{K}$. The thermodynamic requirement becomes

$$\mathbf{E}_1 \cdot \mathbf{K}\mathbf{E}_1 + \mathbf{E}_2 \cdot \mathbf{K}\mathbf{E}_2 + \mathbf{E}_1 \cdot (\mathbf{H} - \mathbf{H}^T)\mathbf{E}_2 < 0 \tag{16}$$

for positive ω and non-zero $\mathbf{E}_1, \mathbf{E}_2 \in \mathrm{Sym}$. This implies that $\mathbf{K}$ is negative definite.

We look for solutions to the equation of motion in the form of inhomogeneous waves. In particular, we have

$$\mathbf{E} = i\,\mathrm{sym}(\mathbf{k}\otimes\mathbf{u}).$$

Substitution yields the propagation condition

$$[\mathbf{Q}(\omega,\mathbf{k}) - \rho\omega^2\mathbf{1}]\tilde{\mathbf{u}} = 0 \tag{17}$$

where

$$\mathbf{Q}(\omega,\mathbf{k}) = \mathbf{k}\mathbf{G}_\omega\mathbf{k},$$

the contraction of $\mathbf{G}_\omega$ with $\mathbf{k}$ being on the second and third subscripts of $\mathbf{G}_\omega$. The system (17) has non-trivial solutions in the unknown vector $\tilde{\mathbf{u}}$ if and only if

$$\det[\mathbf{Q}(\omega,\mathbf{k}) - \rho\omega^2\mathbf{1}] = 0. \tag{18}$$

The determinantal equation (18) is the characteristic condition for the admissible wave vectors $\mathbf{k}$.

Still let $\mathbf{k} = \mathbf{l} + i\mathbf{m}$. The real and imaginary parts $\mathbf{R}$ and $\mathbf{S}$ of $\mathbf{Q}$ are then given by

$$\mathbf{R} = \mathbf{lHl} - \mathbf{mHm} - \mathbf{lKm} - \mathbf{mKl},$$

$$\mathbf{S} = \mathbf{lKl} - \mathbf{mKm} + \mathbf{lHm} + \mathbf{mHl}.$$

Once $\mathbf{k}$ is found, we determine the amplitude vector $\tilde{\mathbf{u}}$ as a solution to (17), or to

$$(\mathbf{A} - \lambda\mathbf{1})\tilde{\mathbf{u}} = 0,$$

by the classical results of linear algebra.

Though, in principle, we are able to determine the wave vector $\mathbf{k}$ and the amplitude vector $\tilde{\mathbf{u}}$, for practical purposes it is often sufficient to consider a consequence of (17). Let $\mathbf{v},\mathbf{w}$ be the real and imaginary parts of $\tilde{\mathbf{u}}$, viz. $\tilde{\mathbf{u}} = \mathbf{v} + i\mathbf{w}$. Inner multiplication of (17) by the complex conjugate $\tilde{\mathbf{u}}^* = \mathbf{v} - i\mathbf{w}$ yields

$$\mathbf{v}\cdot\mathbf{R}\mathbf{v} + \mathbf{w}\cdot\mathbf{R}\mathbf{w} + \mathbf{w}\cdot(\mathbf{S} - \mathbf{S}^T)\mathbf{v} = \rho\omega^2(\mathbf{v}^2 + \mathbf{w}^2), \tag{19}$$

$$\mathbf{v}\cdot\mathbf{S}\mathbf{v} + \mathbf{w}\cdot\mathbf{S}\mathbf{w} + \mathbf{v}\cdot(\mathbf{R} - \mathbf{R}^T)\mathbf{w} = 0,$$

which express general relations between the amplitude vectors $\mathbf{v},\mathbf{w}$ and the wave vectors $\mathbf{l},\mathbf{m}$ occurring in $\mathbf{R},\mathbf{S}$.

If the solid is elastic then

$$\mathbf{R} = \mathbf{lHl} - \mathbf{mHm}, \qquad \mathbf{S} = \mathbf{lHm} + \mathbf{mHl}.$$

Regard $\mathbf{l}$ as fixed. If $\mathbf{m}$ is replaced by $-\mathbf{m}$ then $\mathbf{R}$ remains unchanged while $\mathbf{S}$ is replaced by $-\mathbf{S}$. The determinantal equation results in

$$\begin{aligned}\epsilon_{jhk}\{&(R_{1j} - \rho\omega^2\delta_{1j})(R_{2h} - \rho\omega^2\delta_{2h})(R_{3k} - \rho\omega^2\delta_{3k}) - (R_{1j} - \rho\omega^2\delta_{1j})S_{2h}S_{3k}\\ &- S_{1j}S_{2h}(R_{3k} - \rho\omega^2\delta_{3k}) - S_{1j}(R_{2h} - \rho\omega^2\delta_{2h})S_{3k}\} = 0,\end{aligned} \tag{20}$$

$$\epsilon_{jhk}\{(R_{1j}-\rho\omega^2\delta_{1j})S_{2h}(R_{3k}-\rho\omega^2\delta_{3k})+(R_{1j}-\rho\omega^2\delta_{1j})(R_{2h}-\rho\omega^2\delta_{2h})S_{3k}$$
$$+S_{1j}(R_{2h}-\rho\omega^2\delta_{2h})(R_{3k}-\rho\omega^2\delta_{3k})-S_{1j}S_{2h}S_{3k}\}=0. \tag{21}$$

Each term in (20) is of even order in $\mathbf{m}$ while each term in (21) is of odd order. Accordingly, if $\mathbf{l},\mathbf{m}$ is a solution then $\mathbf{l},-\mathbf{m}$ is a solution as well. This in turn implies that, if $\mathbf{l}\cdot\mathbf{m}\neq 0$, pairs of solutions in $\mathbf{l},\mathbf{m}$ occur with both $\mathbf{l}\cdot\mathbf{m}>0$ and $\mathbf{l}\cdot\mathbf{m}<0$.

Because of dissipation, one expects that the wave amplitude decays in the direction of propagation. While it is so in isotropic solids, the description of attenuation in anisotropic solids is significantly more involved. Let

$$\mathbf{u}=(\mathbf{v}+i\mathbf{w})\exp(-\mathbf{m}\cdot\mathbf{x})\exp[i(\mathbf{l}\cdot\mathbf{x}-\omega t)]. \tag{22}$$

As we show in a moment, the modulus $|\mathbf{u}|$ may increase in the direction of phase propagation but, nevertheless, this is consistent with the dissipative nature of the model.

Let S be a closed surface with outward normal $\mathbf{n}$. For complex-valued velocity fields $\dot{\mathbf{u}}$, the time-averaged power $\mathcal{P}$ delivered by the surface tractions over a time period, $2\pi/\omega$, is given by

$$\mathcal{P}=\tfrac{1}{2}\mathrm{Re}\int_S(\mathbf{T}\mathbf{n})\cdot\dot{\mathbf{u}}^*da.$$

This allows

$$\mathbf{J}=-\tfrac{1}{2}\mathrm{Re}(\mathbf{T}\dot{\mathbf{u}}^*)$$

to be viewed as the mean energy flux. Since $\dot{\mathbf{u}}^*=i\omega\mathbf{u}^*$, we can write

$$\mathbf{J}=\tfrac{1}{2}\omega\mathrm{Im}(\mathbf{T}\mathbf{u}^*).$$

Owing to dissipation, one expects that if $\mathbf{J}$ really represents an energy flux then

$$\mathcal{P}=-\int_S\mathbf{J}\cdot\mathbf{n}\,da>0.$$

If it is so, the divergence theorem, the arbitrariness of the region enclosed by S, and the continuity of $\nabla\cdot\mathbf{J}$ imply that

$$\nabla\cdot\mathbf{J}<0. \tag{23}$$

We now show that the condition (23) is satisfied for each (inhomogeneous) wave in anisotropic viscoelastic solids.

By the equation of motion, inner multiplication by $\dot{\mathbf{u}}^*$ yields

$$\rho\ddot{\mathbf{u}}\cdot\dot{\mathbf{u}}^*+\mathbf{T}\cdot\nabla\dot{\mathbf{u}}^*=\nabla\cdot(\mathbf{T}\dot{\mathbf{u}}^*). \tag{24}$$

For inhomogeneous waves of the form (22) we have

$$\rho\ddot{\mathbf{u}}\cdot\dot{\mathbf{u}}^*=-i\rho\omega^3(\mathbf{v}^2+\mathbf{w}^2)\exp(-2\mathbf{m}\cdot\mathbf{x}),$$

and, since $\mathbf{T}=(\mathbf{H}+i\mathbf{K})\nabla\mathbf{u}$,

$$\begin{aligned}&\mathbf{T}\cdot\nabla\dot{\mathbf{u}}^*\\
=\;&-\omega[(\mathbf{m}\otimes\mathbf{v}+\mathbf{l}\otimes\mathbf{w})\cdot\mathbf{K}(\mathbf{m}\otimes\mathbf{v}+\mathbf{l}\otimes\mathbf{w})+(\mathbf{l}\otimes\mathbf{v}-\mathbf{m}\otimes\mathbf{w})\cdot\mathbf{K}(\mathbf{l}\otimes\mathbf{v}-\mathbf{m}\otimes\mathbf{w})\\
&\quad+(\mathbf{l}\otimes\mathbf{v}-\mathbf{m}\otimes\mathbf{w})\cdot(\mathbf{H}-\mathbf{H}^T)(\mathbf{l}\otimes\mathbf{w}+\mathbf{m}\otimes\mathbf{v})]\exp(-2\mathbf{m}\cdot\mathbf{x})\\
&+i\omega[(\mathbf{m}\otimes\mathbf{v}+\mathbf{l}\otimes\mathbf{w})\cdot\mathbf{H}(\mathbf{m}\otimes\mathbf{v}+\mathbf{l}\otimes\mathbf{w})+(\mathbf{l}\otimes\mathbf{v}-\mathbf{m}\otimes\mathbf{w})\cdot\mathbf{H}(\mathbf{l}\otimes\mathbf{v}-\mathbf{m}\otimes\mathbf{w})\\
&\quad+(\mathbf{m}\otimes\mathbf{v}+\mathbf{l}\otimes\mathbf{w})\cdot(\mathbf{K}-\mathbf{K}^T)(\mathbf{l}\otimes\mathbf{v}-\mathbf{m}\otimes\mathbf{w})]\exp(-2\mathbf{m}\cdot\mathbf{x}).\end{aligned}$$

Substitution in (24), taking the real part, and the observation that

$$\mathrm{Re}[\nabla\cdot(\mathbf{T}\dot{\mathbf{u}}^*)] = \nabla\cdot[\mathrm{Re}(\mathbf{T}\dot{\mathbf{u}}^*)] = -2\nabla\cdot\mathbf{J},$$

yield

$$\begin{aligned}\nabla\cdot\mathbf{J} &= \tfrac{1}{2}\omega[(\mathbf{m}\otimes\mathbf{v}+\mathbf{l}\otimes\mathbf{w})\cdot\mathbf{K}(\mathbf{m}\otimes\mathbf{v}+\mathbf{l}\otimes\mathbf{w})+(\mathbf{l}\otimes\mathbf{v}-\mathbf{m}\otimes\mathbf{w})\cdot\mathbf{K}(\mathbf{l}\otimes\mathbf{v}-\mathbf{m}\otimes\mathbf{w})\\ &\quad +(\mathbf{l}\otimes\mathbf{v}-\mathbf{m}\otimes\mathbf{w})\cdot(\mathbf{H}-\mathbf{H}^T)(\mathbf{l}\otimes\mathbf{w}+\mathbf{m}\otimes\mathbf{v})]\exp(-2\mathbf{m}\cdot\mathbf{x}).\end{aligned}$$

The inequality (16) implies (23) unless

$$\mathrm{sym}(\mathbf{m}\otimes\mathbf{v}+\mathbf{l}\otimes\mathbf{w})=0, \qquad \mathrm{sym}(\mathbf{l}\otimes\mathbf{v}-\mathbf{m}\otimes\mathbf{w})=0,$$

namely,

$$\mathrm{sym}((\mathbf{l}+i\mathbf{m})\otimes(\mathbf{v}+i\mathbf{w}))=0.$$

Owing to (22), this condition is equivalent to

$$\mathrm{sym}(\nabla\mathbf{u})=0,$$

which in turn means that the strain tensor $\mathbf{E}$ vanishes. Accordingly, $\nabla\cdot\mathbf{J}=0$ if and only if the body undergoes a rigid motion.

It is of interest to remark that the negativity of $\nabla\cdot\mathbf{J}$ is proved by means only of the restriction, of thermodynamic character, (16).

Observe that

$$\begin{aligned}\mathrm{Im}(\mathbf{T}\mathbf{u}^*) &= \{[\mathbf{H}(\mathbf{l}\otimes\mathbf{v})-\mathbf{K}(\mathbf{m}\otimes\mathbf{v})-\mathbf{K}(\mathbf{l}\otimes\mathbf{w})-\mathbf{H}(\mathbf{m}\otimes\mathbf{w})]\mathbf{v}\\ &\quad +[\mathbf{K}(\mathbf{l}\otimes\mathbf{v})+\mathbf{H}(\mathbf{m}\otimes\mathbf{v})+\mathbf{H}(\mathbf{l}\otimes\mathbf{w})-\mathbf{K}(\mathbf{m}\otimes\mathbf{w})]\mathbf{w}\}\exp(-2\mathbf{m}\cdot\mathbf{x}). \quad (25)\end{aligned}$$

The dependence of $\mathbf{J}=\frac{1}{2}\omega\,\mathrm{Im}(\mathbf{T}\mathbf{u}^*)$ on $\mathbf{x}$ is through $\exp(-2\mathbf{m}\cdot\mathbf{x})$ only. Hence

$$\nabla\cdot\mathbf{J}=-2\mathbf{m}\cdot\mathbf{J}.$$

Because of (23) it follows that

$$\mathbf{J}\cdot\mathbf{m}>0. \qquad (26)$$

The inequality (26) signifies that the modulus of $\mathbf{u}$ decays in the direction of $\mathbf{J}$. For, if $\mathbf{x}=x\mathbf{J}/|\mathbf{J}|$, the modulus of $\mathbf{u}$ varies as

$$\exp(-\mathbf{m}\cdot\mathbf{J}x/|\mathbf{J}|)$$

and hence decreases as x increases.

The inequality (26), along with (25), resembles the result $\mathbf{J}\cdot\mathbf{m}=\langle\mathcal{D}\rangle/2$, where $\langle\mathcal{D}\rangle$ is the (the mean value of the) rate of dissipation, for isotropic solids [15] or that for anisotropic solids (cf. Hayes in [30]); incidentally, in Hayes' notation, $\tilde{D}=\langle\mathcal{D}\rangle/\omega$. Of course we expect that $\langle\mathcal{D}\rangle$ and $\tilde{D}$ are positive. Here we obtain

$$\begin{aligned}\mathbf{J}\cdot\mathbf{m} &= -\frac{\omega}{4}[(\mathbf{m}\otimes\mathbf{v}+\mathbf{l}\otimes\mathbf{w})\cdot\mathbf{K}(\mathbf{m}\otimes\mathbf{v}+\mathbf{l}\otimes\mathbf{w})+(\mathbf{l}\otimes\mathbf{v}-\mathbf{m}\otimes\mathbf{w})\cdot\mathbf{K}(\mathbf{l}\otimes\mathbf{v}-\mathbf{m}\otimes\mathbf{w})\\ &\quad +(\mathbf{l}\otimes\mathbf{v}-\mathbf{m}\otimes\mathbf{w})\cdot(\mathbf{H}-\mathbf{H}^T)(\mathbf{l}\otimes\mathbf{w}+\mathbf{m}\otimes\mathbf{v})]\exp(-2\mathbf{m}\cdot\mathbf{x})\end{aligned}$$

and hence $\mathbf{J}\cdot\mathbf{m}$ is strictly positive for non-rigid motions. Elastic solids correspond to $\mathbf{K}=0$, $\mathbf{H}=\mathbf{G}_0$. Hence we have

$$\mathbf{J}\cdot\mathbf{m}=[(\mathbf{l}\otimes\mathbf{v}-\mathbf{m}\otimes\mathbf{w})\cdot(\mathbf{G}_0-\mathbf{G}_0^T)(\mathbf{l}\otimes\mathbf{w}+\mathbf{m}\otimes\mathbf{v})]\exp(-2\mathbf{m}\cdot\mathbf{x}). \tag{27}$$

The symmetry of $\mathbf{G}_0$ makes $\mathbf{J}\cdot\mathbf{m}$ vanish and the amplitude $|\mathbf{u}|$ be constant in the direction of the energy flux $\mathbf{J}$.

Though (26) always holds for dissipative solids, as shown above for elastic anisotropic solids, the inner product $\mathbf{l}\cdot\mathbf{m}$ may equally well take positive and negative values. Since

$$\mathbf{u}=\tilde{\mathbf{u}}\exp(-\mathbf{m}\cdot\mathbf{x})\exp[i(\mathbf{l}\cdot\mathbf{x}-\omega t)],$$

in the direction of (phase) propagation, viz. $\mathbf{x}=x\mathbf{l}/|\mathbf{l}|$, we have

$$|\mathbf{u}|=|\tilde{\mathbf{u}}|\exp(-\mathbf{m}\cdot\mathbf{l}x/|\mathbf{l}|).$$

Accordingly, if $\mathbf{l}\cdot\mathbf{m}>0$ then $|\mathbf{u}|$ decreases as x increases and viceversa. The first instance seems to be consistent with the dissipativity of the solid while the second one does not. In fact there is no strict relation between dissipativity and the sign of $\mathbf{l}\cdot\mathbf{m}$. The modulus $|\mathbf{u}|$ of $\mathbf{u}$ may even increase in the direction of $\mathbf{l}$ but, nonetheless, for any region of the body the overall inward power flow is positive.

It is of interest to evaluate $\mathbf{J}\cdot\mathbf{l}$. Inner multiplication of (25) by $\mathbf{l}$ yields

$$\begin{aligned}\mathbf{J}\cdot\mathbf{l} \;=\; & \tfrac{1}{2}\omega[(\mathbf{v}\otimes\mathbf{l})\cdot\mathbf{H}(\mathbf{l}\otimes\mathbf{v})+(\mathbf{w}\otimes\mathbf{l})\cdot\mathbf{H}(\mathbf{l}\otimes\mathbf{w})-(\mathbf{v}\otimes\mathbf{l})\cdot\mathbf{H}(\mathbf{m}\otimes\mathbf{w})\\ &+(\mathbf{w}\otimes\mathbf{l})\cdot\mathbf{H}(\mathbf{m}\otimes\mathbf{v})+(\mathbf{l}\otimes\mathbf{w})\cdot(\mathbf{K}-\mathbf{K}^T)(\mathbf{l}\otimes\mathbf{v})\\ &-(\mathbf{v}\otimes\mathbf{l})\cdot\mathbf{K}(\mathbf{m}\otimes\mathbf{v})-(\mathbf{w}\otimes\mathbf{l})\cdot\mathbf{K}(\mathbf{m}\otimes\mathbf{w})]\exp(-2\mathbf{m}\cdot\mathbf{x}).\end{aligned}$$

Hence it seems that in general $\mathbf{J}\cdot\mathbf{l}$ is not definite. It is then worth looking at the case when $\mathbf{H},\mathbf{K}$ are symmetric and $\mathbf{H}$ is positive definite. A direct calculation yields

$$\begin{aligned}\mathbf{v}\cdot\mathbf{R}\mathbf{v}+\mathbf{w}\cdot\mathbf{R}\mathbf{w} \;=\; & (\mathbf{v}\otimes\mathbf{l})\cdot\mathbf{H}(\mathbf{l}\otimes\mathbf{v})-(\mathbf{v}\otimes\mathbf{m})\cdot\mathbf{H}(\mathbf{m}\otimes\mathbf{v})+(\mathbf{w}\otimes\mathbf{l})\cdot\mathbf{H}(\mathbf{l}\otimes\mathbf{w})\\ &-(\mathbf{w}\otimes\mathbf{m})\cdot\mathbf{H}(\mathbf{m}\otimes\mathbf{w})-2(\mathbf{v}\otimes\mathbf{l})\cdot\mathbf{K}(\mathbf{m}\otimes\mathbf{v})-2(\mathbf{w}\otimes\mathbf{l})\cdot\mathbf{K}(\mathbf{m}\otimes\mathbf{w}).\end{aligned}$$

Hence we derive an expression for $(\mathbf{v}\otimes\mathbf{l})\cdot\mathbf{K}(\mathbf{m}\otimes\mathbf{v})+(\mathbf{w}\otimes\mathbf{l})\cdot\mathbf{K}(\mathbf{m}\otimes\mathbf{w})$, make use of (19) where now $\mathbf{S}-\mathbf{S}^T=0$ - and substitute to obtain

$$\begin{aligned}\mathbf{J}\cdot\mathbf{l}=\frac{1}{4}\omega[\rho\omega^2(\mathbf{v}^2+\mathbf{w}^2) \;&+\; (\mathbf{v}\otimes\mathbf{l}-\mathbf{w}\otimes\mathbf{m})\cdot\mathbf{H}(\mathbf{v}\otimes\mathbf{l}-\mathbf{w}\otimes\mathbf{m})\\ &+\; (\mathbf{w}\otimes\mathbf{l}+\mathbf{v}\otimes\mathbf{m})\cdot\mathbf{H}(\mathbf{w}\otimes\mathbf{l}+\mathbf{v}\otimes\mathbf{m})]\exp(-2\mathbf{m}\cdot\mathbf{x}).\end{aligned}$$

The positive definiteness of $\mathbf{H}$ implies that

$$\mathbf{J}\cdot\mathbf{l}>0$$

for any non-rigid motion. This inequality is a generalization of that obtained by Hayes for anisotropic elasticity and that obtained by Buchen for isotropic viscoelastic solids [15].

For isotropic solids, in addition to (26) we have the restriction

$$\mathbf{l}\cdot\mathbf{m}>0$$

which is a feature of the propagation modes in isotropic solids and is a consequence of thermodynamics. The dissipative character of the solid is expressed by (23) and is equivalent to (26).

Some of the calculations exhibited in this section can be performed equally well in $\mathbb{R}^6$ by representing symmetric tensors as six-tuples and complex moduli as 6×6 matrices. The present notation, to our mind, allows a more direct interpretation. Analogous results hold for fluids [31].

7. Rays in pre-stressed solids

Consider a solid which occupies an unstressed placement $\mathcal{B}$, in the absence of gravity, and is at equilibrium in the placement $\mathcal{B}_i$ under the action of the body force (gravity) $\mathbf{g}$ per unit mass. Denote by $\mathbf{X}$ and $\mathbf{x}$ the vector position in $\mathcal{B}$ and $\mathcal{B}_i$. Denote by $\bar{\mathbf{S}}$ the first Piola-Kirchhoff stress tensor in the equilibrium configuration. Hence we write the equilibrium equation as

$$\nabla_{\mathbf{X}} \cdot \bar{\mathbf{S}} + \rho_0 \mathbf{g} = 0. \tag{28}$$

Now let any point $\mathbf{x}(\mathbf{X})$ of $\mathcal{B}_i$ undergo a motion. Denote by $\mathbf{u}$ the displacement and $\tilde{\mathbf{x}} = \mathbf{x} + \mathbf{u}$ the present position; $\tilde{\mathbf{x}} = \mathbf{x}(\mathbf{X}) + \mathbf{u}(\mathbf{x}(\mathbf{X}), t)$. Letting $\tilde{\mathbf{S}}$ be the present Piola-Kirchhoff stress, we write the equation of motion as

$$\rho_0 \ddot{\mathbf{u}} = \nabla_{\mathbf{X}} \cdot \tilde{\mathbf{S}} + \rho_0 \mathbf{g}. \tag{29}$$

Subtraction of (28) from (29) yields

$$\rho_0 \ddot{\mathbf{u}} = \nabla_{\mathbf{X}} \cdot (\tilde{\mathbf{S}} - \bar{\mathbf{S}}). \tag{30}$$

Now we assume that the motion $\mathbf{u}(\mathbf{x}(\mathbf{X}), t)$ superposed on the equilibrium configuration is small. Let ∇ be the gradient with respect to the position $\mathbf{x}$ in the equilibrium configuration and define the displacement gradient as $\mathbf{H} = \nabla \mathbf{u}^T$. Also, if $\mathbf{F}$ and $\tilde{\mathbf{F}}$ denote the deformation gradient in $\mathcal{B}_i$ and $\mathcal{B}_t$ then $\tilde{\mathbf{F}} = (\mathbf{1} + \mathbf{H})\mathbf{F}$. By the smallness of the motion, second-order terms in $\mathbf{u}, \mathbf{H}$ and higher are neglected.

Consider the second Piola-Kirchhoff stresses $\bar{\mathbf{Y}}$ and $\bar{\mathbf{Y}} + \hat{\mathbf{Y}}$ corresponding to the equilibrium deformation $\mathbf{x}(\mathbf{X})$ and the motion $\tilde{\mathbf{x}}(\mathbf{X}, t)$. To within the quadratic term $\mathbf{HF}\hat{\mathbf{Y}}$,

$$\tilde{\mathbf{S}} = \tilde{\mathbf{F}}(\bar{\mathbf{Y}} + \hat{\mathbf{Y}}) = \mathbf{F}\bar{\mathbf{Y}} + \mathbf{HF}\bar{\mathbf{Y}} + \mathbf{F}\hat{\mathbf{Y}}.$$

Hence (30) becomes

$$\rho_0 \ddot{\mathbf{u}} = \nabla_{\mathbf{X}} \cdot (\mathbf{HF}\bar{\mathbf{Y}} + \mathbf{F}\hat{\mathbf{Y}}). \tag{31}$$

Let $J = \det \mathbf{F}$. Multiplication of (31) by $1/J$ and use of the identity $\nabla \cdot (\mathbf{F}/J) = 0$ yields

$$\rho \ddot{\mathbf{u}} = \nabla \cdot (\frac{1}{J}\mathbf{HF}\bar{\mathbf{Y}}\mathbf{F}^T + \frac{1}{J}\mathbf{F}\hat{\mathbf{Y}}\mathbf{F}^T)$$

where $\rho = \rho_0/J$ is the mass density in $\mathcal{B}_i$. Further, $\mathbf{F}\bar{\mathbf{Y}}\mathbf{F}^T/J$ is the Cauchy stress $\bar{\mathbf{T}}$ in the equilibrium configuration $\mathcal{B}_i$ and hence we write

$$\rho \ddot{\mathbf{u}} = \nabla \cdot (\mathbf{H}\bar{\mathbf{T}} + \frac{1}{J}\mathbf{F}\hat{\mathbf{Y}}\mathbf{F}^T). \tag{32}$$

In (32), $\mathbf{u}$ is the unknown function while $\bar{\mathbf{T}}$ and $\mathbf{F}$, and hence J, are regarded as given once the solution $\mathbf{x}(\mathbf{X})$ to the equilibrium equation (28) is determined. Now, to solve (28) we have to specify the constitutive equation for $\mathbf{S}^0$ in terms of $\mathbf{F}$ and the boundary conditions. In the realistic nonlinear case, solving (28) is a formidable problem. Furthermore, the result is related to the chosen model. Accordingly, here we prefer to develop general considerations which are not affected by the choice of the constitutive model.

Now we consider the term $\hat{\mathbf{Y}}$. Let $\mathbf{Y}$ be given by a nonlinear functional of the history of $\mathbf{C} = \tilde{\mathbf{F}}^T\tilde{\mathbf{F}}$ or of the present value $\mathbf{C}(t)$ and the restricted history ${}_r\mathbf{C}^t$, viz.

$$\mathbf{Y}(t) = \mathcal{Y}(\mathbf{C}(t), {}_r\mathbf{C}^t)).$$

Relative to the standard fading memory norm (cf. [1], §1.3) assume that $\mathcal{Y}$ is Fréchet differentiable at any constant history $\mathbf{C}^\dagger$, with value $\mathbf{C}^0$. Hence the approximation of $\hat{\mathbf{Y}}$ by $d\mathcal{Y}(\mathbf{C}^0, {}_r\mathbf{C}^\dagger | \hat{\mathbf{C}}(t), {}_r\hat{\mathbf{C}}^t)$ yields the representation

$$\hat{Y}_{AB}(t) = \Xi_{ABHL}\hat{C}_{HL}(t) + \int_0^\infty \Gamma_{ABHL}(\xi)\hat{C}_{HL}(t-\xi)d\xi,$$

the tensors $\Xi_{ABHL} = \partial\mathcal{Y}_{AB}/\partial C_{HL}$ and $\Gamma_{ABHL}(s)$ being parameterized by $\mathbf{C}^0$. The tensor function $\boldsymbol{\Gamma}$ on $\mathbb{R}^+$ is required to be integrable.

It is a consequence of thermodynamics that the half-range Fourier-sine transform

$$\boldsymbol{\Gamma}_s(\omega) = \int_0^\infty \boldsymbol{\Gamma}(\xi)\sin\omega\xi \, d\xi$$

is negative definite in Sym. To prove this result we apply the second law in the form

$$\int_0^d \frac{1}{\tilde{\rho}}\tilde{\mathbf{T}}\cdot\mathbf{L}\,dt > 0 \tag{33}$$

for all non-trivial cycles of duration d; $\tilde{\rho}, \tilde{T}, \mathbf{L}$ are the mass density, the Cauchy stress and the velocity gradient in $\mathcal{B}_t$. First observe that $\tilde{\mathbf{T}}\cdot\mathbf{L} = \mathbf{Y}\cdot\dot{\mathbf{C}}/2\mathcal{J}$ where $\mathcal{J}$ is the effective deformation gradient. Since $\tilde{\mathbf{F}} = (\mathbf{1}+\mathbf{H})\mathbf{F}$, in the linear approximation we have

$$\mathcal{J} = \det\tilde{\mathbf{F}} \simeq J(1+\nabla\cdot\mathbf{u}).$$

Also,

$$\frac{1}{\tilde{\rho}}\tilde{\mathbf{T}}\cdot\mathbf{L} = \frac{1}{2\rho_0}\bar{\mathbf{Y}}\cdot\dot{\mathbf{C}} + \frac{1}{2\rho_0}\hat{\mathbf{Y}}\cdot\dot{\mathbf{C}},$$

where ρ_0 is the reference mass density. Upon integration on any cycle, the contribution of the first term is zero. Now take $\hat{\mathbf{C}}$ as a symmetric tensor $\mathbf{A}$ times $\sin\omega t$ and then $\dot{\mathbf{C}} = \omega\mathbf{A}\cos\omega t$. Substitution in (33) and integration with respect to t yield

$$0 < \int_0^d \frac{1}{\tilde{\rho}}\tilde{\mathbf{T}}\cdot\mathbf{L}\,dt = -\frac{\pi}{2\rho_0}\mathbf{A}\cdot\boldsymbol{\Gamma}_s(\omega)\mathbf{A}, \qquad \forall\omega > 0,$$

whence the desired result.

Still up to linear terms in $\mathbf{H}$,

$$\hat{\mathbf{C}} = \tilde{\mathbf{F}}^T\tilde{\mathbf{F}} - \mathbf{F}^T\mathbf{F} = 2\mathbf{F}^T\mathbf{E}\mathbf{F},$$

where $\mathbf{E} = \mathrm{sym}\mathbf{H}$, and

$$\Gamma_{ABMN}(\xi)\hat{C}_{MN}(t-\xi) = 2\Gamma_{ABMN}(\xi)F_{jM}F_{hN}E_{jh}(t-\xi).$$

Hence

$$\frac{1}{J}\mathbf{F}\hat{\mathbf{Y}}(t)\mathbf{F}^T = \bar{\mathbf{G}}\mathbf{E}(t) + \int_0^\infty \boldsymbol{\Lambda}(\xi)\mathbf{E}(t-\xi)d\xi$$

where

$$\bar{G}_{ikjh} = \frac{2}{J}F_{iA}F_{kB}F_{jM}F_{hN}\Xi_{ABMN}, \qquad \Lambda_{ikjh}(\xi) = \frac{2}{J}F_{iA}F_{kB}F_{jM}F_{hN}\Gamma_{ABMN}(\xi).$$

Now, for any symmetric tensor $\mathbf{S}$

$$\Lambda_{ikjh}S_{ik}S_{jh} = \Gamma_{ABMN}\Sigma_{AB}\Sigma_{MN}$$

where $\Sigma_{AB} = S_{ik}F_{iA}F_{kB}\sqrt{2/J}$. Of course Σ_{AB} is symmetric too and vanishes if and only if S_{ik} vanishes. Hence $\boldsymbol{\Gamma}_s < 0$ in Sym implies that $\boldsymbol{\Lambda}_s < 0$ in Sym.

By (32) we can write the equation of motion as

$$\rho\ddot{\mathbf{u}}(t) = \nabla\cdot\left(\mathbf{H}(t)\bar{\mathbf{T}} + \bar{\mathbf{G}}\mathbf{E}(t) + \int_0^\infty \boldsymbol{\Lambda}(\xi)\mathbf{E}(t-\xi)d\xi\right).$$

If attention is restricted to the time-harmonic dependence $\exp(-i\omega t)$ the equation of motion becomes

$$-\rho\omega^2\mathbf{u} = \nabla\cdot(\mathbf{H}\bar{\mathbf{T}} + \bar{\mathbf{G}}\mathbf{E} + \boldsymbol{\Lambda}_F\mathbf{E})$$

where $\boldsymbol{\Lambda}_F$ is the Fourier transform of $\boldsymbol{\Lambda}$, namely

$$\boldsymbol{\Lambda}_F = \int_0^\infty \boldsymbol{\Lambda}(\xi)\exp(i\omega\xi)d\xi.$$

The fourth-order tensor $\boldsymbol{\Lambda}_F$ is complex valued and parameterized by ω. For later convenience we observe that if $\boldsymbol{\Lambda} \in W^{2,1}(\mathbb{R}^+)$ then two integrations by parts show that

$$\boldsymbol{\Lambda}_F = -(i\omega)^{-1}\boldsymbol{\Pi} + O(\omega^{-2})$$

where $\boldsymbol{\Pi} = \boldsymbol{\Lambda}(0)$.

Let the subscript c denote the cosine transform so that

$$\boldsymbol{\Lambda}_F = \boldsymbol{\Lambda}_c(\omega) + i\boldsymbol{\Lambda}_s(\omega).$$

Hence, for any $\mathbf{S} \in \mathrm{Sym}$, it follows that

$$i\omega\mathbf{S}\cdot\boldsymbol{\Lambda}_c(\omega)\mathbf{S} - \omega\mathbf{S}\cdot\boldsymbol{\Lambda}_s(\omega)\mathbf{S} = -\mathbf{S}\cdot\boldsymbol{\Pi}\mathbf{S} + O(\omega^{-1}).$$

For large enough ω we have

$$\operatorname{sgn}\omega \mathbf{S} \cdot \boldsymbol{\Lambda}_s(\omega)\mathbf{S} = \operatorname{sgn}\mathbf{S} \cdot \boldsymbol{\Pi}\mathbf{S}.$$

This implies that $\boldsymbol{\Pi}$ is negative definite.

By generalizing standard asymptotic analyses, take the displacement $\mathbf{u}$ as the series

$$\mathbf{u} = \exp[i\omega\tau(\mathbf{x})] \sum_{\alpha=0}^{\infty} \mathbf{U}^{\alpha}(i\omega)^{-\alpha}.$$

Evaluation of the pertinent derivatives and substitution in the equation of motion yields a series of powers of ω in the form $\sum_{\alpha=0}^{\infty} c_\alpha \omega^{2-\alpha}$. We determine the series solution by setting the coefficient of each power separately equal to zero. Let

$$M_{phlk} = G_{phlk} + \bar{T}_{hk}\delta_{lp}.$$

The vanishing of each coefficient results in the relations

$$[\rho\delta_{pl} - M_{phlk}\tau_{,h}\tau_{,k}]U_l^0 = 0, \tag{34}$$

$$\begin{aligned}&[-\rho\delta_{pl} + M_{phlk}\tau_{,h}\tau_{,k}]U_l^1 \\ &+M_{phlk}(\tau_{,hk}U_l^0 + \tau_{,k}U_{l,h}^0 + \tau_{,h}U_{l,k}^0) + M_{phlk,h}\tau_{,k}U_l^0 - \Pi_{phlk}\tau_{,k}\tau_{,h}U_l^0 = 0,\end{aligned} \tag{35}$$

along with higher-order equations which are inessential to the present purposes. Equation (34) allows for non-trivial solutions only if

$$\det(\rho\delta_{pl} - M_{phlk}\tau_{,h}\tau_{,k}) = 0. \tag{36}$$

Equation (36) is the eikonal equation for the phase function τ. For each solution $\tau(\mathbf{x})$, the matrix applied to U_l^1 in (35) is singular. Let $\mathbf{v}_1, \mathbf{v}_2, \mathbf{v}_3$ be the three column vectors of the symmetric matrix

$$\Upsilon_{pl} = \rho\delta_{pl} - M_{phlk}\tau_{,h}\tau_{,k}.$$

By (36), $\mathbf{v}_1, \mathbf{v}_2, \mathbf{v}_3$ are coplanar and, by (34), $\mathbf{U}^0$ is orthogonal to the plane $L(\mathbf{v}_1, \mathbf{v}_2, \mathbf{v}_3)$. Accordingly, U_p^0 times (4.2) yields an equation for $\mathbf{U}^0$, free from $\mathbf{U}^1$. By this equation and (34) we determine $\mathbf{U}^0(\mathbf{x})$. Once $\mathbf{U}^0$ is known, we might proceed in a similar way to determine $\mathbf{U}^\alpha$, $\alpha = 1, 2, \ldots$. Now we consider the pertinent families of rays.

Represent by $\mathbf{x} = \mathbf{x}(\sigma, \xi_1, \xi_2)$ the position vector along the ray; ξ_1, ξ_2 label the rays and σ is a coordinate along the ray. If

$$\mathcal{H}(\mathbf{h}, \mathbf{x}) = 0$$

is the partial differential equation satisfied by $\tau(\mathbf{x})$, where $\mathbf{h} = \nabla\tau$, rays are a family of curves, parameterized by ξ_1, ξ_2, such that

$$\dot{\mathbf{x}} = \phi\frac{\partial\mathcal{H}}{\partial\mathbf{h}}, \tag{37}$$

where ϕ is an arbitrary function of $\mathbf{x}$ and the superposed dot denotes $d/d\sigma$. Meanwhile $\mathbf{h}$ varies along any ray according to the equation

$$\dot{\mathbf{h}} = -\phi \frac{\partial \mathcal{H}}{\partial \mathbf{x}}. \tag{38}$$

Owing to the arbitrariness of ϕ we choose $\phi = 1/2$. Now, (37) and (38) constitute a system of six ordinary differential equations in the six unknowns $\mathbf{x}(\sigma), \mathbf{h}(\sigma)$, parameterized by ξ_1, ξ_2. The ray vector $\partial\mathcal{H}/\partial\mathbf{h}$ is also called group-velocity vector.

By analogy with [32] and borrowing from a joint research with Caviglia [33] we can say that the function $\mathcal{H}(\mathbf{h}, \mathbf{x})$ is in fact the product of three functions $\mathcal{H}_r(\mathbf{h}, \mathbf{x})$, $r = 1, 2, 3$, and that

$$\frac{\partial \mathcal{H}_r}{\partial h_i} = w_{i(r)} := M_{pijk} q_{p(r)} q_{j(r)} h_k.$$

Here $\mathbf{q}_{(1)}, \mathbf{q}_{(3)}, \mathbf{q}_{(3)}$ are eigenvectors of

$$Q_{pj} = M_{pijk} h_i h_k$$

and are an orthonormal basis for the vector space V. Correspondingly, the polarization vector $\mathbf{U}^0$ is found to be parallel to $\mathbf{q}_{(r)}$. Let $\mathbf{U}^0 = \nu_r \mathbf{q}_{(r)}$, where ν_r is an undetermined scalar, possibly dependent on $\mathbf{x}$.

The amplitude ν_r of $\mathbf{U}^0$ is determined by a projection of (35). Let $\mathbf{h}_{(r)} = \nabla\tau_r$ and $B_{pr} = \nu_r q_{p(r)}$. In view of (34) we have

$$B_{pr}[Q_{pl}(\mathbf{h}_{(r)}) - \rho\delta_{pl}] = \nu_r\{[\mathbf{Q}(\mathbf{h}_{(r)}) - \rho\mathbf{1}]\mathbf{q}_{(r)}\}_l = 0, \tag{39}$$

for any $r = 1, 2, 3$. Consider (35) in connection with $\tau = \tau_r$, $\mathbf{U}^0 = \nu_r \mathbf{q}_{(r)}$. Hence, multiplication of (35) by B_{pr} and use of (39) yields

$$B_{pr}[M_{pjlk}(h_k B_{lk,j} + h_j B_{lr\,k} + h_{j,k} B_{lr}) + M_{pjlk,j} h_k B_{lr} - \Pi_{pjlk} h_j h_k B_{lr}] = 0$$

whence, by use of the symmetry of M, we have

$$(M_{pjlk} h_j B_{lr} B_{pr})_{,k} = \Pi_{pjlk} h_j h_k B_{pr} B_{lr}. \tag{40}$$

The definition of B implies that

$$\Pi_{pjlk} h_j h_k B_{pr} B_{lr} = \nu_r^2 f_r$$

where

$$f_r = \Pi_{pjlk} h_j h_k q_{p(r)} q_{l(r)}.$$

Comparison allows (40) to be written as

$$\nabla \cdot (\nu_r^2 \mathbf{w}_{(r)}) = \nu_r^2 f_r. \tag{41}$$

For any regular region D, integration of (41) over D yields

$$\int_{\partial D} \mathbf{w}_{(r)} \cdot \mathbf{n}\, da = \int_D \nu_r^2 f_r\, dv. \tag{42}$$

Equations (41) and (42) are in the form of a balance equation and involve the vector $\mathbf{w}_{(r)}$ that generates the rays associated with the r-th solution to (36). In elasticity, $\boldsymbol{\Pi}$ vanishes and hence the vector $\nu_r^2\mathbf{w}_{(r)}$ turns out to be divergence-free or conserved. In viscoelasticity, $f_r \neq 0$. Indeed, by the negative definiteness of $\boldsymbol{\Pi}$ it follows that $f_r < 0$, for non-zero vectors $\mathbf{h}, \mathbf{q}$. Hence dissipativity results in a negative value of the divergence or the net flux through a closed surface of $\nu_r^2\mathbf{w}_{(r)}$.

To save writing we now omit the label r for quantities pertaining to the r-th solution. Choose any point $\mathbf{x}_0$ and let W_0 be a plane, bounded surface that passes through $\mathbf{x}_0$ and is orthogonal to $\mathbf{w}(\mathbf{x}_0)$. Now consider the tube constituted by the rays that pass through W_0. Let $\bar{\mathbf{x}}$ be a point on the ray passing through $\mathbf{x}_0$ and W be the intersection, with the tube, of the plane that passes through $\bar{\mathbf{x}}$ and is orthogonal to $\mathbf{w}(\bar{\mathbf{x}})$. Denote by d_0 the diameter of W_0. Letting w be the length of $\mathbf{w}$, we have

$$\mathbf{w}\cdot\mathbf{n} = \begin{cases} w(\bar{\mathbf{x}}) + O(d_0), & \text{on } W \\ -w(\mathbf{x}_0) + O(d_0), & \text{on } W_0. \end{cases}$$

Moreover, $\mathbf{w}\cdot\mathbf{n} = 0$ on the lateral surface of the tube (between W_0 and W). Hence, by (42) we have

$$(w\nu^2)(\bar{\mathbf{x}})a - (w\nu^2)(\mathbf{x}_0)a_0 + a_0 O(d_0) = \int_D f\nu^2\, dv$$

where a_0, a are the areas of W_0, W. The $O(d_0)$ term suggests that we pass to the limit as $d_0 \to 0$, which corresponds to an infintesimal tube around the pertinent ray. The ratio a/a_0 goes as the ratio of mean values of the Jacobian J of mapping via rays (cf. [34]; [2], ch. 9). Moreover, at the limit of arbitrarily small surfaces W_0, W we can view $\mathbf{x}$ as a function of the arclength σ only. Hence, by dividing throughout by a_0 and letting $d_0 \to 0$ we have

$$(wJ\nu^2)(\sigma) - (wJ\nu^2)(\sigma_0) = \int_{\sigma_0}^{\sigma} f(s)J(s)\nu^2(s)ds.$$

Differentiation with respect to σ, some rearrangement and a straighforward integration yield

$$\nu^2(\sigma) = \frac{(wJ)(\sigma_0)}{(wJ)(\sigma)}\nu^2(\sigma_0)\exp\Big[\int_{\sigma_0}^{\sigma}\frac{f}{w}(s)ds\Big]. \tag{43}$$

According to (43), the amplitude ν of $\mathbf{U}^0$ varies along the rays because of two causes. One is the heterogeneity of the material and is reflected in the wJ term and in the denominator of the integrand. The other one is the dissipativity and is expressed by the quantity f in the integrand. Because $f < 0$, as expected dissipativity results in the decay of the amplitude ν along the ray tube.

References

1. Fabrizio, M. and A. Morro: *Mathematical Problems in Linear Viscoelasticity,* SIAM, Philadelphia, 1992.

2. Caviglia, G. and A. Morro: *Inhomogeneous Waves in Solids and Fluids,* World Scientific, Singapore, 1992.

3. Boltzmann, L.: *Zur Theorie der elastichen Nachwirkung,* Sitzber. Kaiserl. Akad. Wiss. Wien, Math.-Naturw. Kl. 70 (1874), 275-300.

4. Lamb, J. and J. Richter: *Anisotropic acoustic attenuation with new measurements for quartz at room temperatures,* Proc. R. Soc. London A 293 (1966), 479-492.

5. Truesdell, C.: *Rational Thermodynamics,* Springer, Berlin, 1984.

6. Morro, A. and M. Vianello: *Minimal and maximal free energy for materials with memory,* Boll. Un. Mat. Ital. A 4 (1990), 45-55.

7. Day, W. A.: *An objection to using entropy as a primitive concept in continuum thermodynamics,* Acta Mech. 27 (1977), 251-255.

8. Caviglia, G. and A. Morro: *Dissipative effects on wave propagation in viscoelastic solids,* Eur. J. Mech. A/Solids 12 (1993), 387-402.

9. Leitman, M. J. and G. M. C. Fisher: *The linear theory of viscoelasticity,* in: Encyclopedia of Physics, vol. VIa/3 (Ed. C. Truesdell), Springer, Berlin, 1973.

10. Chen, P. and A. Morro: *On induced discontinuities in a class of linear materials with internal state parameters,* Meccanica 22 (1987), 14-18.

11. Fisher, G. M. C. and M. E. Gurtin: *Wave propagation in the linear theory of viscoelasticity,* Quart. Appl. Math. 23 (1965), 257-263.

12. Morro, A.: *Temperature waves in rigid materials with memory,* Meccanica 12 (1977), 73-77.

13. Fisher, G. M. C.: *The decay of plane waves in the linear theory of viscoelasticity,* Brown University, Report NONR 562 (40)/2, Providence, RI, 1965.

14. Lockett, F. J.: *The reflection and refraction of waves at an interface between viscelastic materials,* J. Mech. Phys. Solids 10 (1962), 58-64.

15. Buchen, P. W.: *Plane waves in linear viscoelastic media,* Geophys. J. R. Astr. Soc. 23 (1971), 531-542.

16. Currie, P. K., M. A. Hayes and P. M. O'Leary: *Viscoelastic Rayleigh waves,* Quart. Appl. Math. 35 (1977), 35-53.

17. Currie, P. K. and P. M. O'Leary: *Viscoelastic Rayleigh waves II,* Quart. Appl. Math. 35 (1978), 445-454.

18. Boulanger, Ph. and M. Hayes: *Inhomogeneous plane waves in viscous fluids,* Continuum Mech. Thermodyn. 2 (1990), 1-9.

19. Poirée, B.: *Complex harmonic plane waves,* in: Physical Acoustics (Eds. O. Leroy and M. A. Breazeale), Plenum, New York, 1991.

20. Aki, K. and P. G. Richards: *Quantitative Seismology I* Freeman, San Francisco, 1980; p. 170.

21. Morro, A.: *Wave propagation in dissipative solids,* Proc. Workshop "Advanced mathematical tools in metrology", Turin, 1993.

22. Hayes, M.: *Inhomogeneous plane waves,* Arch. Rational Mech. Anal. 85 (1984), 41-79.

23. Fedorov, F. I.: *Theory of Elastic Waves in Crystals,* Plenum, New York, 1968.

24. Musgrave, M. J. P.: *Crystal Acoustics,* Holden Day, San Francisco, 1970.

25. Crampin, S.: *A review of wave motion in anisotropic and cracked elastic media,* Wave Motion 3 (1981), 343-391.

26. Kerner, C., B. Dyer and M. Worthington: *Wave propagation in a vertical transversely isotropic medium: field experiment and model study,* Geophys. J. 97 (1989), 295-305.

27. Chadwick, P.: *Wave propagation in transversely isotropic elastic media I. Homogenous plane waves,* Proc. Roy. Soc. London A 422 (1989), 23-66.

28. Hayes, M. A. and R. S. Rivlin: *Longitudinal waves in a linear viscoelastic material,* ZAMP 23 (1972), 153-156.

29. Hayes, M. A. and R. S. Rivlin: *Plane waves in linear viscoelastic materials,* Q. Appl. Math. 32 (1974), 113-121.

30. Mainardi, F.: *Wave propagation in viscoelastic media,* Pitman, Boston, 1982.

31. Caviglia, G. and A. Morro: *Inhomogeneous waves in anisotropic dissipative solids,* Continuum Mech. Thermodyn., to appear.

32. Caviglia, G. and A. Morro: *Asymptotic ray theory in heterogeneous viscoelastic solids,* Q. Jl Mech. appl. Math. 46 (1993), 569-582.

33. Caviglia, G. and A. Morro: *Asymptotic rays in pre-stressed, anisotropic, dissipative solids,* in preparation.

34. Bleistein, N.: *Mathematical Methods for Wave Phenomena,* Academic Press, London, 1984.

DYNAMIC VISCOELASTIC FRACTURE

J.R. Walton
Texas A&M University, College Station, TX, USA

ABSTRACT

A semi-infinite crack in an infinite material body provides a canonical model with which to study the local crack-tip behavior of propagating cracks that avoids complicated analytical details arising in the consideration of finite length cracks in bounded bodies. Moreover, semi-infinite crack problems frequently admit analytical solutions with which one can examine precisely the behavior of relevant fracture parameters and calibrate alternative numerical schemes which must be employed in more complicated settings for which analytical solutions are not obtainable. These five chapters present analytical solution methods and results for a variety of dynamic, semi-infinite crack models within the context of linear viscoelasticity theory. In the introductory chapter, the general framework for constructing solutions to these semi-infinite crack problems is presented along with a discussion of the physical settings in which classical quasi-static analyses for elastic material models prove inadequate and which motivate the consideration of inertial effects and material viscoelasticity. The remaining chapters present detailed solutions to various canonical problems for both anti-plane shear and plane strain modes of deformation, and under both steady-state and transient crack propagation conditions. In particular, chapters 4 and 5 contain a description of a recently developed method for constructing solutions to dynamically accelerating, semi-infinite, anti-plane shear crack problems in viscoelastic material, and the result of applying the method to the case of an Achenbach-Chao material model.

1. INTRODUCTION

1.1 Introduction to Dynamic Viscoelastic Fracture

Due to its vital importance in technological applications, fracture mechanics is an important and diverse branch of both applied mechanics and applied mathematics which was given a firm foundation more than fifty years ago through the seminal work of Griffith on brittle elastic fracture. Kanninen and Popelar [1] give an excellent survey of fracture mechanics with myriad references to the many milestones achieved throughout its long history. In spite of the fact that brittle elastic fracture has been an enormously successful theory, there are serious limitations to its applicability, prime examples being furnished by materials that exhibit significant time dependent mechanical response, such as metal plasticity and polymer viscoelasticity. It is to the latter class of material behavior, which occurs in a wide variety of materials with important engineering applications, that the present discussion is addressed. (Ferry [2] and Lockett [3] give extensive accounts of the effective viscoelastic characterization of many materials.)

The consideration of material inertia in modelling fracture (*dynamic analyses*) becomes important whenever either the crack-speed or loading rate is sufficiently high. For brittle elastic fracture, this has traditionally meant that either the crack propagates at a significant fraction of a characteristic material wave speed or the loading is of an impact nature. For viscoelastic material, the situation is much more complicated due to the competing mechanisms of energy transport to a crack-tip through stress wave propagation versus viscous energy dissipation. In particular, one must decide whether the glassy or equilibrium wave speeds (which can differ by three orders of magnitude) are the relevant ones for determining the appropriateness of a *quasi-static analysis* (inertial effects ignored) versus a dynamic analysis. It is engineering custom to consider the glassy wave-speed as playing the role for viscoelastic material of the elastic wave-speed. However, as will be shown in later chapters, there are many situations in which significant dynamic effects occur in viscoelastic material for crack-speeds near the relevant equibrium wave-speed. Thus, for modelling propagating cracks in viscoelastic material, consideration of material inertia can be warranted (depending upon material mean relaxation times) for crack-speeds considerably slower than would be required for elastic material with wave speeds comparable to the glassy viscoelastic values.

There is an extensive literature on dynamic elastic crack problems owing to the many different analytical methods available for their solutions. Among the most important techniques that have been successfully applied to these problems are integral-transform/complex-variable based methods, asymptotic/scaling methods, and homogeneous function methods. An excellent account of this subject is given by Freund [4] and an important more recent contribution is given by Willis [5]. For viscoelastic material, there is an extensive literature on quasi-static analyses owing largely to the applicability of generalized correspondence principles such as those of Graham [6] and Schapery [7] which permit one to construct solutions to viscoelastic problems from reference elastic ones by the use of convolution integrals. The book by Golden and Graham [8] is the definitive reference for this development. However, compared to these rather mature subjects, there is a relative paucity of solutions available for dynamic crack analyses in viscoelasticity theory. This is due largely to the fact that the very powerful correspondence principle based methods of quasi-static viscoelasticity no longer apply, and of the direct methods of elasticity theory, only the integral-transform/complex-variable based methods are adaptable to the viscoelastic setting. Due to the analytical complexities encountered in applying direct transform methods to growing, finite length crack problems, attention has focused on semi-infinite crack geometries in either infinite bodies or infinitely long layers of finite thickness. Moreover, except until very recently, only either dynamic, steady-state or constant crack-speed, transient solutions have been constructed for viscoelastic media. In particular, it has been widely accepted that direct transform methods could not be applied to dynamic, viscoelastic, accelerating crack problems since relative to a fixed coordinate frame, the boundary conditions have incomplete time histories whereas relative to a coordinate frame moving with the crack-tip, the governing equations of motion have time dependent coefficients. In either case, it would seem that the application of the Laplace transform in time, which is needed in order to apply the standard integral-transform/complex-variable based machinery, leads to an intractable problem. It will be shown in chapters 4 and 5 how the the standard integral-transform/complex-variable methodology can be modified to circumvent this impediment.

The first solution to a dynamic viscoelastic crack problem was presented by Willis [9] who considered the steady-state propagation of a semi-infinite, anti-plane shear crack in an infinite viscoelastic body modelled as a standard linear solid. Willis utilized an integral-transform/complex-variable methodology, which was subsequently used by a number of researchers in generalizations of this pioneering paper, to construct a simple expression for the stress intensity factor (SIF), K, the coefficient multiplying the square-root singular dominant term in the asymptotic expansion of the stress in a neighborhood of the crack-tip. In particular, the essence of the complex-variable portion of Willis' analysis is the classical Wiener-Hopf technique, as presented for example in the book by Noble [10]. The principle result of this paper is that as a function of the crack-speed, V, $K = K(V)$ is a constant for $0 < V < c^*$ equal to the corresponding static elastic SIF, whereas for $c^* < V < c$ it is a function of both v and material properties that decreases monotonically to zero as $v \to c$, where c^* and c denote the equilibrium and glassy shear wave speeds of the standard linear solid. Walton [11] later generalized Willis' result to a very general class of viscoelastic materials, essentially those for which the second law of thermodynamics is valid in the sense considered by Fabrizio and Morro [12], and showed Willis' result to be generally true, even to the elegantly simple form for the SIF.

Prior and subsequent to Walton's paper [11], there have appeared several analyses by Atkinson and co-workers and Walton and co-workers of a wide variety of dynamic, constant crack-speed problems for semi-infinite cracks in viscoelastic media assuming either mode III (anti-plane shear) or mode I (plane strain) deformation and for a variety of material models. The principle references for this work are [13]-[22]. Chapters 2 and 3 present an overview of the solution methods and kinds of results contained in these papers for mode III and mode I, respectively. In the next section, the general framework for constructing such solutions is presented.

1.2 General Solution Method for Steady-State, Semi-Infinite Crack Problems

The basic problem considered in this section is the dynamic steady-state propagation of a semi-infinite (2-dimensional) crack in an infinite, isotropic, homogeneous, linear viscoelastic body. The presentation will be generic in that it applies both to anti-plane shear and plane strain modes of deformation. Also, only minor modifications are needed to treat semi-infinite cracks in finitely thick layers. The crack is assumed to travel to the right with speed, V, along the x_1-axis in a Cartesian frame, (x_1, x_2), under the action of a travelling distribution of applied tractions ($\sigma_{23}(x_1, 0, t) = L_e l((x_1 - Vt)/a_e)$ or $\sigma_{22}(x_1, 0, t) = L_e l((x_1 - Vt)/a_e)$ in mode III or mode I, respectively) on the crack-faces. The constant L_e is a load parameter, $l(r)$ a dimensionless function and a_e is a length parameter associated with the applied traction that plays the role of the crack length in semi-infinite crack problems. The crack occupies the ray $-\infty < x_1 < Vt$ across which the appropriate displacement (u_3 or u_2 in mode III or mode I, respectively) has a discontinuity and in front of which it is continuous. The governing differential equations express the balance of linear momentum

$$\text{Div}\Sigma = \rho\ddot{\mathbf{u}}, \tag{1.1}$$

where Σ and $\mathbf{u}$ denote the (Cauchy) stress tensor and displacement field, respectively.

It is convenient to write the constitutive equations for a homogeneous, isotropic, linear viscoelastic material in terms of the shear and bulk relaxation functions, $\mu(t)$ and $\kappa(t)$, respectively, as

$$\begin{aligned}
\sigma_{ij} &= 2\mu * d\epsilon_{ij} + \delta_{ij}\lambda * d\epsilon_{kk} \\
\lambda(t) &= \kappa(t) + \frac{4}{3}\mu(t) \\
\epsilon_{ij} &= \frac{1}{2}(u_{i,j} + u_{j,i}) \\
\epsilon_{kk} &= \epsilon_{11} + \epsilon_{22} + \epsilon_{33}
\end{aligned} \tag{1.2}$$

where

$$\mu * d\epsilon = \int_{-\infty}^{t} \mu(t-\tau)d\epsilon(\tau) \tag{1.3}$$

defines a Riemann-Stieltjes convolution. In all subsequent analyses, the time derivatives, $\dot{\mu}(t)$ and $\dot{\kappa}(t)$, are assumed to have negative, semi-definite Fourier sine transforms. (c.f. Fabrizio and Morro [12].) In particular, this property holds whenever $\mu(t)$

and $\kappa(t)$ are non-negative, decreasing, and convex. Poisson's ratio, $\nu(t)$, is defined through its s-multiplied Laplace transform, $\bar{\nu}(s)$, by

$$\bar{\nu}(s) = \frac{(\bar{\kappa}(s) - \frac{2}{3}\bar{\mu}(s))}{2(\bar{\kappa}(s) + \frac{1}{3}\bar{\mu}(s))} \tag{1.4}$$

with $\bar{\nu}(s)$ given by

$$\bar{\nu}(s) = \nu(0) + \int_0^\infty e^{-ts}\dot{\nu}(t)\, dt. \tag{1.5}$$

The functions $\bar{\mu}(s)$ and $\bar{\kappa}(s)$ are defined analogously. It is evident from (1.4) and (1.5) that $\nu(t)$ need not be monotone. The material is said to have *synchronous moduli* whenever Poisson's ratio is constant.

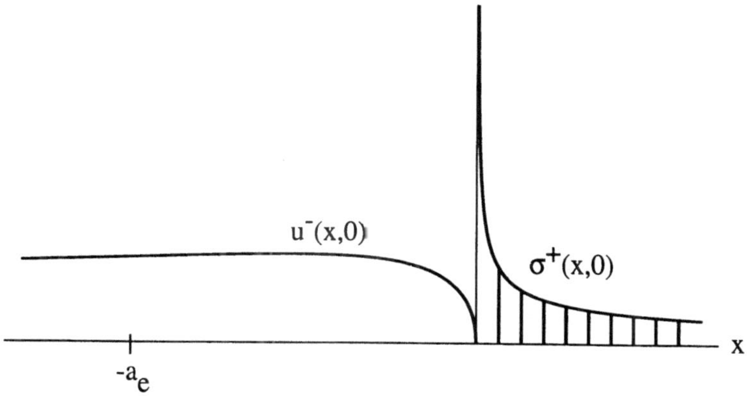

Fig. 1. Semi-infinite crack-face separation profile, $u^-(x,0)$, and singular stress field, $\sigma^+(x,0)$, in the Galilean coordinate frame.

For steady-state crack propagation, it is useful to adopt a coordinate frame attached to the steadily moving crack-tip. (See Fig. 1.) Introducing the Galilean variables $x = x_1 - Vt$, $y = x_2$ and the associated displacement field $\mathbf{w}(x,y) = \mathbf{u}(x_1,x_2,t)$ with components $u(x,y)$, $v(x,y)$ and $w(x,y)$, (1.1) becomes

$$\mathrm{Div}\Sigma = \rho V^2 \mathbf{w}_{xx}, \tag{1.6}$$

where the x-subscript on $\mathbf{w}$ denotes partial differentiation. The form of the applied tractions and considerations of symmetry lead to mixed boundary conditions for the semi-infinite crack (which occupies the ray $(-\infty,0)$ in the moving frame) which in mode I become

$$\begin{aligned} \sigma_{xy}(x,0) &= 0, \quad -\infty < x < \infty \\ \sigma_{yy}(x,0) &= L_e l(x/a_e), \quad x < 0 \\ v(x,0) &= 0, \quad x > 0 \\ \sigma_{yy}(x,y) &\to 0 \quad \text{as} \quad x^2 + y^2 \to \infty \end{aligned} \tag{1.7}$$

and in mode III become

$$\begin{aligned} \sigma_{yz}(x,0) &= L_{el}(x/a_e), \quad x<0 \\ w(x,0) &= 0, \quad x>0 \\ \sigma_{yz}(x,y) &\to 0 \quad \text{as} \quad x^2+y^2\to\infty. \end{aligned} \tag{1.8}$$

The boundary value problems (1.2), (1.6) and (1.7) or (1.8) are solved by first applying the Fourier transform defined by

$$F(p,y) = \hat{f}(p,y) = \int_{-\infty}^{\infty} e^{ipx} f(x,y)\,dx = F_+(p,y) + F_-(p,y)$$

where

$$F_+(p,y) = \int_0^{\infty} e^{ipx} f^+(x,y)\,dx \qquad F_-(p,y) = \int_{-\infty}^{0} e^{ipx} f^-(x,y)\,dx.$$

The inverse Fourier transform is then given by

$$f(x,y) = \check{F}(x,y) = \frac{1}{2\pi}\int_{-\infty}^{\infty} e^{-ipx} F(p,y)\,dp.$$

Under suitable restrictions on $f(x,y)$, $F_+(z,y)$ and $F_-(z,y)$ are analytic functions of z for $\mathrm{Im}(z) > 0$ and $\mathrm{Im}(z) < 0$, respectively. Briefly summarizing the next series of steps, one applies the Fourier transform to (1.2) (after adopting the Galilean variables by which the convolution in t becomes a convolution in x), (1.6), (1.7) and (1.8), solves the resulting ordinary differential equations in y, appeals to the appropriate transformed boundary conditions and obtains an equation for the transformed boundary values of the form

$$T^I(p)\widehat{v_x}(p,0) = \widehat{\sigma_{yy}}(p,0) \tag{1.9}$$

in mode I and

$$T^{III}(p)\widehat{w_x}(p,0) = \widehat{\sigma_{yz}}(p,0) \tag{1.10}$$

in mode III where v_x and w_x denote partial differentiation with respect to x, and $T^I(p)$ and $T^{III}(p)$ are the mode I and mode III *transfer functions*, respectively, which relate the transformed boundary stress and strain. These are the viscoelastic analogs of the Dirichlet to Neumann map (in transformed variables) from potential theory. The explicit form for the transfer functions will be exhibited in later chapters. Here what is emphasized is the generic nature of the solution method. Consequently, what follows is a formal summary of the principle steps common to the detailed derivations to follow in the succeeding chapters. To that end, a generic form of (1.9) and (1.10) will be considered in the remainder of this chapter

$$T(p)\widehat{u_x^-}(p,0) = \hat{\sigma}(p,0) = \widehat{\sigma^+}(p,0) + \widehat{\sigma^-}(p,0) \tag{1.11}$$

where $f^+(x)$ and $f^-(x)$ denote the restrictions of a function $f(x)$ to the positive and negative portions of the x-axis, respectively.

Appealing to the fact that $\widehat{\sigma^+}(p,0)$ and $\widehat{\sigma^-}(p,0)$ have analytic extensions $\widehat{\sigma^+}(z,0)$ and $\widehat{\sigma^-}(z,0)$ to the upper and lower halves of the complex plane, respectively, that vanish as $\mathrm{Im}(z) \to \pm\infty$, one may regard (1.11) as defining the following Riemann-Hilbert boundary value problem: Find $F^+(z)$ and $F^-(z)$ analytic for $\mathrm{Im}(z)\gtrless 0$, respectively, such that

$$\begin{aligned} &F^+(p) = T(p)F^-(p) + g(p), \quad \text{on} \quad -\infty < p < \infty \\ &F^\pm(z) \to 0 \quad \text{as} \quad \mathrm{Im}(z) \to \pm\infty \end{aligned} \tag{1.12}$$

where

$$F^+(z) = \widehat{\sigma^+}(z,0) \tag{1.13}$$

$$F^-(z) = \widehat{v_x^-}(z,0) \tag{1.14}$$

$$g(p) = -\widehat{\sigma^-}(p,0) = -L_e a_e \hat{l}(a_e p) \tag{1.15}$$

$$F^\pm(p) = \lim_{\mathrm{Im}(z)\to 0\pm} F^\pm(z).$$

Alternatively, (1.11) may be cast as a Wiener-Hopf problem (c.f. [9], [10], [13]–[17]). The choice of which point of view to adopt, Riemann-Hilbert or Wiener-Hopf, is largely one of personal preference. Traditionally the Wiener-Hopf problem has been approached as one requiring clever tricks to produce the Wiener-Hopf factorization required to construct solutions, whereas the Riemann-Hilbert problem has been approached by appealing to the well-developed, general, rigorous theory as presented by Gakov [23], for example. For that reason, the Riemann-Hilbert theory has been utilized in what follows below, and in [11] and [18]—[22]. A formal procedure for solving the Riemann-Hilbert problem (1.12) and extracting useful information for fracture mechanics will now be presented; rigorous justification for these constructions depends upon the particular forms for the transfer function, $T(p)$, and applied traction, $l(r)$, occuring in each application discussed in the remaining chapters.

A formal procedure for solving (1.12) can be derived from the classical Plemelj formulae

$$\Phi^\pm(p) = \pm\frac{1}{2}\phi(p) + \frac{1}{2\pi i}\int_{-\infty}^{\infty} \phi(\tau)\frac{d\tau}{(\tau - p)} \tag{1.16}$$

or equivalently

$$\Phi^{+}(p) - \Phi^{-}(p) = \phi(p) \tag{1.17}$$

$$\Phi^{+}(p) + \Phi^{-}(p) = \frac{1}{\pi i}\int_{-\infty}^{\infty} \phi(\tau)\frac{d\tau}{(\tau - p)} \tag{1.18}$$

where

$$\Phi(z) = \frac{1}{2\pi i}\int_{-\infty}^{\infty} \phi(\tau)\frac{d\tau}{(\tau - z)} \tag{1.19}$$

$$\Phi^{\pm}(p) = \lim_{q\to 0\pm} \Phi(p + iq).$$

The integral in (1.18) and (1.19) is interpreted in the Cauchy principal value sense.

To solve (1.12), one first considers the associated homogeneous Riemann-Hilbert problem

$$X^{+}(p) = T(p)X^{-}(p). \tag{1.20}$$

Substitution of (1.20) into (1.12) yields after rearrangement

$$\frac{F^{+}(p)}{X^{+}(p)} - \frac{F^{-}(p)}{X^{-}(p)} = g(p)$$

which, from (1.17), leads to the unique solution to (1.12)

$$F^{\pm}(z) = X^{\pm}(z)\frac{1}{2\pi i}\int_{-\infty}^{\infty} \frac{g(\tau)}{X^{+}(\tau)}\frac{d\tau}{\tau - z}. \qquad (1.21))$$

Applying the Plemelj formula (1.16) followed by Fourier inversion yields the following expression for the stress ahead and in the plane of the propagating crack

$$\sigma^{+}(x) = \frac{1}{2\pi}\int_{-\infty}^{\infty} e^{-ixp}X^{+}(p)\, dp\frac{1}{2\pi i}\int_{-\infty}^{\infty} \frac{g(\tau)}{X^{+}(\tau)}\frac{d\tau}{\tau - p} \tag{1.22}$$

for all $x > 0$

To solve the homogeneous Riemann-Hilbert problem (1.20), one makes use of the fact that for all of the crack problems to be considered here, $T(p)$ can be factored as

$$T(p) = \operatorname{sgn}(p)T_0(p) \tag{1.23}$$

where $\operatorname{sgn}(p) = p/|p|$ and $T_0(p)$ has a finite, non-zero limit, $T_0(\infty) = \lim_{p\to\pm\infty} T_0(p)$. A solution to (1.20) can then be constructed in the form

$$X^{\pm}(z) = \omega^{\pm}(z)X_0^{\pm}(z) \tag{1.24}$$

with

$$\omega^+(p) = \text{sgn}(p)\omega^-(p) \tag{1.25}$$

$$X_0^+(p) = T_0(p)X_0^-(p). \tag{1.26}$$

Let $\omega^\pm(z)$ denote branches of $z^{1/2}$ whose branch cuts lie along the negative and positive imaginary axes, respectively. Then $\omega^\pm(z)$ is analytic for $\text{Im}(z)\gtrless 0$, respectively, and for real p

$$\omega^+(p)/\omega^-(p) = \text{sgn}(p). \tag{1.27}$$

One next shows that $T_0(p)$ has a single-valued logarithm from which it follows that the canonical solution to (1.26) that is bounded at infinity (See [23].) may be constructed in the form

$$X_0^\pm(z) = \exp(\Gamma_0^\pm(z)) \tag{1.28}$$

$$\Gamma_0^\pm(z) = \frac{1}{2\pi i}\int_{-\infty}^{\infty} \log(T_0(\tau))\frac{d\tau}{\tau - z}. \tag{1.29}$$

The construction of $X_0^\pm(z)$ is then completed by a careful analysis of the integral in (1.29). For some applications, the integral can be evaluated in closed form, whereas in others a numerical quadrature scheme is required.

1.3 General Stress Analysis

Solving the Riemann-Hilbert problem (1.12) produces expressions for the Fourier transformed crack-face separation and crack-plane stress. However, it is highly desirable to have direct expressions for these physical quantities in terms of the applied loading and material constitutive properties. A general procedure for constructing such an expression for the stress field in front of the crack will be presented below. Obtaining a suitable expression for the crack-face separation is more complicated and will be addressed for specific cases considered in later chapters.

Returning to (1.22), one can show that, under suitable assumptions concerning the form for the applied traction $l(r)$,

$$\int_{-\infty}^{\infty} \frac{g(\tau)}{X^+(\tau)}\frac{d\tau}{\tau - p} = -\frac{1}{p}\int_{-\infty}^{\infty} \frac{g(\tau)}{X^+(\tau)}d\tau + O(|p|^{-3/2}) \quad \text{as} \quad |p| \to \infty$$

from which it follows that for $x > 0$ and near zero,

$$\sigma^+(x) \sim -\frac{1}{2\pi}\int_{-\infty}^{\infty} \frac{g(\tau)}{X^+(\tau)}d\tau \frac{1}{2\pi i}\int_{-\infty}^{\infty} \frac{X^+(p)}{p}e^{-ixp}dp. \tag{1.30}$$

From (1.24), (1.28) and (1.29), one sees that $X^+(p) = O(|p|^{1/2})$ as $|p| \to \infty$ which together with (1.30) and an application of the Abelian theorems for the Fourier transform [24], yields the conclusion that

$$\sigma^+(x) \sim \frac{K}{\sqrt{\pi x}} \tag{1.31}$$

with the Stress Intensity Factor, K, given by

$$K = \frac{X_\infty^+}{2\pi} \int_{-\infty}^{\infty} g(p)/X^+(p)\, dp$$
$$X_\infty^+ = \lim_{p\to\infty} X_0^+(p). \tag{1.32}$$

Since $X_\infty^+/X^+(p)$ has a natural analytic extension $X_\infty^+/X^+(z)$ to the upper half-plane that vanishes as $\mathrm{Im}(z) \to +\infty$, it follows that $X_\infty^+/X^+(p)$ is the Fourier transform of a function with support on $(0,\infty)$. Construction of this function is facilitated by first writing

$$\frac{X_\infty^+}{X^+(p)} = \frac{1}{\omega^+(p)} \frac{X_\infty^+}{X_0^+(p)}. \tag{1.33}$$

An elementary calculation shows that

$$\frac{1}{\omega^+(p)} = \frac{\widehat{x_+^{-1/2}}}{\sqrt{\pi i}} \tag{1.34}$$

with the function x_+ defined by

$$x_+ = \begin{cases} x & \text{if} \quad 0 < x \\ 0 & \text{if} \quad x < 0. \end{cases}$$

Combining (1.33) and (1.34), one has after appealing to the convolution theorem for the Fourier transform [24]

$$[X_\infty^+/X^+(p)\check{]} = h_+(x) = x_+^{-1/2} * k_+(x) = \int_0^x k_+(r) \frac{dr}{\sqrt{x-r}} \tag{1.35}$$

where

$$\widehat{k_+}(p) = \frac{X_\infty^+}{X_0^+(p)\sqrt{\pi i}}.$$

Applying the Plancherel theorem [24] to (1.32) and making use of (1.15) and (1.35), one obtains for the Stress Intensity Factor the result

$$\begin{aligned} K &= \frac{X_\infty^+}{2\pi} \int_{-\infty}^{\infty} g(p)/X^+(p)\, dp \\ &= -\frac{1}{2\pi} \int_{-\infty}^{\infty} \widehat{\sigma^-}(p) \hat{h}_+(p)\, dp \\ &= -\int_{-\infty}^{\infty} \widehat{\sigma^-}(p) \check{h}_-(p)\, dp \\ &= -\int_{-\infty}^{0} \sigma^-(x) h_-(x)\, dx \end{aligned} \tag{1.36}$$

where $h_-(x) = h_+(-x)$ and use has been made of the obvious identity $2\pi \check{f}(p) = [\widehat{f(-x)}](p)$. The function $h_-(x)$, which vanishes for $x > 0$, depends upon material constitutive properties and crack-speed and will be constructed for anti-plane shear and plane strain crack problems in chapters 2 and 3, respectively.

It is often useful to have knowledge of $\sigma^+(x)$ for all $x > 0$, rather than merely its dominant asymptotic form for $x \ll 1$. A (formal) procedure for constructing an expression for $\sigma^+(x)$ from (1.22) now follows. Use will be made of the following well-known identity relating the Fourier and Hilbert transforms [24]

$$\mathcal{H}(f)(\tau) = [i \operatorname{sgn}(p) \check{f}(p)\,\hat{]}(\tau) \tag{1.37}$$

where the Hilbert transform, $\mathcal{H}(f)$, is defined by

$$\mathcal{H}(f)(\tau) = \frac{1}{\pi}\int_{-\infty}^{\infty} f(t)\frac{dt}{t-\tau}. \tag{1.38}$$

The integral in (1.38) is to be interpreted in the Cauchy Principal Value sense [24].

Applying (1.35), (1.37) and (1.38) to the inner integral in (1.22), one has

$$\begin{aligned} \frac{1}{2\pi i}\int_{-\infty}^{\infty} g(\tau)\frac{1}{X^+(\tau)}\frac{d\tau}{\tau - p} &= \frac{1}{2X_\infty^+}[\operatorname{sgn}(x)[g(\tau)X_\infty^+/X^+(\tau)\check{]}\,\hat{]} \\ &= \frac{-1}{2X_\infty^+}[\operatorname{sgn}(x)\,\sigma^- * h_+(x)\hat{]}. \end{aligned} \tag{1.39}$$

By a similar calculation to (1.33)–(1.35), there is

$$\begin{aligned} \frac{X^+(p)}{ipX_\infty^+} &= \frac{\omega^+(p)}{ip}\frac{X_0^+(p)}{X_\infty^+} \\ &= \frac{-i}{\omega^+(p)}\frac{X_0^+(p)}{X_\infty^+} \\ &= -i\,\widehat{x_+^{-1/2}}(p)\widehat{k_+^{-1*}}(p)/\sqrt{\pi i} \end{aligned} \tag{1.40}$$

where $k_+^{-1*}(x)$ denotes the convolution inverse of $k_+(x)$ defined by

$$k_+^{-1*} * k_+(x) = \delta(x)$$

with $\delta(x)$ equal to the delta-function point mass measure, or equivalently

$$\widehat{k_+^{-1*}}(p)\,\widehat{k_+}(p) = 1.$$

Finally, combining (1.22), (1.39) and (1.40) one has

$$\sigma^+(x) = \frac{-1}{2\pi}\frac{d}{dx}[\widehat{x_+^{-1/2}}(p)\widehat{k_+^{-1*}}(p)[\mathrm{sgn}(x)\,\sigma^- * h_+(x)]\hat{\,}(p)$$
$$= \frac{-1}{2\pi}\frac{d}{dx}[x_+^{-1/2} * k_+^{-1*}(x)] * [\mathrm{sgn}(x)\,\sigma^- * h_+(x)]. \qquad (1.41)$$

It is not difficult to construct the asymptotic form for $\sigma^+(x)$ as $x \to 0+$ from (1.41) to recover the Stress Intensity Factor result (1.36). What is most interesting from (1.41) is that the function $k_+(x)$, defined in (1.35), that is necessary for constructing the Stress Intensity Factor is also sufficient for constructing $\sigma^+(x)$ for all $x > 0$.

In the next chapter, these general ideas are applied to specific anti-plane shear and plane strain steady-state problems for which careful justification of the key steps in the argument can be provided. Chapter 3 then addresses the corresponding transient initial value problems under the restriction that the crack-speed in constant. In Chapters 4 and 5 this last restriction is abandoned when consideration is given to dynamically accelerating cracks.

2. DYNAMIC STEADY-STATE CRACK PROPAGATION

2.1 Introduction

In section 1.2, a general procedure for constructing solutions to dynamic, steady-state, semi-infinite crack propagation problems was outlined. The presentation was formal and generic. In the present chapter, more careful analyses will be presented for the specific cases of anti-plane shear and plane-strain cracks. In the former case, the analysis will be shown to be valid for a quite general class of shear relaxation functions. The analysis for the latter class of problems involves significantly greater technical difficulties. In particular, it will first be assumed that the Poisson's ratio, defined by (1.4), is constant as a function of time, in which case rigorous justification of the analysis is presented. Materials for which this assumption is valid are said to have *synchronous moduli.* Such is certainly the case for the important class of incompressible materials (rubber, for example), but most viscoelastic materials exhibit a Poisson's ratio that is monotonically increasing with time. The solution will be presented for a very general class of materials with non-constant Poisson's ratio modulo verification of a certain technical fact concerning the winding number of the complex Rayleigh function for the material. This fact must be proved, or demonstrated numerically, for specific applications. Furthermore, it is shown that if this fact is not true for a particular material, then physically unacceptable stress and displacement solutions occur; specifically the energy flow into the crack-tip must either be zero or infinite. However, as yet no "nice theorem" exists demonstrating that the desired property for the complex Rayleigh function must hold for the general class of materials considered here. Certain issues related to the usefulness of these solutions for fracture mechanics are also discussed.

2.2 Anti-Plane Shear Crack Propagation

For a semi-infinite, anti-plane shear crack propagating in a steady-state fashion through an infinite viscoelastic plane, the relevant transformed crack-plane, stress-displacement relation is given by (1.10) in which the transfer function has the form (1.23) with (See, [11].)

$$T_0(p) = -i\bar{\mu}(iVp)\beta(iVp) \tag{2.1}$$

with $\bar{\mu}(s)$ defined as in (1.5) and $\beta(s)$ given by

$$\beta(s) = \sqrt{1 - \rho V^2/\bar{\mu}(s)}. \tag{2.2}$$

As discussed in [11], the square-root in (2.2) must be chosen to have positive real part in order for the stresses to vanish at $y \to \infty$. As shown in the previous chapter, the solution of the Riemann-Hilbert problem (1.12) and the subsequent construction of the stress and displacement fields hinge upon solving the homogeneous Riemann-Hilbert problem (1.24) which in turn rests upon solving the auxiliary problem (1.26) which can be rewritten as

$$X_0^+(p)/X_0^-(p) = T_0(p). \tag{2.3}$$

To that end, it is useful to appeal to the obvious observation that if $T_0(p)$ is factored as

$$T_0(p) = T_1(p)T_2(p) \tag{2.4}$$

and solutions are found for the separate problems

$$X_i^+(p)/X_i^-(p) = T_i, \quad i = 1, 2 \tag{2.5}$$

then the solution to (2.3) can be constructed in the factored form

$$X_0^\pm(z) = X_1^\pm(z)X_2^\pm(z). \tag{2.6}$$

A convenient factorization of (2.1) of the form (2.4) has

$$T_1(p) = -i\bar{\mu}(iVp) \quad \text{and} \quad T_2(p) = \beta(iVp). \tag{2.7}$$

It is in constructing the functions $X_i^\pm(z)$ that the constitutive assumptions concerning the shear relaxation function $\mu(t)$ play a crucial role. Specifically, under suitable assumptions $\bar{\mu}(iVp)$ has a natural analytic extension, $\bar{\mu}(iVz)$, to the lower-half of the complex plane that approaches $\mu(0)$ as $\mathrm{Im}(z) \to -\infty$. To see this, one can write $z = p - iq$ and consider

$$\bar{\mu}(iVz) = \mu(0) + \int_0^\infty \cos(Vpt)\dot{\mu}(t)e^{-Vqt}dt$$
$$- \int_0^\infty \sin(Vpt)\dot{\mu}(t)e^{-Vqt}dt. \tag{2.8}$$

If one assumes that $\dot{\mu}(t)$ is integrable on $(0,\infty)$ then clearly one has that $\bar{\mu}(iV(p-iq))$ is analytic for $q>0$ with $\lim_{q\to\infty}\bar{\mu}(iV(p-iq))=\mu(0)$. It is then natural to define the functions $X_1^{\pm}(z)$ by

$$X_1^+(z)=1 \quad \text{and} \quad X_1^-(z)=1/T_1(z). \tag{2.9}$$

Clearly, for (2.9) to be a valid solution, $T_1(z)$ must have no zeros for $\mathrm{Im}(z)<0$. Appealing to the Argument Principle, such will be the case provided the winding number of $T_1(p)$ about the origin for $-\infty<p<\infty$ is zero. From chapter 1, the principal constitutive assumption for the shear relaxation function is that its Fourier sine transform be negative semi-definite. It then follows that

$$\mathrm{Im}(\bar{\mu}(iVp))=\begin{cases} >0 & \text{if } p>0 \\ <0 & \text{if } p<0. \end{cases}$$

Moreover, $\mu(0)=\bar{\mu}(\infty)\geq\bar{\mu}(0)=\mu(\infty)\geq 0$ and it is easily seen that $\mathrm{Re}(\bar{\mu}(iVp))$ and $\mathrm{Im}(\bar{\mu}(iVp))$ are even and odd functions of p, respectively. From these facts one concludes that the winding number of $T_1(p)$ about the origin is zero and hence that (2.9) is a valid solution to the homogeneous Riemann-Hilbert problem (2.5).

The construction of $X_2^{\pm}(z)$ requires a bit more care. In particular, it proves useful to reduce the analysis to the consideration of two separate cases:

Case 1. $0\leq V\leq c^*$ and Case 2. $c^*<V<c$

where

$$\begin{aligned} c^* &= \sqrt{\mu(\infty)/\rho} \\ c &= \sqrt{\mu(0)/\rho} \end{aligned} \tag{2.10}$$

are the equilibrium (long time) and glassy (short time) shear wave speeds, respectively.

For case 1, one shows easily that $\beta(iVp)^2$ has similar properties to $\bar{\mu}(iVp)$, namely that $0\leq\beta(0)^2=(1-(V/c^*)^2)\leq(1-(V/c)^2)=\beta(\infty)^2$, $\mathrm{Im}(\beta(iVp)^2)>0$ for $p>0$, and $\mathrm{Im}(\beta(iVp)^2)$ and $\mathrm{Re}(\beta(iVp)^2)$ are odd and even functions of p, respectively. Thus, by arguments similar to those above for $\bar{\mu}(iVp)$, one shows that $\beta(iVp)^2$ has a natural analytic extension, without zeros, to the lower-half of the complex plane with $\lim_{q\to\infty}\beta(iV(p-iq))^2=(1-(V/c)^2)^2$. Furthermore, if $T_2(p)=\beta(iVp)$ is defined by taking the branch of the square-root with branch cut lying along the negative real axis, then $T_2(p)$ has similar properties to $\beta(iVp)^2$ and one can chose for $X_2^{\pm}(z)$ the solution

$$\begin{aligned} X_2^+(z) &= 1 \\ X_2^-(z) &= 1/T_2(z). \end{aligned} \tag{2.11}$$

For case 2, one has $\beta(0)^2=(1-(V/c^*)^2)<0<(1-(V/c)^2)=\beta(\infty)^2$, and hence the winding number of $\beta(iVp)^2$ about zero is 1. Taking the square-root as before produces a function $T_2(z)$ that is analytic in the lower-half of the complex plane except for a branch cut along part of the negative imaginary axis. To proceed from here, an additional constitutive assumption is made; namely, that $\bar{\mu}(s)$ is an increasing function for $0 < s$. Focusing attention to horizontal lines $p - iq$ for fixed $q > 0$, reveals that there is a unique nonnegative number $q^* = q^*(V)$ such that for $0 \leq q < q^*(V)$, $\beta(V(ip+q))$ has winding number 1 about the origin whereas for $q^*(V) < q$ the winding number is zero. Indeed, $q^*(V)$ is defined to be the unique solution to

$$v^2 = \bar{\mu}(vq^*(V))/\rho, \tag{2.12}$$

uniqueness being assured by the assumed monotonicity of $\bar{\mu}(s)$. One can now construct $X_2^{\pm}(z)$ by appealing to the formulas (1.28), (1.29). The corresponding Cauchy integral defining $\Gamma_2^{\pm}(z)$ has density $\log(T_2(p))$ which has an analytic extension, $\log(T_2(p-iq))$, to the lower-half of the complex plane with a branch cut along the interval $p = 0$, $0 < q < q^*(V)$. In particular, $\log(T_2(p-iq))$ is analytic for $q > q^*(V)$. One may then deform the contour of integration in (1.29) down horizontal lines leaving the sum of an integral along the horizontal line $q = q^*(V)$, which can be evaluated by Cauchy's theorem, and the vertical segment $p = 0$, $0 < q < q^*(V)$. Across the vertical segment, $\log(T_2(p-iq))$ has a jump discontinuity equal to the constant πi independent of q producing an integral which is easily evaluated. In this way, one obtains for $X_2^{\pm}(z)$

$$\begin{aligned} X_2^+(z) &= \sqrt{z + iq^*(V)/z} \\ X_2^-(z) &= (1/T_2(z))\sqrt{z + iq^*(V)/z} \end{aligned} \tag{2.13}$$

where the branch cut for $\sqrt{z - iq^*(V)/z}$ is chosen to be the interval Re(z)=0, $0 <$Im$(z) < q^*(V)$.

Combining(1.25), (1.26), (2.6), (2.9) and (2.13) one obtains

$$X^+(z) = \begin{cases} \sqrt{z} & \text{for } 0 \leq V \leq c^* \\ \sqrt{z + iq^*(V)} & \text{for } c^* < V < c \end{cases} \tag{2.14}$$

and

$$X^-(z) = \begin{cases} \omega^-(z)/T_0(z) & \text{for } 0 \leq V \leq c^* \\ (\omega^-(z)/T_0(z))\sqrt{z + iq^*(V)/z} & \text{for } c^* < V < c. \end{cases} \tag{2.15}$$

In view of (2.14), the Fourier inversion required by (1.35) and (1.36) is seen to be elementary and yields

$$K = -\int_{-\infty}^{0} \sigma^-(x) h_-(x) dx$$

with

$$h_-(x) = \frac{1}{\sqrt{|x|}} \begin{cases} 1 & \text{for } 0 \leq V \leq c^* \\ e^{xq^*(V)} & \text{for } c^* < V < c. \end{cases} \tag{2.16}$$

One important observation to be made from (2.16) is that for $0 \leq V \leq c^*$, the viscoelastic stress intensity factor equals that for the corresponding elastic problem and is independent of crack speed and material properties, whereas for $c^* < V < c$, it depends upon both crack speed and material properties and tends to zero as $V \to c$, usually monotonically. Additionally, the influence of crack speed and material properties on the stress field occurs only through the single parameter $q^*(V)$ defined by (2.12). Finally, the constitutive assumptions upon which the analysis is predicated are that the stress relaxation function $\mu(t)$ satisfy

$$\text{a. } \operatorname{Im}(\bar{\mu}(is)) > 0 \text{ for } s > 0 \qquad \text{b. } \bar{\mu}(s) \text{ is increasing for } s > 0. \tag{2.17}$$

Both of these restrictions are mild. Indeed, Fabrizio and Morro [12] have shown (2.17a) to be necessary for the second law of thermodynamics to hold and (2.17b) to be a consequence of (2.17a)†.

Combining (1.41) and (2.16), one can construct the full stress field in front of the propagating crack, $\sigma^+(x)$, $0 < x$. In particular, it can be shown without much difficulty that

$$\sigma^+(x) = \frac{1}{2\pi x}\int_{-\infty}^{0} \sigma^-(r) l(r/x)(r/x) n(q^*(V)(x-r))dr$$

where

$$l(r) = \int_0^1 (1-t)^{-1/2}(t-r)^{-3/2}dt \quad \text{and} \quad n(r) = \begin{cases} 1 & \text{if } 0 \leq V \leq c^* \\ e^{-r} & \text{if } c^* < V < c. \end{cases}$$

Again one sees that for $0 \leq V \leq c^*$, the viscoelastic stress ahead of the crack is the same as for elastic material, whereas for $c^* < V < c$ the viscoelastic stress depends upon material properties and crack-speed, but only through the single parameter $q^*(V)$.

It is also of interest to determine the angular dependence of the near crack-tip stress field. For elastic material, standard asymptotic methods were used by Achenbach and Bazant [26] to construct the near crack-tip stress and displacement fields with respect to a local polar coordinate system centered at the crack-tip. Unfortunately, these asymptotic methods do not seem applicable to the viscoelastic wave equation (1.6). However, one can construct the near crack-tip stress field through an asymptotic analysis of the integral representation for $\sigma(x,y)$

$$\sigma(x,y) = \frac{1}{2\pi}\int_{-\infty}^{\infty} \sigma(s,0)ds \int_{-\infty}^{\infty} e^{-|p|y\beta(iVp)+ip(s-x)}dp \tag{2.18}$$

† This author is grateful to Prof. M. Fabrizio for pointing out to him that (2.17a) implies (2.17b). See the Appendix of [12].

derived in [11] with $\beta(s)$ given by (2.2). Adopting a local polar coordinate system, (r, θ), centered at the crack-tip (which is located at the origin of the Gálilean coordinate system), one concludes after an obvious change of variables that

$$\sigma(x, y) = \sigma^*(r, \theta) = \frac{1}{2\pi} \int_{-\infty}^{\infty} \sigma(rt, 0) dt \int_{-\infty}^{\infty} e^{-|\tau| \sin(\theta) \beta(iV\tau/r) + i\tau(t - \cos(\theta))} d\tau. \quad (2.19)$$

It is straight forward, albeit somewhat tedious, to show that the dominant term in the asymptotic expansion of (2.19) in powers of r for $r \to 0+$ is given by

$$\sigma^*(r, \theta) \sim \frac{K}{\sqrt{r}} f(V, \theta)$$

with

$$f(V, \theta) = \frac{1}{2\pi} \int_0^{\infty} \frac{dt}{\sqrt{t}} \int_{-\infty}^{\infty} e^{-|\tau| \sin(\theta) \sqrt{1-(V/c)^2} + i\tau(t - \cos(\theta))} d\tau. \quad (2.20)$$

The integral in (2.20) is easily evaluated and yields the result

$$f(v, \theta) = \mathrm{Re}[\cos(\theta) - i \sin(\theta) \sqrt{1 - (V/c)^2}]^{-1/2}. \quad (2.21)$$

The function $f(V, \theta)$ in (2.21) is the same as occurs in elastic material with shear modulus equal to the glassy viscoelastic value, $\mu(0)$. Thus, the viscoelastic near crack-tip stress field shows the same angular dependence as occurs in elastic material; the two stress fields differ only in the stress intensity factor, K.

2.3 Plane-Strain Crack Propagation

The analysis of the steady-state propagation of a semi-infinite, plane-strain crack in an infinite viscoelastic body is considerably more complicated than the anti-plane shear analysis described in the previous section. The plane-strain transfer function from (1.9) is given by (1.23) in which

$$T_0(p) = T_1(p) T_2(p) T_3(p) \quad (2.22)$$

with

$$T_1(p) = i\bar{\mu}(iVp)[\bar{c}_T(iVp)/V]^2 \quad (2.23)$$

$$T_2(p) = 1/\beta_L(iVp) \quad (2.24)$$

$$T_3(p) = 4\beta_L(iVp, V)\beta_T(iVp, V) - (1 + \beta_T^2(iVp, V))^2 \quad (2.25)$$

where

$$\beta_L(z, V) = \sqrt{1 - [V/\bar{c}_L(z)]^2} \quad \text{and} \quad \beta_T(z, V) = \sqrt{1 - [V/\bar{c}_T(z)]^2}. \quad (2.26)$$

In (2.23)-(2.26), $\bar{c}_L(z)$ and $\bar{c}_T(z)$ denote the complex longintudinal and transverse wave speeds, respectively, given by

$$[\bar{c}_L(s)]^2 = \frac{\bar{\kappa}(s) + \frac{4}{3}\bar{\mu}(s)}{\rho} \quad \text{and} \quad [\bar{c}_T(s)]^2 = \frac{\bar{\mu}(s)}{\rho} \tag{2.27}$$

and $T_3(p)$ is the complex, viscoelastic Rayleigh function. Moreover, branches for the square-roots in (2.26) must be chosen to have positive real parts. The solution to the homogeneous Riemann-Hilbert problem (1.26) can then be constructed in the form

$$X_0^{\pm}(z) = X_1^{\pm}(z)X_2^{\pm}(z)X_3^{\pm}(z) \tag{2.28}$$

with

$$X_j^{+}(p) = T_j(p)X_j^{-}(p), \quad j = 1, 2, 3. \tag{2.29}$$

The constitutive assumptions adopted here are that both $\bar{\mu}(s)$ and $\bar{\kappa}(s)$ satisfy (2.17) from which it follows that the complex wave speeds $\bar{c}_L(s)$ and $\bar{c}_T(s)$ inherit the same properties. The auxiliary Riemann-Hilbert problems (2.29) for $i = 1, 2$ may then solved by the arguments employed in the anti-plane shear analysis. Specifically, $T_1(p)$ has a natural analytic extension, $T_1(z)$, to the lower half of the complex plane that does not vanish and that is bounded as $\mathrm{Im}(z) \to -\infty$. Therefore, for $X_1^{\pm}(z)$ one may take

$$X_1^{+}(z) = 1 \quad \text{and} \quad X_1^{-}(z) = \frac{1}{T_1(z)}. \tag{2.30}$$

For $X_2^{\pm}(z)$, there are two cases: $0 < V < c_L^*$ and $c_L^* < V < c_L$ in which $c_L^* = \bar{c}_L(0)$ and $c_L = \bar{c}_L(\infty)$ are the equibrium (long time) and glassy (short time) longitudinal wave speeds, respectively. For $0 < V < c_L^*$, one has

$$X_2^{+}(z) = 1 \quad \text{and} \quad X_2^{-}(z) = \frac{1}{T_2(z)} = \beta_L(iVz, V) \tag{2.31}$$

whereas for $c_L^* < V < c_L$, one has

$$X_2^{+}(z) = \frac{1}{\sqrt{1 + iq_L^*(V)/z}} \quad \text{and} \quad X_2^{-}(z) = \frac{\beta_L(iVz, V)}{\sqrt{1 + iq_L^*(V)/z}} \tag{2.32}$$

where $q_L^*(V)$ is the unique positive real root of

$$\beta_L(Vq_L^*(V), V) = 0. \tag{2.33}$$

The interval $-q_L^*(V) < \mathrm{Im}(z) < 0$, $\mathrm{Re}(z) = 0$ is taken for the branch cut for $\sqrt{1 + iq_L^*(V)/z}$. Also, $q_L^*(V)$ is an increasing function with $q_L^*(c_L^*) = 0$ and $\lim_{V \nearrow c_L} q_L^*(V) = \infty$.

The solution of (2.29) for $j = 3$, requires a careful analysis of the complex mapping properties of the Rayleigh function $T_3(p)$. It is useful to indicate by $T_3 = T_3(z,V)$ that T_3 depends independently upon V and p through $T_3(z,V) = 4\beta_L(z,V)\beta_T(z,V) - (1+\beta_T^2(z,V))^2$. One sees now, for example, that $T_3(0,V)$ and $T_3(\infty,V)$ are the (real-valued) Rayleigh functions corresponding to an elastic material with the equilibrium and instantaneous viscoelastic properties, respectively. The equilibrium and instantaneous Rayleigh wave speeds for the viscoelastic material, c_R^* and c_R, are the unique positive numbers for which $T_3(0,c_R^*) = T_3(\infty,c_R) = 0$. As is well-known (See [27], for example.), the equilibrium and instantaneous wave speeds must satisfy $c_R^* < c_T^* < c_L^*$ and $c_R < c_T < c_L$, and the constitutive assumptions imply that $c_j^* \leq c_j$ for $j = R,T,L$. However, it is certainly to be expected that for many viscoelastic materials one would have $c_L^* < c_R$. From these observations and the obvious fact that the analysis breaks down for $V \geq c_R$, it follows that consideration must be given to the four possible cases: Case 1. $0 \leq V \leq c_R^*$; Case 2. $c_R^* < V \leq c_T^*$; Case 3. $c_T^* < V \leq c_L^*$; Case 4. $c_L^* < V < c_R$. If $c_T^* > c_R$ there are only two cases to be considered and if $c_T^* < c_R < c_L^*$ there are only three.

The construction of $X_3^{\pm}(z)$ requires careful analysis of the analytic continuation of $T_3(p,V)$ to the lower half of the complex plane and in particular the mapping properties of $T_3(p-iq,V)$ on fixed horizontal lines, $q = constant$. As shown above for $\beta_L(Vz,V)$ and in the previous section for $\beta_T(Vz,V)$, there exist unique positive numbers, $q_L^*(V)$ and $q_T^*(V)$ defined by (2.33) and

$$\beta_T(Vq_T^*(V),V) = 0 \tag{2.34}$$

such $\beta_j(iV(p-iq),V)$ is analytic for $q > 0$ whenever $0 \leq V \leq c_j^*$ and for $c_j^* < V < c_j$ it is analytic for $q > 0$ except for a branch cut along the interval $p = 0$, $0 < q < q_j^*(V)$, $j = T,L$. Moreover, $q_j^*(V)$ are increasing functions of V with $q_j^*(V) = 0$ if $V \leq c_j^*$ and $q_L^*(V) < q_T^*(V)$ for $V \geq c_T^*$. One can also show easily that for $c_L^* < V < c_R$, the product $\beta_L(iV(p-iq),V)\beta_T(iV(p-iq),V)$ is analytic for $q > 0$ except for a branch cut along $p = 0$, $q_L^*(V) < q < q_T^*(V)$.

Consideration is given first to the case of material for which Poisson's ratio is constant. Specifically, it is assumed that

$$(\bar{c}_T(s)/\bar{c}_L(s))^2 = \frac{(1-2\nu)}{2(1-\nu)} \tag{2.35}$$

in which the Poisson's ratio, ν, is a constant. Introducing $\bar{d}(s) = \bar{c}_T(s)/c_T = \bar{c}_L(s)/c_L$, one may then rewrite $T_3(z,V))$ in (2.25) as

$$T_3(z,V) = R(V/\bar{d}(iVz)) \tag{2.36}$$

where $R(s)$ denotes the Rayleigh function

$$R(s) = 4\sqrt{1-(s/c_T)^2}\sqrt{1-(s/c_L)^2} - (2-(s/c_T)^2)^2. \tag{2.37}$$

It then follows that for Case 1, $0 \le V \le c_R^*$, $T_3(p - iq, V)$ is analytic for $q > 0$ and one may define

$$X_3^+(z) = 1 \quad \text{and} \quad X_3^-(z) = 1/T_3(z, V) \tag{2.38}$$

provided $T_3(p - iq, V) = R(V/\bar{d}(iV(p - iq)))$ has no zeros for $q > 0$. However, it is well-known (See [27], for example.) that the only zeros in the complex s-plane of the Rayleigh function (2.37) occur at $s = \pm c_R$. Thus, for any zeros of $T_3(p - iq, V)$, $\bar{d}(iV(p-iq))$ must be real. But by the constitutive assumptions, $\bar{d}(iV(p-iq))$ is real only for $p = 0$ and in the limit as $p, q \to \infty$. Moreover, $T_3(-iq, V)$ is never zero for $q > 0$ and $0 \le V < c_R$ and one concludes that (2.38) is valid.

For Case 2, $c_R^* \le V \le c_T^*$, $T_3(-iq, V)$ is real and there exists a unique $q = q_R^*(V)$ such that

$$T_3(-iq_R^*(V), V) = 0 \tag{2.39}$$

and satisfying the conditions: i. $q_R^*(c_R^*) = 0$; ii. $q_R^*(V) \nearrow \infty$ as $V \nearrow c_R$; iii. $T_3(-iq, V) < 0$ for $0 < q < q_R^*(V)$ and $T_3(-iq, V) > 0$ for $q_R^*(V) < q$. It follows by arguments similar to those for Case 1, that $T_3(p - iq, V)$ has one zero (counting multiplicity) for $q > 0$, occurring for $p = 0$ and $q = q_R^*(V) > 0$. Moreover, as argued in [22], $X^{\pm}(z)$ will have the correct asymptotic form as $|z| \to \infty$ and $z \to 0$ provided one takes

$$X_3^+(z) = (z + iq_R^*(V))/z \quad \text{and} \quad X_3^-(z) = (z + iq_R^*(V))/(zT_3(z)). \tag{2.40}$$

For Case 3, $c_T^* < V < c_L^*$, and Case 4, $c_L^* < V < c_R$, simple expression for $X_3^{\pm}(z)$ corresponding to (2.38) and (2.40) do not exist because $T_3(p - iq, V)$ has a branch cut along the interval $p = 0$, $0 < q < q_T^*(V)$ (Case 3) or $q_L^*(V) < q < q_T^*(V)$ (Case 4). Moreover, as in the anti-plane shear case, the discontinuity across the branch cut is only in $\arg(T_3(p - iq, V))$; $|T_3(p - iq, V)|$ is continuous across the cut. However, unlike in the anti-plane shear case, the jump in $\arg(T_3(p-iq, V))$ is not constant, but rather is a complicated function of q. Nevertheless, in [11], the following expressions for $X_3^{\pm}(z)$ were derived:

$$X_3^+(z) = \begin{cases} \left[\frac{z+iq_R^*(V)}{z+iq_T^*(V)}\right] \exp(J_3(z)) & \text{for } c_T^*(V) < V < c_L^*(V) \\ \left[\frac{z+iq_L^*(V)}{z}\right]\left[\frac{z+iq_R^*(V)}{z+iq_T^*(V)}\right] \exp(J_4(z)) & \text{for } c_L^*(V) < V < c_R \end{cases} \tag{2.41}$$

with

$$J_3(z) = \frac{1}{\pi}\int_0^{q_T^*(V)} \theta(q)\frac{dq}{q - iz} \quad \text{and} \quad J_4(z) = \frac{1}{\pi}\int_{q_L^*(V)}^{q_T^*(V)} \theta(q)\frac{dq}{q - iz}$$

in which $\theta(q) = \lim_{p\to 0+} \arg(T_3(p - iq, V))$. In both Cases 3 and 4, $X_3^-(z) = X_3^+(z)/T_3(z, V)$.

Combining (2.28), (2.30)-(2.32), (2.37), (2.40)-(2.41) one constructs $X^{\pm}(z)$ and subsequently from (1.32) one derives an expresson for the stress intensity factor K.

An important observation is that for Case 1, $0 \leq V \leq c_R^*$, K is independent of crack-speed and material properties, as is the case for elastic material. Moreover, since $q_R^*(V) \to \infty$ as $V \to c_R$, K must vanish as $V \to c_R$. Also, numerical calculations of K for a variety of material models suggest that K is a decreasing function of crack-speed, V.

For materials with non-constant Poisson's ratio, the arguments outlined above and proved in [11] also apply provided the Rayleigh function $T_3(z, V)$ has the mapping properties proved above, the essential features being that: i. the equation $T_3(-iq, V) = 0$ has a unique solution $q = q_R^*(V) > 0$ for $c_R^* \leq V < c_R$; ii. that the winding number of $T_3(p - iq, V)$ about the origin for q fixed and $-\infty < p < \infty$ be either zero or one. However, the proof of the required mapping properties presented in [11] relies heavily upon the representation (2.36) which is valid only when Poisson's ratio is constant. It is easy to see that if i. and ii. do not hold, then $X^+(p)$ will not be asymptotic to $\sqrt{|p|}$ as $|p| \to \infty$ and the crack-tip stress singularity will not be of the square-root type. It would follow that the energy flux into the crack-tip would be either zero or infinite, neither of which is physically reasonable. Moreover, for every specific material model examined by the author, properties i. and ii. held true. However, it would be desireable to have a direct proof that i. and ii. hold for the constitutive class considered here, namely materials for which $\bar{\mu}(s)$ and $\bar{\kappa}(s)$ satisfy (2.17). At present this seems to be an open question. Finally, [11] contains an analysis of the energy release rate for a plane strain crack and in particular the importance of its definition through the inclusion of a crack-tip failure-zone by which the singular stresses are cancelled. However, for the sake of brevity, the energy release rate analysis will be deferred until the next section in which the steady-state problems examined in this section are reconsidered with the inclusion of transient effects.

3. TRANSIENT CONSTANT CRACK-SPEED CRACK PROPAGATION

3.1 Introduction

In the previous two chapters, a general method for constructing solutions to dynamic steady-state semi-infinite crack problems in the context of linear viscoelasticity was presented, at first formally in chapter 1 and then more carefully in chapter 2 with proper attention given to the serious analytical issues arising in the rigorous justification of the essential steps of the construction for both anti-plane shear and plane strain modes of deformation. In the present chapter, consideration is given to the corresponding transient problems. A number of papers have appeared in which dynamic, transient analyses (either approximate or exact) for moving cracks in viscoelastic material are presented under either anti-plane shear or plane strain modes of deformation. The principal references to this literature are [13]-[17], [20], [25], and [28]-[30]. With the exception of the approximate analysis in [25] and the exact analysis in [30] that is valid only for a very special material constitutive law, all of these transient analyses have assumed that when the crack begins to propagate due to the application of tractions, it does so at a constant speed. The reason for this emphasis given to constant crack speed problems is one of mathematical necessity; the integral transform/complex variable methodology described in chapters 1 and 2 for solving steady-state problems is readily adaptable to transient problems if the crack speed is assumed constant, whereas, for accelerating cracks in viscoelastic material neither these methods nor the ones utilized to develop the theory of dynamically accelerating cracks in elastic material (As presented in [4], for example.) seem to be applicable. A new method for treating dynamically accelerating crack problems in viscoelastic material, recently developed in [30] and [31], will be described subsequently in chapters 4 and 5.

Though studied primarily as a compomise to mathematical convenience, the dynamic, transient, constant crack speed problems addressed in the literature have nonetheless provided valuable insight into dynamic fracture mechanics. For example, through these problems one can study the combined influence of material viscoelasticity, inertia and crack speed on important fracture mechanics parameters, such as the stress intensity factor and energy release rate, and one can prove theorems describing the rate at which these parameters converge to their steady-state values thereby shedding light on the question of how long initial transients persist as important effects. It should also be mentioned in this regard, that in several important experimental investigations of dynamic fracture in polymeric materials (See [32]–[35], for example.) it has been seen that, within the limit of the observable time scales, the cracks propagated at a constant speed until either crack branching or arrest occurred or the specimen failed completely. In section 2, the problem is considered for an anti-plane shear crack and expressions for the stress and displacement fields are derived. Section 3 contains a procedure for constructing the energy release rate within the context of a crack-tip failure zone and some results on the influence of crack speed and material properties on the rate of convergence of the transient energy release rate to the limiting steady-state value.

3.2 Constant Speed Anti-Plane Shear Crack

The problem to be solved is that of a semi-infinite anti-plane shear crack that begins to propagate at time $t = 0$ at a constant speed, V, in an infinite viscoelastic body due to the sudden application of crack-face tractions. As in chapter 2, the governing field equations are (1.1) in which only the out-of-plane displacement, $u_3(x_1, x_2, t)$ is non-zero, but here one must impose the initial conditions $u_3(x_1, x_2, 0) = \dot{u}_3(x_1, x_2, 0) = 0$. Again as in chapter 2, it proves useful to adopt a coordinate system attached to the crack-tip given by $x = x_1 - Vt$, $y = x_2$, and to define $w(x, y, t) = w(x_1 - Vt, x_2, t) = u_3(x_1, x_2, t)$ and $\sigma_{yz}(x, y, t) = \sigma_{23}(x_1, x_2, t)$. In the moving coordinate system, the governing equation of motion becomes (See [20] for more details.)

$$\rho\left[\frac{\partial}{\partial t}-V\frac{\partial}{\partial x}\right]w(x,y,t)=\Delta\left[\mu(0)w(x,y,t)+\int_0^t w(x+V(t-r),y,r)\mu'(t-r)dr\right] \quad (3.1)$$

where Δ denotes the Laplacian operator. The boundary conditions, valid for all $t>0$, are (1.8) modified appropriately to accommodate transient behavior

$$\begin{aligned} \sigma_{yz}(x,0,t) &= L_e l(x/a_e,t), \quad x<0 \\ w(x,0,t) &= 0, \quad x>0 \\ \sigma_{yz}(x,y,t) &\to 0 \quad \text{as} \quad x^2+y^2\to\infty. \end{aligned} \quad (3.2)$$

In (3.2), L_e and a_e have the dimensions of stress and length, respectively, whereas $l(\cdot,\cdot)$ is dimensionless. Application of the Fourier transform to (3.1) results in

$$\rho\left[\frac{\partial}{\partial t}+iVp\right]^2\hat{w}(p,y,t)=\left[(-ip)^2+\frac{\partial^2}{\partial y^2}\right] \times\left[\mu(0)\hat{w}(p,y,t)+\int_0^t \hat{w}(p,y,r)e^{-iVp(t-r)}\mu'(t-r)dr\right]. \quad (3.3)$$

Subsequent application of the Laplace transform, defined by

$$\bar{g}(p,y,s)=\int_0^\infty g(p,y,t)e^{-ts}dt,$$

to (3.3) produces the relation

$$\rho(s+iVp)^2\hat{\bar{w}}(p,y,s)=\left[(-ip)^2+\frac{\partial^2}{\partial y^2}\right]\bar{\mu}(s+iVp)\hat{\bar{w}}(p,y,s) \quad (3.4)$$

in which $\bar{\mu}(s)$ is given by (1.5). For much of the subsequent analysis, s will be assumed to be real and positive; analytic continuation arguments will be invoked as appropriate when Laplace inversion of resulting expressions necessitates the consideration of s as lying along some Bronwich path in the complex s-plane. Regarding (3.3) as an ordinary differential equation in y with p and s as fixed parameters and in consideration of the boundary condition (3.2c), one obtains the solution

$$\hat{\bar{w}}(p,y,s)=A(p,s)e^{-\beta(s,p)|y|} \quad (3.5)$$

where

$$\beta(s,p)=\left[p^2+\rho(s+iVp)^2/\bar{\mu}(s+iVp)\right]^{1/2} \quad (3.6)$$

must be chosen so that $\mathrm{Re}\beta(s,p) \geq 0$ for all s and p, restricted appropriately.

In a similar manner, applying the Fourier and Laplace transforms to the constitutive equations (1.2) yields

$$\hat{\bar{\sigma}}_{yz}(p,y,s) = \bar{\mu}(s+iVp)\frac{\partial}{\partial y}\hat{\bar{w}}(p,y,s). \tag{3.7}$$

Combining (3.5)–(3.7), one obtains the fundamental boundary relation

$$\begin{aligned}\hat{\bar{\sigma}}_{yz}(p,0,s) &= -\beta(s,p)\bar{\mu}(s+iVp)\hat{\bar{w}}(p,0,s)\\ &= T(p,s)\hat{\bar{w}}(p,0,s).\end{aligned} \tag{3.8}$$

Equation (3.8) is the transient analog of (1.10) with the important distinction that (1.10) relates the transformed boundary stress and displacement x-partial derivative, whereas in (3.8) the displacement, not its x-partial derivative, appears.

At this point, one is now in the position to apply to (3.8) the (formal) Riemann-Hilbert methodology developed in chapter 1. In particular, one obtains an expression like (1.22) for the Laplace transform of the stress ahead of the crack

$$\bar{\sigma}^+(x,s) = \frac{1}{2\pi}\int_{-\infty}^{\infty} e^{-ixp}X^+(p,s)\,dp\frac{1}{2\pi i}\int_{-\infty}^{\infty}\frac{g(\tau,s)}{X^+(\tau,s)}\frac{d\tau}{\tau-p} \tag{3.9}$$

for all $x > 0$ and like (1.31) and (1.32) for the Laplace transformed stress intensity factor:

$$\bar{\sigma}^+(x,s) \sim \frac{\bar{K}(s)}{\sqrt{\pi x}} \quad \text{with} \quad \bar{K}(s) = \frac{X_\infty^+}{2\pi}\int_{-\infty}^{\infty} g(p,s)/X^+(p,s)\,dp \tag{3.10}$$

and with

$$X_\infty^+ = \lim_{p\to\infty} X_0^+(p,s).$$

As in chapter 2, the formal manipulations of chapter 1 leading to (3.9) and (3.10) must be proved rigorously by appealing to a careful analysis of the analyticity properties of the transfer function $T(p,s)$ in (3.8). To this end, it is useful to consider a factorization analogous to (1.23)–(1.26) and (2.6)–(2.7) given by

$$T_1(p,s) = -\bar{\mu}(s+iVp)(p-is/V) \quad \text{and} \quad T_2(p,s) = \left[\frac{p^2}{(p-is/V)^2} - \frac{\rho V^2}{\bar{\mu}(s+iVp)}\right]^{1/2}. \tag{3.11}$$

One argues easily in the manner of chapter 2 that for the constitutive class under consideration, $\bar{\mu}(s+iVp)$ has an analytic continuation $\bar{\mu}(s+iVz)$ for $\mathrm{Im}(z) < 0$ without zeros, and hence for $X_1^\pm(z,s)$ one must take

$$X_1^+(z,s) = 1 \quad \text{and} \quad X_1^-(z,s) = \frac{-1}{\bar{\mu}(s+iVz)(z-is/V)}. \tag{3.12}$$

The analysis of $T_2(p,s)$ is more delicate. It can be shown ([20]) for the constitutive class under consideration and for $s > 0$, that $T_2(p,s)$ has a continuation $T_2(z,s)$ analytic for $\mathrm{Im}(z) < 0$ except for a branch cut along the interval $\mathrm{Re}(z) = 0$, $0 \geq \mathrm{Im}(z) > -q^*(s,V)$, where $z = -iq^*(s,V)$ is the unique root of $T_2^2(z,s)$ in the half-plane $\mathrm{Im}(z) < 0$, i.e.

$$T_2^2(-iq^*(s,V),s) = 0. \tag{3.13}$$

As in the steady-state analysis of chapter 2, the jump in $\log(T_2(z,s))$ across the branch cut is a constant equal to $i\pi$ from which it can be shown that for $X_2^{\pm}(z,s)$ one may take

$$X_2^+(z,s) = \sqrt{(z - iq^*(s,V))/z} \tag{3.14}$$

and $X_2^-(z,s) = X_2^+(z,s)/T_2(z,s)$. One may now show, in the manner of chapter 2, that the Laplace transformed stress intensity factor in (3.10) has the form

$$\bar{K}(s) = \frac{-1}{\sqrt{t\pi}} L_e \sqrt{a_e} \int_{-\infty}^{0} l(x,s) e^{xq^*(s,V)a_e} \frac{dx}{\sqrt{|x|}}. \tag{3.15}$$

Performing the Laplace inversion of (3.15) in closed form would seem to be a formidable task owing to the implicit nature in which $q^*(s,V)$ is defined through (3.13). However, it can also be shown ([29]) that equations (3.9), (3.10), (3.12), (3.14) and (3.15) hold for each fixed s in the half-plane $\mathrm{Re}(s) > 0$, though the argument is somewhat trickier than for $s > 0$. Thus these expressions (slightly modified) are seen to be valid on a Bromwich path thereby allowing their numerical Laplace inversion. A convenient and accurate algorithm for performing the numerical Laplace inversion is described and illustrated in [29]. In the next section, the issue of defining and constructing the energy release rate is considered.

3.3 Crack-Tip Failure-Zone and Energy Release Rate

Energy criteria for fracture are based upon the notion that fracture can occur only if the energy that is removed from the body during the fracture process (required energy for fracture) can be supplied by the external energy sources (available energy). The rate at which energy is absorbed from the body by a moving crack-tip, E_c, is defined through an overall energy balance (ignoring thermal effects) given by

$$P = \dot{U} + \dot{K} + D + E_c \tag{3.16}$$

in which P is the power supplied by external traction, U is the stored (reversible) elastic energy, K is the kinetic energy and D represents energy dissipation due to viscous effects. It has been shown (See [36] and [37], for example.) that for viscoelastic material, an energy fracture criterion that is based upon the singular stress solution will not fully reflect the viscoelastic properties of the material. Moreover, there are good physical arguments ([36], [37]) for modifying the classical crack boundary conditions (1.8) to include a small failure-zone of length a_f behind the crack-tip in which cohesive stresses σ_f act to resist the fracture process and which have the effect of cancelling the crack-tip stress singularity. The idea to introduce such a cohesive-zone dates from the seminal paper of Barenblatt [38]. Ironically, Barenblatt was interested primarily in brittle elastic fracture for which is was subsequently shown by a number of people that fracture criteria based upon the stress intensity factor, the energy release rate calculated using the singular crack-tip stress field and the energy release rate calculated using a failure-zone model with bounded crack-tip stresses must be equivalent. However, for viscoelastic material one obtains qualitatively different fracture criteria from these three parameters.

Adopting the Barenblatt model and assuming the crack speed is constant, one can show (See [39] or [40] for example.) that the energy flux into the crack-tip region (the energy available for fracture) is given by

$$E = \int_{Vt-a_f}^{Vt} \sigma_f(x_1 - Vt, t)\dot{u}^-(x_1 - Vt, 0, t)\, dx_1 \tag{3.17}$$

where $u(x_1, 0, t)$ is the relative crack-face displacement corresponding to the combined crack-face loading $\sigma_e + \sigma_f$. $\sigma_e(x_1, t)$ denotes the external crack-face traction for which σ_f is the corresponding cohesive-zone response. From (3.16) it follows that a necessary condition for fracture is that $E = E_c$, i.e. the available energy must equal the required energy. For steady-state crack propagation, $\dot{u}^- = -u^-_{,x}V$ and if the energy release rate, G, is defined by $E = GV$, one has that

$$G = -\int_{-a_f}^{0} \sigma_f(x)u^-_{,x}(x)\, dx. \tag{3.18}$$

Thus, for the cohesive-zone model, G is seen from (3.18) to be the work done by the cohesive forces per unit length of crack in the failure-zone. For general loadings, the integral in (3.18) is difficult to calculate directly. However, for a variety of special loading forms computationally convenient expressions for (3.18) can be derived. One such class of functions for σ_e and σ_f have the forms

$$\sigma_e(x,t) = L_e l_e(t)e^{x/a_e}, \quad \sigma_f(x,t) = -L_f l_f(t)e^{x/a_f}, \quad -\infty < x < 0 \tag{3.19}$$

in which L_e andL_f have the dimensions of stress, a_e and a_f are length scales associated with the external loading and cohesive stress, respectively , and $l_e(t)$ and $l_f(t)$ are dimensionless functions of time. In these semi-infinite crack problems, a_e and a_f play the role of crack length and cohesive-zone length, respectively. The forms (3.19) are clearly not intended to model physical reality; they are adopted for mathematical convenience and, because σ_f is not compactly supported, they necessitate the slight modification of (3.18)

$$G = -\int_{-\infty}^{0} \sigma_f(x)u^-_{,x}(x)\, dx. \tag{3.20}$$

Nevertheless, they can be made to reflect the essential features of the Barenblatt model. Namely, that

$$K_e + K_f = 0 \quad \text{and} \quad a_f/a_e \ll 1 \tag{3.21}$$

where K_e and K_f are the stress intensity factors resulting from boundary value problems in which only σ_e or σ_f act on the crack-faces, respectively. Because the governing equations of motion and the boundary conditions are linear, (3.21a) guarantees that the cohesive-zone response serves to cancel the stress singularity produced by the external loading. The condition (3.21b) means that the cohesive-zone is small relative to the crack length parameter, a_e, and serves to diminish in significance the unphysical feature of σ_f in (3.19b) that it is not compactly supported.

What makes the functions (3.19) convenient for evaluating the expression (3.20) is that their Fourier transforms have the simple forms

$$\hat{\sigma}_f(p) = \frac{-a_f L_f}{1 + i a_f p}, \qquad \check{\sigma}_f(p) = \frac{-a_f L_f}{2\pi(1 - i a_f p)} \tag{3.22}$$

which are meromorphic with simple poles in the upper/lower half of the complex p-plane, respectively. Similar expressions hold for $\hat{\sigma}_e(p)$. The steady-state G in (3.20) can now be evaluated by first applying the Parseval relation obtaining

$$G = -\int_{-\infty}^{\infty} \check{\sigma}_f(p)\hat{u}^-_{,x}(p)\, dp. \tag{3.23}$$

Since $\hat{u}^-_{,x}(p)$ has the analytic extension $F^-(z)$ to the lower half-plane $\mathrm{Im}(z) < 0$ and $\check{\sigma}_f(z)$ is meromorphic with a simple pole at $z = -i/a_f$, (3.23) can be evaluated by residues, yielding

$$G = L_f F^-(-i/a_f). \tag{3.24}$$

Moreover, the stress intensity factors K_e and K_f can be evaluated similarly from (1.36)

$$K_e = \frac{-X^+_\infty L_e}{X^+(i/a_e)}, \qquad K_f = \frac{X^+_\infty L_f}{X^+(i/a_f)}. \tag{3.25}$$

The cancellation of stress singularities condition (3.21a) requires

$$L_e X^+(i/a_f) = L_f X^+(i/a_e). \tag{3.26}$$

In particular, for the steady-state anti-plane shear crack, substitution of (2.14) into (3.26) yields the relation

$$L_e\sqrt{\frac{a_e}{1 + a_e q^*(V)}} = L_f\sqrt{\frac{a_f}{1 + a_f q^*(V)}}. \tag{3.27}$$

A general method for evaluating $F^-(i/a_f)$ in (3.24) was developed in [19] and simplified in [41] with the result that

$$G = \left(\frac{a_e - a_f}{a_e + a_f}\right) \frac{a_f L_e^2}{2} \frac{X^-(-i/a_f)X^+(i/a_f)}{[X^+(i/a_e)]^2}. \tag{3.28}$$

The result (3.28) is general in that it holds for all problems in which the methodology of chapter 1 applies and for which the special forms (3.19) are employed. Naturally, the functions $X^\pm(z)$ in (3.28) must be constructed separately for each application. In particular for a steady-state anti-plane shear crack, one can show that (3.28) becomes

$$G = \frac{K^2(V)}{2\bar{\mu}(V\tau/a_f)} \left(\frac{a_e - a_f}{a_e + a_f}\right) \sqrt{\frac{1 - (a_f q^*(V))^2}{1 - (V/\bar{c}_T(V\tau/a_f))^2}} \tag{3.29}$$

in which $\bar{c}_T(s)$ is the viscoelastic transverse wave speed function defined in (2.27) and τ is a characteristic shear relaxation time introduced through

$$\mu(t) = \mu_0 m(t/\tau) \tag{3.30}$$

with $m(t)$ a dimensionless relaxation function. The energy release rate defined in the absence of a failure-zone, G_{nf}, results from taking the limit of (3.28) as $a_f \to 0$. For an anti-plane shear crack, one obtains from (3.29) the limiting result

$$G_{nf} = \frac{K^2(V)}{2\mu_0\sqrt{1 - (V/c_T)^2}}. \tag{3.31}$$

It is clear from (3.29) and (3.31) that without a failure-zone in the model, G depends upon material properties only through $K(V)$ and the glassy viscoelastic modulus, whereas with the additional length scale provided by a failure-zone, G depends fully on $\mu(t)$. Moreover, one sees from (3.29) the fundamental role played by the dimensionless parameter $V\tau/a_f$.

The formulas (3.24)–(3.31) are readily modified (See [19] for the details.) for generalizations of the loading (3.19) given by

$$\sigma_e(x) = L_e \int_0^\infty e^{tx/a_e}\, dh_e(t) \qquad \sigma_f(x) = -L_f \int_0^\infty e^{tx/a_f}\, dh_f(t) \tag{3.32}$$

in which $h_e(t)$ and $h_f(t)$ are arbitrary signed (i.e. not necessarily positive) measures restricted only to the extent that the required integrals converge. For example, the forms (3.19) correspond to $dh_e(t) = dh_f(t) = \delta(t-1)$, the Dirac point measure concentrated at $t = 1$. Corresponding to (3.22) one then has

$$\hat{\sigma}_f(p) = -a_f L_f \int_0^\infty \frac{dh_f(t)}{t + ipa_f} \qquad \check{\sigma}_f(p) = \frac{-a_f L_f}{\pi} \int_0^\infty \frac{dh_f(t)}{t - ipa_f}$$

from which one shows that the generalization of (3.27) becomes

$$|K_e| = \frac{L_e\sqrt{a_e}}{\sqrt{\pi}} \int_0^\infty \frac{dh_e(t)}{\sqrt{t + a_e q^*(V)}} = \frac{L_f\sqrt{a_f}}{\sqrt{\pi}} \int_0^\infty \frac{dh_f(t)}{\sqrt{t + a_f q^*(V)}} = |K_f|.$$

In a similar fashion, one constructs the natural generalization of (3.29)

$$G = -L_f \int_0^\infty \sqrt{\frac{1 - t/(a_f q^*(V))}{1 - (V/\bar{c}_T(V\tau/a_f))^2} \frac{H(-it/a_f)}{\bar{\mu}(V\tau t/a_f)}}\, dh_f(t)$$

with

$$\begin{aligned} H(p) = c_e L_e &\int_0^\infty \sqrt{a_e q^*(V)/(t + a_e q^*(V))} \frac{dh_e(t)}{t + ia_e p)} \\ - a_f L_f &\int_0^\infty \sqrt{a_f q^*(V)/(t + a_f q^*(V))} \frac{dh_f(t)}{t + ia_f p)}. \end{aligned}$$

Returning to the transient problem and using the forms (3.19) in (3.17) one has after an obvious change of variables

$$G(t) = \frac{-L_f l_f(t)}{V} \int_{-\infty}^0 \exp^{x/a_f} \left[\frac{\partial}{\partial t} - V\frac{\partial}{\partial x}\right] u^-(x,0,t)dx \tag{3.33}$$

which may be written $G(t) = I(t) + W(t)$ with

$$\begin{aligned} I(t) &= \frac{-L_f l_f(t)}{V} \int_{-\infty}^0 \exp^{x/a_f} \frac{\partial}{\partial t} u^-(x,0,t)dx \\ W(t) &= L_f l_f(t) \int_{-\infty}^0 \exp^{x/a_f} \frac{\partial}{\partial x} u^-(x,0,t)dx. \end{aligned} \tag{3.34}$$

$I(t)$ and $W(t)$ individually do not have direct physical significance, but the decomposition (3.33) does provide insight into the behavior of $G(t)$. Specifically, one can show that $\lim_{t\to 0} G(t) = \lim_{t\to 0} I(t)$ and $\lim_{t\to\infty} G(t) = \lim_{t\to\infty} W(t)$ whereas $\lim_{t\to 0} W(t) = \lim_{t\to\infty} I(t) = 0$. Thus, the short time behavior of $G(t)$ is governed primarily by $I(t)$ and the long time behavior by $W(t)$. Because the model assumes that the crack speed instantaneously takes and maintains a constant value when the tractions are applied and does not account for an initial crack acceleration phase, the short time asymptotic analysis of $I(t)$ presented in [20] exhibits unphysical (and singular as $V \to 0$) behavior. On the other hand, the long time asymptotic analysis presented for $W(t)$ is physically meaningful. With similar arguments to those employed above for the steady-state analysis, one can show that for the transient anti-plane shear crack problem considered in this chapter, the energy release rate $G(t) = -L_f l_f(t) g(t)$ with the Laplace transform of $g(t)$ given by

$$\bar{g}(s) = \frac{-L_e \bar{l}_e(s)}{2\bar{\mu}(s + V/a_f)} \frac{(a_e - a_f)}{(a_e + a_f)} \sqrt{a_e a_f} \left[\frac{1 + a_f q^*(V,s)}{1 - a_e q^*(V,s)}\right]^2$$
$$\times \left[\frac{1}{(1 + s a_f/V)^2} - (V/\bar{c}_T(s + V/a_f))^2\right]^{-1/2} . \tag{3.35}$$

It should be remarked that implicit in the result (3.35) is the failure-zone constitutive assumption that a_f is constant and $l_f(t)$ is the (dimensionless) cohesive stress response to the applied traction that cancels the crack-tip stress singularity. In particular, one easily shows that the Laplace transform of $l_f(t)$ must satisfy

$$\bar{l}_f(s) = \bar{l}_e(s) \frac{L_e}{L_f} \sqrt{\frac{a_e}{a_f}} \left[\frac{1 + a_f q^*(V,s)}{1 + a_e q^*(V,s)}\right] . \tag{3.36}$$

This cohesive-zone constitutive assumption is objectionable on physical grounds since, for example, it requires that the stress level in the cohesive-zone adjust in a spatially uniform manner to transient effects which in turn suggests that information propagates throughout the cohesive-zone with infinite speed. A far more reasonable model has $l_f(t) = 1$ and $a_f = a_f(V, t)$, as occurs in the Dugdale model, for example. However, the analysis of this model requires consideration of a dynamically accelerating crack-tip, which is a much more difficult problem and the subject of the next two chapters.

These physical objections to the model leading to (3.35) and (3.36) notwithstanding, the long time asymptotic analysis of $W(t)$ is useful for studying questions such as the rate of convergence to the steady-state solution. By way of comparison, it is shown in [22] that for the corresponding problem in elastic material,

$$W_{EL}(t) = W(\infty) + o(e^{-t/b}) \quad \text{as} \quad t \to \infty \quad \text{with} \quad b = \frac{a_e}{c - V} . \tag{3.37}$$

Thus, $W_{EL}(t)$ converges exponentially to its steady-state limit with a rate that tends to zero as the crack speed approaches the shear wave speed.

For viscoelastic material the situation is more complicated and the asymptotic behavior of $W(t)$ as $t \to \infty$ clearly depends upon the asymptotic behavior of the relaxation function $\mu(t)$. However, the question was studied in [22] for two important special cases: i. power-law material for which $\mu(t) \sim t^{-\alpha}$ as $t \to \infty$; and ii. exponential material for which $\mu(t) \sim e^{-ta}$ as $t \to \infty$. The analysis makes crucial use of the general property of the Laplace transform $F(s)$ of a function $f(t)$ that $f(t) \sim e^{-ta}$ as $t \to \infty$ with $a > 0$ if and only if $F(s)$ is analytic in the halfplane $\mathrm{Re}(s) > -a$.

The heart of the analysis concerns determining the analyticity properties of $q^*(V,s)$ defined implicitly through (3.13).

For power-law material, one shows that for $0 \leq V < c^*$, $q^*(V,s)$ is not analytic at $a = 0$ from which it follows that $G(t)$ cannot decay exponentially to its steady-state value. On the other hand, for $c^* < V < c$, $q^*(V,s)$ is analytic in a half-plane $\mathrm{Re}(s) > -a$ for some $a > 0$. Since, $\bar{\mu}(s + V/a_f)$ must be analytic in the half-plane $\mathrm{Re}(s) > -V/a_f$ it follows from (3.35) that for $l_e(t) = 1$, $G(t)$ must decay exponentially to its steady-state value. In the remaining case $V = c^*$, one can show that $q^*(V,s)$ is not analytic in any half-plane $\mathrm{Re}(s) > -a$ with $a > 0$ and hence that $G(t)$ cannot converge exponentially fast to equilibrium.

For material with an exponentially decaying relaxation function, such as a standard linear solid for example, the long time asymptotic results when $c^* \leq V < c$ are the same as for power-law material. However, such is not the case when $0 \leq V < c^*$. In particular, it can be shown that now $q^*(V,s)$ is analytic in a half-plane $\mathrm{Re}(s) > -a$ for some $a > 0$. Thus, one concludes that $G(t)$ converges exponentially fast to equilibrium for all crack speeds V except for the unique value $V = c^*$.

In [29], an algorithm for the efficient numerical Laplace inversion of (3.35) and (3.36) is derived and illustrated. The corresponding theory for a plane-strain crack in presented in [28], but for the sake of brevity these results will not be described here. The next chapter begins the analysis of dynamically accelerating cracks in viscoelastic material by presenting a recently developed method for their solution in the simpler context of elastic material.

4. AN ACCELERATING ELASTIC ANTI-PLANE SHEAR CRACK

4.1 Introduction

The previous three chapters have focused on constructing solutions to dynamic, semi-infinite crack problems in viscoelastic material under the assumption that the crack speed is constant. Clearly, many important issues in fracture mechanics, such as crack initiation, arrest and other more complicated propagation events, cannot be fully explored by considering only constant crack speed solutions. Indeed, one of the milestones in the history of dynamic fracture was the solution presented by Kostrov [42] in 1966 for an accelerating, semi-infinite, anti-plane shear crack in an infinite, homogeneous, isotropic, linear elastic body. Subsequently, Kostrov's approach, which is based upon the construction of a Green's function for the wave equation in a half-plane satisfying certain boundary conditions, was generalized and augmented by asymptotic arguments through the work of Achenbach, Freund and many others (See [4] for an excellent account of the subject.) to create a mature theory encompasing both opening and shearing modes of deformation, curved as well as straight cracks, anisotropy and various complicating geometric and loading effects. However, the Green's function based methods, and more recent alternative approaches presented by Burridge [43] and Willis [5] that are based upon certain results on homogeneous function solutions for the elastic wave equations, do not seem to generalize to viscoelastic material.

In [31] another approach to solving accelerating semi-infinite crack problems was introduced through a reconsideration of the problem originally studied by Kostrov [42]. The method in [31] utilizes a modification of the integral transform and complex variable technique described in chapters 1–3 for steady-state and constant crack speed problems. It is certainly far from evident that such a technique would prove useful in constructing solutions to dynamically accelerating crack problems. Indeed, when formulated relative to a coordinate system moving with the crack-tip, the governing viscoelastic wave equations have a time dependent coefficient ($\dot{a}(t)$) in the acceleration term , whereas when formulated relative to a fixed inertial frame, the displacement boundary condition for the crack-plane at all points ahead of the initial crack-tip location and through which the crack-tip eventually passes has incomplete time histories. It was shown in [31] how these apparent impediments to applying the integral transform/complex variable machinery to accelerating crack problems can be surmounted to produce closed form solutions for the stress and displacement fields. This chapter contains a summary of the key ideas underlying the method and its application to the elastic problem considered in [31].

4.2 Problem Formulation and General Solution Method

The specific initial/boundary value problem to be solved concerns the anti-plane shear elastic wave equation

$$\ddot{u}_3 = c^2 \Delta u_3 \tag{4.1}$$

where Δ denotes the two dimensional Laplacian and c is the elastic shear wave speed. The initial conditions are

$$u_3(x_1, x_2, 0) = \dot{u}_3(x_1, x_2, 0) = 0 \tag{4.2}$$

and the boundary conditions

$$\begin{aligned} u_3(x_1, 0, t) &= 0 \qquad \text{for} \qquad x_1 > a(t) \\ \sigma_{23}(x_1, 0, t) &= \sigma_e(x_1, t) \qquad \text{for} \qquad x_1 < a(t) \\ u_3(x_1, x_2, t) &\to 0 \qquad \text{as} \qquad x_2 \to \infty. \end{aligned} \tag{4.3}$$

The relevant elastic constitutive equation is

$$\sigma_{23}(x_1, x_2, t) = \mu \frac{\partial u_3}{\partial x_2} \tag{4.4}$$

and the crack-tip motion $a(t)$ is assumed to be a non-negative, smooth function for which

$$0 \le \dot{a}(t) < c. \tag{4.5}$$

Thus, crack closure and supersonic crack speeds are not permitted.

Following the methodology presented in chapter 3 for the constant crack speed problem, one may apply the Fourier and Laplace transforms to (4.1) and (4.4), solve the resulting ordinary differential equation in x_2 and show that

$$\hat{\bar{u}}_3(p, x_2, s) = A(p, s)e^{-\beta(p,s)x_2} \tag{4.6}$$

with $A(p, s) = \hat{\bar{u}}_3(p, 0, s)$ and $\beta(p, s)$ given by

$$\beta(p, s) = \sqrt{p^2 + s^2/c^2}. \tag{4.7}$$

In (4.7), the square root must be chosen so that $\mathrm{Re}(\beta(p, s)) > 0$. From (4.4), (4.6) and (4.7) one next derives the basic crack plane stress/displacement relation (in transformed variables)

$$\hat{\bar{\sigma}}_{23}(p, 0, s) = -\mu\beta(p, s)\hat{\bar{u}}_3(p, 0, s). \tag{4.8}$$

It is useful at this point to formulate (4.6) relative to a coordinate system (x, y) moving with the crack-tip. Adopting the notation

$$\begin{aligned} x = x_1 - a(t), \qquad & y = x_2, \\ w(x, y, t) = w(x_1 - a(t), x_2, t) &= u_3(x_1, x_2, t) \\ \sigma(x, y, t) = \sigma(x_1 - a(t), x_2, t) &= \sigma_{23}(x_1, x_2, t), \end{aligned} \tag{4.9}$$

the relation (4.8) becomes

$$\eta(p, s) = -\mu\beta(p, s)\xi^-(p, s) \tag{4.10}$$

where $\eta(p, s)$ and $\xi^-(p, s)$ are defined by

$$\begin{aligned} \eta(p, s) &= \int_0^\infty e^{-ts} e^{ipa(t)} \hat{\sigma}(p, 0, t)dt \\ \xi^-(p, s) &= \int_0^\infty e^{-ts} e^{ipa(t)} \hat{w}(p, 0, t)dt. \end{aligned} \tag{4.11}$$

The canonical methodology of chapters 1–3 applied to (4.11) leads to the Riemann-Hilbert linear jump relation

$$\eta^+(p, s) = -\mu\beta(p, s)\xi^-(p, s) - \eta^-(p, s) \tag{4.12}$$

in which $\eta^\pm(p, s)$ are given by

$$\begin{aligned} \eta^+(p, s) &= \int_0^\infty e^{-ts} e^{ipa(t)} \hat{\sigma}^+(p, 0, t)dt \\ &= \int_0^\infty e^{-ts}dt \int_{a(t)}^\infty e^{ipx}\sigma^+(x - a(t), 0, t)dx \end{aligned} \tag{4.13}$$

$$\eta^-(p, s) = \int_0^\infty e^{-ts}dt \int_{-\infty}^{a(t)} e^{ipx}\sigma^-(x - a(t), 0, t)dx. \tag{4.14}$$

In (4.12), $\eta^-(p,s)$ is known from (4.14) and $\eta^+(p,s)$ and $\xi^-(p,s)$ have natural analytic extensions $\eta^+(z,s)$ and $\xi^-(z,s)$ for $\mathrm{Im}(z)\gtrless 0$, respectively. Therefore, equation (4.12) would define a canonical Riemann-Hilbert boundary value problem if one were permitted to assume á priori that $\eta^+(z,s)$ and $\xi^-(z,s)$ decayed to zero in the appropriate half-plane as $\mathrm{Im}(z) \to \pm\infty$. Such is case for $\eta^+(z,s)$. However, the method breaks down since, in fact, $\xi^-(z,s)$ grows exponentially as $\mathrm{Im}(z) \to -\infty$. The source of this exponential growth in $\xi^-(z,s)$ is revealed by writing the transform $\hat{w}(p,0,t)$ in (4.11) as an integral in the fixed inertial frame and then introducing the decomposition

$$\begin{aligned}\xi^-(p,s) &= \int_0^\infty e^{-ts}dt \int_{-\infty}^{a(t)} e^{ipx_1}u_3(x_1,0,t)dx_1 \\ &= \hat{\hat{d}}_L(p,s) + \hat{\hat{d}}_R(p,s)\end{aligned} \tag{4.15}$$

with

$$\begin{aligned} d_L(x_1,t) &= \begin{cases} u_3(x_1,0,t), & \text{if} \quad -\infty < x_1 < 0 \\ 0, & \text{if} \quad 0 < x_1 \end{cases} \\ d_R(x_1,t) &= \begin{cases} 0, & \text{if} \quad -\infty < x_1 < 0, \text{ or } \ a(t) < x_1 \\ u_3(x_1,0,t), & \text{if} \quad 0 < x_1 < a(t) \end{cases}\end{aligned} \tag{4.16}$$

and

$$\hat{\hat{d}}_L(p,s) = \int_0^\infty e^{-ts}dt \int_{-\infty}^{0} e^{ipx_1}u_3(x_1,0,t)dx_1 \tag{4.17}$$

$$\hat{\hat{d}}_R(p,s) = \int_0^\infty e^{-ts}dt \int_{0}^{a(t)} e^{ipx_1}u_3(x_1,0,t)dx_1. \tag{4.18}$$

It is now apparent that the term $\hat{\hat{d}}_L(p,s)$ has a natural analytic extension $\hat{\hat{d}}_L(z,s)$ to the lower complex half-plane that decays as $\mathrm{Im}(z) \to -\infty$, whereas the analytic continuation $\hat{\hat{d}}_R(z,s)$ of $\hat{\hat{d}}_R(p,s)$ grows exponentially as $\mathrm{Im}(z) \to -\infty$. These observations suggest posing the following Riemann-Hilbert problem: Find $F^\pm(z,s)$ analytic for $\mathrm{Im}(z)\gtrless 0$, respectively, and such that

$$\begin{aligned} F^+(p,s) &= -\mu\beta(p,s)F^-(p,s) + g(p,s) \\ \lim_{\mathrm{Im}(z)\to\pm\infty} F^\pm(z,s) &= 0, \end{aligned} \tag{4.19}$$

where

$$\begin{aligned} F^+(p,s) &= \eta^+(p,s) \\ F^-(p,s) &= \hat{\hat{d}}_L(p,s) \\ g(p,s) &= -\eta^-(p,s) - \mu\beta(p,s)\hat{\hat{d}}_R(p,s). \end{aligned} \tag{4.20}$$

Since both $d_R(x_1,t)$ and $d_L(x_1,t)$ are unknown, the inhomogeneous term $g(p,s)$ as well as $F^+(p,s)$ and $F^-(p,s)$ in the R-H problem (4.19) are all unknown. Consequently, (4.19) cannot be solved directly for $F^+(p,s)$ and $F^-(p,s)$. Nevertheless, by formally solving (4.19), two operator equations will be constructed relating the three unknown functions $d_R(x_1,t)$, $d_L(x_1,t)$ and $\sigma^+(x,t)$ to the known function $\sigma^-(x,t)$. These operator equations will then be solved successively for $d_R(x_1,t)$, $d_L(x_1,t)$ and $\sigma^+(x,t)$.

The Riemann-Hilbert problem (4.19) is readily solved by applying the machinery of chapter 1. In deriving the following formulas, s is restricted to the positive real line. The extension to complex s then follows by analytic continuation arguments. The central issue is the homogeneous Riemann-Hilbert problem

$$X^+(p,s) = -\mu\beta(p,s)X^-(p,s) \tag{4.21}$$

the appropriate solution of which is easily seen from (4.7) to be

$$\begin{aligned} X^+(z,s) &= \sqrt{z+is/c} \\ X^-(z,s) &= \frac{-1}{\mu\sqrt{z-is/c}} \end{aligned} \tag{4.22}$$

in which the branch cuts for $X^+(z,s)$ and $X^-(z,s)$ lie along the negative and positive imaginary axes, respectively.

The unique solution to the inhomogeneous R-H problem (4.19) is then given by

$$\begin{aligned} F^\pm(z,s) &= X^\pm(z,s)\frac{1}{2\pi i}\int_{-\infty}^{\infty}\frac{g(\tau,s)}{X^+(\tau,s)}\frac{d\tau}{(\tau-z)} \\ &= -X^\pm(z,s)\frac{1}{2\pi i}\int_{-\infty}^{\infty}\frac{\eta^-(\tau,s)}{X^+(\tau,s)}\frac{d\tau}{(\tau-z)} \\ &\quad + X^\pm(z,s)\frac{1}{2\pi i}\int_{-\infty}^{\infty}\frac{\hat{\hat{d}}_R(\tau,s)}{X^-(\tau,s)}\frac{d\tau}{(\tau-z)}. \end{aligned} \tag{4.23}$$

Application of the Plemelj formula to (4.23) leads to

$$\begin{aligned} \hat{\hat{d}}_L(p,s) &= F^-(p,s) \\ &= \hat{\hat{d}}_0(p,s) - \frac{1}{2}\hat{\hat{d}}_R(p,s) \\ &\quad + \frac{1}{\sqrt{p-is/c}}\frac{1}{2\pi i}\int_{-\infty}^{\infty}\hat{\hat{d}}_R(\tau,s)\sqrt{\tau-is/c}\frac{d\tau}{(\tau-p)} \end{aligned} \tag{4.24}$$

and

$$\begin{aligned} \eta^+(p,s) &= F^+(p,s) \\ &= \hat{\hat{t}}_0(p,s) - \frac{\mu}{2}\sqrt{p^2+s^2/c^2}\,\hat{\hat{d}}_R(p,s) \\ &\quad - \mu\sqrt{p+is/c}\frac{1}{2\pi i}\int_{-\infty}^{\infty}\hat{\hat{d}}_R(\tau,s)\sqrt{\tau-is/c}\frac{d\tau}{(\tau-p)} \end{aligned} \tag{4.25}$$

in which $\hat{\hat{d}}_0(p,s)$ and $\hat{\hat{t}}_0(p,s)$ are known functions given by

$$\hat{\hat{d}}_0(p,s) = \frac{-\eta^-(p,s)}{2\mu\sqrt{p^2+s^2/c^2}} + \frac{1}{\mu\sqrt{p-is/c}}\frac{1}{2\pi i}\int_{-\infty}^{\infty}\frac{\eta^-(\tau,s)}{\sqrt{\tau+is/c}}\frac{d\tau}{(\tau-p)} \tag{4.26}$$

$$\hat{\hat{t}}_0(p,s) = -\frac{1}{2}\eta^-(p,s) - \sqrt{p+is/c}\frac{1}{2\pi i}\int_{-\infty}^{\infty}\frac{\eta^-(\tau,s)}{\sqrt{\tau+is/c}}\frac{d\tau}{(\tau-p)}. \tag{4.27}$$

Equations (4.24) and (4.25) are the fundamental operator equations, in the transform (p,s)-domain, relating the two pieces of the crack-face displacement, $d_R(x_1,t)$, $d_L(x_1,t)$, the crack-face traction, $\sigma^-(x,t)$, and the crack-plane stress, $\sigma^+(x,t)$, ahead of the crack-tip. Fourier and Laplace inversion is required to construct the corresponding operator equations in the physical (x_1,t)-domain which are then solved directly for the three unknown functions $d_R(x_1,t)$, $d_L(x_1,t)$ and $\sigma^+(x,t)$. To that end, consideration is given first to constructing the operator in the (x,t)-domain corresponding to the multiplier operator $\sqrt{s+ipc}\,\hat{\hat{d}}_R(p,s)$ in the (p,s)-domain. In this and subsequent calculations it proves useful to introduce the retarded and advanced time scales

$$\begin{aligned} b_0(t) &= t - a(t)/c \\ b_1(t) &= t + a(t)/c \end{aligned} \tag{4.28}$$

and the auxiliary function $\delta_L^{-1/2}(x,t)$ defined through

$$\hat{\hat{\delta}}_L^{-1/2}(p,s) = \sqrt{s+ipc}\,\hat{\hat{d}}_R(p,s). \tag{4.29}$$

From the assumption (4.5), it follows that $b_0(t)$ and $b_1(t)$ are both increasing functions of t, and hence invertible.

Expanding the right hand side of (4.29), making the change of variables $s_1 = s+ipc$, $x_1 = x+ct$, appealing to the convolution theorem for the Laplace transform and the elementary fact that $1/\sqrt{t}$ and $\sqrt{\pi/s}$ are a Laplace transform pair, one finds that

$$\begin{aligned} \hat{\hat{\delta}}_L^{-1/2}(p,s) &= \sqrt{s+ipc}\int_0^{\infty} e^{-ts}dt\int_{-\infty}^{\infty} e^{ipx}d_R(x,t)dx \\ &= \frac{s_1}{\sqrt{s_1}}\int_0^{\infty} e^{-ts_1}dt\int_{-\infty}^{\infty} e^{ipx_1}d_R(x_1-ct,t)dx_1 \\ &= s_1\int_0^{\infty} e^{-ts_1}dt\int_{-\infty}^{\infty} e^{ipx_1}dx_1\frac{1}{\sqrt{\pi}}\int_0^t d_R(x_1-cq,q)\frac{dq}{\sqrt{t-q}} \\ &= \int_0^{\infty} e^{-ts_1}dt\int_{-\infty}^{\infty} e^{ipx_1}dx_1\frac{\partial}{\partial t}\frac{1}{\sqrt{\pi}}\int_0^t d_R(x_1-cq,q)\frac{dq}{\sqrt{t-q}}. \end{aligned} \tag{4.30}$$

Using the support property of $d_R(x_1,t)$ from (4.16) in (4.30) and changing x_1 and s_1 back to x and t, one arrives at the result

$$\delta_L^{-1/2}(x,t)=\left[\frac{\partial}{\partial t}-c\frac{\partial}{\partial x}\right]\frac{1}{\sqrt{\pi}}\int_{\max\{0,b_1^{-1}(t+x/c)\}}^{\min\{t,t+x/c\}} d_R(x+c(t-q),q)\frac{dq}{\sqrt{t-q}}. \tag{4.31}$$

Substitution of (4.29) into (4.24) produces

$$\begin{aligned}\hat{\hat{d}}_L(p,s)&=\hat{\hat{d}}_0(p,s)-\frac{1}{2\sqrt{s+ipc}}\\&\quad\times\left[\hat{\hat{\delta}}_L^{-1/2}(p,s)-\frac{1}{\pi i}\int_{-\infty}^{\infty}\hat{\hat{\delta}}_L^{-1/2}(\tau,s)\frac{d\tau}{(\tau-p)}\right].\end{aligned} \tag{4.32}$$

Use will now be made of the well-known formula

$$\mathcal{F}[i\operatorname{sgn}(\xi)\mathcal{F}^{-1}(f)(\xi)](x)=\frac{1}{\pi}\int_{-\infty}^{\infty}f(t)\frac{dt}{(t-x)} \tag{4.33}$$

relating the Fourier and Hilbert transforms which, when applied to the Hilbert transform in (4.32), yields the alternative representations

$$\begin{aligned}\frac{1}{\pi i}\int_{-\infty}^{\infty}\hat{\hat{\delta}}_L^{-1/2}(\tau,s)\frac{d\tau}{(\tau-p)}&=\mathcal{F}\left\{\operatorname{sgn}(r)\bar{\delta}_L^{-1/2}(r,s)\right\}\\&=-\hat{\hat{\delta}}_L^{-1/2}(p,s)+2\int_0^{\infty}e^{ixp}\bar{\delta}_L^{-1/2}(x,s)dx \qquad (4.34)\\&=\hat{\hat{\delta}}_L^{-1/2}(p,s)-2\int_{-\infty}^{0}e^{ixp}\bar{\delta}_L^{-1/2}(x,s)dx. \qquad (4.35)\end{aligned}$$

Formula (4.35) is the appropriate alternative to substitute into (4.32) thereby obtaining the result

$$\hat{\hat{d}}_L(p,s)=\hat{\hat{d}}_0(p,s)-\frac{1}{\sqrt{s+icp}}\int_{-\infty}^{0}e^{ixp}\bar{\delta}_L^{-1/2}(x,s)dx. \tag{4.36}$$

With similar calculations to those leading to (4.30) one shows that

$$\begin{aligned}\frac{1}{\sqrt{s+ipc}}\int_{-\infty}^{0}e^{ixp}\bar{\delta}_L^{-1/2}(x,s)dx&=\int_0^{\infty}e^{-ts}dt\int_{-\infty}^{0}e^{ixp}dx\\&\quad\times\frac{1}{\sqrt{\pi}}\int_{\max\{0,t+x/c\}}^{t}\delta_L^{-1/2}(x+c(t-r),r)\frac{dr}{\sqrt{t-r}}.\end{aligned} \tag{4.37}$$

Substitution of (4.37) into (4.36) followed by Fourier and Laplace inversion yields for $x < 0$ the relation

$$d_L(x,t) = d_0(x,t) - \frac{1}{\sqrt{\pi}} \int_{\max\{0,t+x/c\}}^{t} \delta_L^{-1/2}(x + c(t-r), r)\frac{dr}{\sqrt{t-r}}. \qquad (4.38)$$

Combining (4.31) and (4.38) produces finally the equation

$$d_L(x,t) = d_0(x,t) - \frac{1}{\pi} \int_{\max\{0,t+x/c\}}^{t} \frac{dr}{\sqrt{t-r}} \frac{\partial}{\partial r} \int_{\max\{0,b_1^{-1}(t+x/c)\}}^{t+x/c} d_R(x + c(t-q), q)\frac{dq}{\sqrt{r-q}} \qquad (4.39)$$

relating $d_L(x,t)$ and $d_R(x,t)$ for $x < 0$.

Similar calculations applied to (4.25) yield the following equation valid for $x > 0$

$$f_R(x,t) = t_0(x,t) - \frac{\mu}{c\sqrt{\pi}}\left[\frac{\partial}{\partial t} + c\frac{\partial}{\partial x}\right] \int_{\max\{0,t-x/c\}}^{t} \delta_L^{-1/2}(x - c(t-r), r)\frac{dr}{\sqrt{t-r}} \qquad (4.40)$$

in which

$$f_R(x,t) = \begin{cases} 0, & \text{if } 0 < x < a(t) \\ \sigma^+(x - a(t), 0, t), & \text{if } a(t) < x. \end{cases}$$

Equation (4.40) is used in two ways.

First, by restriction to the (x,t)-set $\{(x,t) : 0 < x < a(t)\}$, one generates the operator equation

$$t_0(x,t) = \frac{\mu}{c\sqrt{\pi}}\left[\frac{\partial}{\partial t} + c\frac{\partial}{\partial x}\right] \int_{t-x/c}^{t} \delta_L^{-1/2}(x - c(t-r), r)\frac{dr}{\sqrt{t-r}} \qquad (4.41)$$

which may be inverted to solve for $\delta_L^{-1/2}(x,t)$. Subsequently, equation (4.31) may be inverted to solve for $d_R(x,t)$. Finally, substitution of $d_R(x,t)$ into (4.40) followed by restriction to the (x,t)-set $\{(x,t) : a(t) < x\}$, yields an expression for $f_R(x,t)$, the stress ahead of the crack-tip.

Solving (4.41) is facilitated by recognizing that the right hand side is the composition of fractional derivatives of $d_R(x,t)$ of order 1/2 along each of the two characteristic directions. This is most easily seen by changing the variables in (4.41) to the characteristic coordinates $\xi = t - x/c$, $\eta = t + x/c$ and then noting that since $0 < x < a(t)$, η and ξ must satisfy

$$\xi < \eta < 2b_0^{-1}(\xi) - \xi,$$

or equivalently

$$2b_1^{-1}(\eta) - \eta < \xi < \eta \qquad (4.42)$$

where $b_0^{-1}(\xi)$ and $b_1^{-1}(\eta)$ denote the inverse functions of the retarded and advanced time scales $b_0(q)$ and $b_1(q)$ given by (4.28). One then concludes that

$$t_0(c(\eta-\xi)/2,(\eta+\xi)/2)=\frac{\mu}{c}\sqrt{\frac{2}{\pi}}\frac{\partial}{\partial\eta}\int_\xi^\eta \delta_L^{-1/2}(c(r-\xi)/2,(r+\xi)/2)\frac{dr}{\sqrt{\eta-r}}. \quad (4.43)$$

For fixed ξ, equation (4.43) is an Abel integral equation on the interval $\xi<\eta<2b_0^{-1}(\xi)-\xi$ whose solution is

$$\begin{aligned}\delta_L^{-1/2}(x,t)&=\delta_L^{-1/2}(c(\eta-\xi)/2,(\eta+\xi)/2)\\&=\frac{c}{\mu\sqrt{2\pi}}\int_\xi^\eta t_0(c(q-\xi)/2,(q+\xi)/2)\frac{dq}{\sqrt{\eta-q}}.\end{aligned} \quad (4.44)$$

Similarly, substituting (4.31) into (4.44) and taking $\eta=t+x/c$ to be constant produces an Abel integral equation on the ξ interval, $2b_1^{-1}(\eta)-\eta<\xi<\eta$, whose solution is

$$\begin{aligned}d_R(x,t)&=d_R(c(\eta-\xi)/2,(\eta+\xi)/2)\\&=\frac{c}{2\mu\pi}\int_{2b_1^{-1}(\eta)-\eta}^{\xi}\frac{dr}{\sqrt{\xi-r}}\int_r^\eta t_0(c(q-r)/2,(q+r)/2)\frac{dq}{\sqrt{\eta-q}} \quad (4.45)\\&=\frac{c}{\mu\pi}\int_{b_1^{-1}(t+x/c)}^{t}\frac{dr}{\sqrt{t-r}}\\&\quad\times\int_r^{t+x/c} t_0(c(q-r),q+r-t-x/c)\frac{dq}{\sqrt{t+x/c-q}}. \quad (4.46)\end{aligned}$$

Next one must derive expressions for $d_0(x,t)$ and $t_0(x,t)$ defined through (4.26) and (4.27). It can be shown by arguments similar to those employed previously, that for $x<0$, $d_0(x,t)$ is given by

$$d_0(x,t)=\frac{-c}{\pi\mu}\int_{\max\{0,t+x/c\}}^{t}\frac{dq}{\sqrt{t-q}}\int_0^q\sigma^-(x+c(-2q+t+b_0(r)),0,r)\frac{dr}{\sqrt{q-r}}, \quad (4.47)$$

whereas for $x>0$, $t_0(x,t)$ is given by

$$\begin{aligned}t_0(x,t)&=-\left[\frac{\partial}{\partial t}+c\frac{\partial}{\partial x}\right]\frac{1}{\pi}\int_{\max\{0,t-x/c\}}^{t}\frac{dq}{\sqrt{t-q}}\\&\quad\times\int_0^{\min\{q,b_0^{-1}(t-x/c)\}}\sigma^-(x-ct+cb_0(r)),0,r)\frac{dr}{\sqrt{q-r}}.\end{aligned} \quad (4.48)$$

Substituting (4.48) into (4.46), one can show that the crack-face displacement

$$d_R(x,t) = \frac{-c}{\mu\pi}\int_{b_1^{-1}(t+x/c)}^{t}\frac{dq}{\sqrt{t-q}}\int_0^q \sigma^-(x+c(-2q+t+b_0(r)),0,r)\frac{dr}{\sqrt{q-r}} \tag{4.49}$$

for $0 < x < a(t)$ and subsequently that

$$d_L(x,t) = \frac{-c}{\mu\pi}\int_{\max\{0,b_1^{-1}(t+x/c)\}}^{t}\frac{dq}{\sqrt{t-q}}\int_0^q \sigma^-(x+c(-2q-t+b_0(r)),0,r)\frac{dr}{\sqrt{q-r}} \tag{4.50}$$

for $x < 0$.

The stress ahead of the crack-tip, $f_R(x,t)$, can be calculated by substituting (4.48) and (4.49) into (4.40). Specifically, one shows thereby that for $a(t) < x < ct$,

$$f_R(x,t) = \int_0^{b_0^{-1}(\xi)} \sigma^-(c(b_0(q)-\xi),0,q)h(\xi,t,q)dq \tag{4.51}$$

in which

$$h(\xi,t,q) = \frac{-1}{2\pi}\int_q^{b_0^{-1}(\xi)} (t-r)^{-3/2}(r-q)^{-1/2}dr. \tag{4.52}$$

Finally, from (4.52) and the relation $\sigma^+(x,t) = f_R(x+a(t),t)$ one obtains the result that

$$\sigma^+(x,t) = \int_0^{b_0^{-1}(b_0(t)-x/c)} \sigma^-(c(b_0(q)-b_0(t))+x,0,q)h(b_0(t)-x/c,t,q)dq \tag{4.53}$$

for $x > 0$.

The stress intensity factor may be extracted from (4.53) by rewriting $h(\xi,t,q)$ in (4.52) as

$$h(\xi,t,q) = \frac{-1/\pi}{\sqrt{t-b_0^{-1}(\xi)}\sqrt{b_0^{-1}(\xi)-q}} + h_1(\xi,t,q) \tag{4.54}$$

with $h_1(\xi,t,q)$ given by

$$h_1(\xi,t,q) = \frac{1}{2\pi}\int_{b_0^{-1}(\xi)}^{t} (t-r)^{-1/2}(r-q)^{-3/2}dr.$$

Since $h_1(\xi,t,q)$ produces no contribution to the crack-tip stress singularity, $\sigma^+(x,t)$ is asymptotically represented by

$$\sigma^+(x,t) \sim \frac{1}{\sqrt{t-b_0^{-1}(b_0(t)-x/c)}}\mathcal{K}(x,t) \tag{4.55}$$

as $x \to 0+$ with

$$\mathcal{X}(x,t) = \frac{-1}{\pi} \int_0^{b_0^{-1}(b_0(t)-x/c)} \sigma^-(c(b_0(q) - b_0(t) + x/c), 0, q) \frac{dq}{\sqrt{b_0^{-1}(b_0(t) - x/c) - q}}. \tag{4.56}$$

From (4.18) one has that $\dot{b}_0(t) = 1 - \dot{a}(t)/c$ with which one concludes that

$$\lim_{x \to 0+} \sqrt{\frac{x}{t - b_0^{-1}(b_0(t) - x/c)}} = \lim_{x \to 0+} \sqrt{c\dot{b}_0(b_0^{-1}(b_0(t) - x/c))} = \sqrt{c - \dot{a}(t)}. \tag{4.57}$$

It then follows from(4.55)–(4.57) that

$$\sigma(x,t) \sim \frac{K(t)}{\sqrt{x}} \quad \text{as } x \to 0+$$

where

$$K(t) = -\sqrt{c - \dot{a}(t)} \, \frac{1}{\pi} \int_0^t \sigma^-(c(b_0(q) - b_0(t)), 0, q) \frac{dq}{\sqrt{t-q}}. \tag{4.58}$$

is the stress intensity factor.

In the next chapter, the program outlined above is applied to the corresponding problem in a viscoelastic material. The derivation of the viscoelastic counterparts to the fundamental equations (2.24)–(4.27) proceeds almost exactly as described above and is valid for general shear relaxation functions. However, the inversion of these relations to produce an integral equation for the crack-face displacement is more complicated for viscoelastic material than was shown above for elastic material. In particular, to date the detailed calculations have been performed only for a special case of the Achenbach-Chao material model.

5. AN ACCELERATING VISCOELASTIC ANTI-PLANE SHEAR CRACK

5.1 Introduction

In the previous chapter, a overview was given of a new method for solving dynamically, accelerating, semi-infinite crack problems that was introduced in [31]. That paper considered specifically an anti-plane shear crack in the context of elasticity theory. Even though solutions to that problem had been derived previously by various people using a variety of different techniques, it was utilized in [30] as the vehicle by which to introduce the new solution method in order to enhance understanding through presention in the simplest possible setting. In this chapter, the method is applied to the corresponding problem in a viscoelastic body. Whereas, in principle the method is applicable to a very general viscoelastic constitutive class, a detailed solution has only been constructed to date for the special case of an Achenbach-Chao material with vanishing equilibrium modulus. This solution, derived in [30], is presented here with few details, but then applied to a study of the extent to which the stress intensity factor and crack-face opening displacement exhibit a memory of an initial transient acceleration phase prior to the onset of steady crack motion. This question has important implications for fracture mechanics and serves to illustrate both similarities and differences between elastic and viscoelastic dynamic fracture theory.

5.2 Solution for an Achenbach-Chao Viscoelastic Material

The boundary value problem to be considered is the viscoelastic analog to (4.1)

$$\rho \ddot{u}_3 = \mu * \Delta u_3 \tag{5.1}$$

together with the initial conditions (4.2), boundary conditions (4.3) and viscoelastic constitutive law

$$\sigma_{23}(x_1, x_2, t) = \frac{\partial}{\partial x_2}(\mu * du_3). \tag{5.2}$$

Much of the following analysis is valid for the general constitutive class (2.17), but eventually will be specialized to an Achenbach-Chao material. Also as in chapter 4, the crack-tip motion is restricted only by (4.5).

The general integral transform/complex variable approach of [31] summarized in chapter 4 applies to equation (5.1). Specifically, Fourier and Laplace transformation of (5.1) followed by solution of the resulting ordinary differential equation in x_2 leads to (4.6) where now

$$\beta(p, s) = \sqrt{p^2 + s^2/\bar{c}^2(s)} \tag{5.3}$$

and

$$\bar{c}^2(s) = \bar{\mu}(s)/\rho. \tag{5.4}$$

Corresponding to (4.8) there is

$$\hat{\bar{\sigma}}_{23}(p, 0, s) = -\bar{\mu}(s)\beta(p, s)\hat{\bar{u}}_3(p, 0, s) \tag{5.5}$$

and (4.10) becomes

$$\eta(p, s) = -\bar{\mu}(s)\beta(p, s)\xi^-(p, s). \tag{5.6}$$

Again, in order to derive a canonical Riemann-Hilbert boundary value problem, one must adopt the decomposition (4.15)–(4.18), arriving eventually at the viscoelastic counterpart to (4.19)–(4.20) with $\bar{\mu}(s)$ and (5.3) replacing μ and $\beta(p, s)$, respectively.

The Riemann-Hilbert problem (4.19), modified appropriately, is solved in a similar manner to that in chapter 4 with

$$\begin{aligned} X^+(z, s) &= \sqrt{z + is/\bar{c}(s)} \\ X^-(z, s) &= \frac{-1}{\bar{\mu}(s)\sqrt{z - is/\bar{c}(s)}} \end{aligned} \tag{5.7}$$

as the solution of the associated homogeneous Riemann-Hilbert problem. Again the general Riemann-Hilbert machinery applies and one is lead to the fundamental operator equations in the Fourier/Laplace transform domain

$$\begin{aligned}\hat{\bar{d}}_L(p,s) &= F^-(p,s)\\ &= \hat{\bar{d}}_0(p,s) - \frac{1}{2}\hat{\bar{d}}_R(p,s)\\ &\quad + \frac{1}{\sqrt{p - is/\bar{c}(s)}}\frac{1}{2\pi i}\int_{-\infty}^{\infty}\hat{\bar{d}}_R(\tau,s)\sqrt{\tau - is/\bar{c}(s)}\frac{d\tau}{(\tau - p)}\end{aligned} \tag{5.8}$$

and

$$\begin{aligned}\eta^+(p,s) &= F^+(p,s)\\ &= \hat{\bar{t}}_0(p,s) - \frac{\bar{\mu}(s)}{2}\sqrt{p^2 + s^2/\bar{c}^2(s)}\,\hat{\bar{d}}_R(p,s)\\ &\quad - \bar{\mu}(s)\sqrt{p + is/\bar{c}(s)}\frac{1}{2\pi i}\int_{-\infty}^{\infty}\hat{\bar{d}}_R(\tau,s)\sqrt{\tau - is/\bar{c}(s)}\frac{d\tau}{(\tau - p)}\end{aligned} \tag{5.9}$$

in which the known functions $\hat{\bar{d}}_0(p,s)$ and $\hat{\bar{t}}_0(p,s)$ are given by

$$\begin{aligned}\hat{\bar{d}}_0(p,s) &= \frac{-\eta^-(p,s)}{2\bar{\mu}(s)\sqrt{p^2 + s^2/\bar{c}^2(s)}}\\ &\quad + \frac{1}{\bar{\mu}(s)\sqrt{p - is/\bar{c}(s)}}\frac{1}{2\pi i}\int_{-\infty}^{\infty}\frac{\eta^-(\tau,s)}{\sqrt{\tau + is/\bar{c}(s)}}\frac{d\tau}{(\tau - p)}\end{aligned} \tag{5.10}$$

$$\hat{\bar{t}}_0(p,s) = -\frac{1}{2}\eta^-(p,s) - \sqrt{p + is/\bar{c}(s)}\frac{1}{2\pi i}\int_{-\infty}^{\infty}\frac{\eta^-(\tau,s)}{\sqrt{\tau + is/\bar{c}(s)}}\frac{d\tau}{(\tau - p)} \tag{5.11}$$

which are the viscoelastic counterparts to (4.14)–(4.17).

The derivation of (5.8)–(5.11) is valid for the general constitutive class (2.17). It is in performing the Fourier and Laplace inversion of (5.8)–(5.11) in order to derive the governing integro-differential operator equations relating the unknown crack-face displacement pieces $d_R(x,t)$ and $d_L(x,t)$ to the unknown and known stress functions $f_R(x,t)$ and $\sigma^-(x,t)$,respectively, that additional information about $\mu(t)$ is required. In particular, only for the special case of an Achenbach-Chao material with vanishing equilibrium modulus has the closed form solution to the problem been constructed to date. Achenbach and Chao [45] introduced their model as an approximation to the standard linear solid that is particularly well-suited to constructing solutions to wave propagation problems. Specifically, the Achenbach-Chao, Maxwell and standard linear solid models can be written in the form

$$u(t) = \mu_0 m(t/\tau) \tag{5.12}$$

in which μ_0 is the glassy modulus, τ is a characteristic relaxation time and $m(t)$ is given by

$$m(t) = \delta^2 + (1-\delta^2)e^{-t} \qquad \text{Standard Linear Solid}$$

$$m(t) = \delta^2 + (1-\delta^2)\Big[1 - \Big(\frac{1-\delta}{1+\delta}\Big)t\Big]e^{-t} \qquad \text{Achenbach-Chao}$$

with $0 \le \delta \le 1$. The equilibrium modulus, μ_∞, is given by $\mu_\infty = \mu_0\,\delta^2$ and the Maxwell model is obtained from the standard linear solid in the limit as $\delta \to 0$. The model adopted in [30] has

$$m(t) = (1-t)e^{-t}, \tag{5.13}$$

which is the limit of the Achenbach-Chao model as $\delta \to 0$. The Achenbach-Chao model has the unphysical feature that $m(t)$ is neither monotonically decreasing nor convex, as is usually assumed to be the case for the shear relaxation function and is satisfied by both a standard linear solid and a Maxwell material. Nevertheless, it conforms to the constitutive restrictions (2.17) and as shown in [45] and [30] it gives physically reasonable, at least qualitatively, solutions to wave propagation and dynamic fracture problems. Moreover, because of the form taken by its transform $\bar{\mu}(s)$, it is very convenient for constructing solutions to the viscoelastic momentum equation (5.1). Specifically, one has that $\bar{\mu}(s) = \mu_0\bar{m}(\tau s)$ with

$$\bar{m}(s) = \Big[\frac{\delta^2+s}{1+s}\Big] \qquad \text{Standard Linear Solid}$$

$$\bar{m}(s) = \Big[\frac{\delta+s}{1+s}\Big]^2 \qquad \text{Achenbach-Chao.} \tag{5.14}$$

From (5.14), it is apparent that the Achenbach-Chao model satisfies (2.17) for all $0 \le \delta \le 1$. Moreover, its corresponding complex viscoelastic wave speed function

$$\bar{c}(s) = c\sqrt{\bar{m}(\tau s)} = c\Big[\frac{\delta+s}{1+s}\Big] \tag{5.15}$$

is seen to be analytic in all of the complex s-plane except for a simple pole at $s=-1$. In contrast, the wave speed function $\bar{c}(s)$ for a standard linear solid has branch points at $s=-1$ and $s=\delta$. Also as shown in [30], a useful simplification in the analysis occurs for the special case $\delta = 0$ in (5.15). Consequently, from here on it is assumed that

$$\bar{m}(s) = \Big[\frac{s}{1+s}\Big]^2 \quad \text{and} \quad \bar{c}(s) = c\sqrt{\bar{m}(\tau s)} = c\Big[\frac{s}{1+s}\Big].$$

Many of the manipulations and arguments required to perform the inverse Fourier and Laplace transforms to (5.8)–(5.11) mirror those described in chapter 4 for elastic material. In particular, corresponding to (4.32) one now has

$$\hat{\bar{d}}_L(p,s) = \hat{\bar{d}}_0(p,s) - \frac{1}{2\sqrt{s+1/\tau+ipc}} \times \left[\hat{\bar{\delta}}_L^{-1/2}(p,s) - \frac{1}{\pi i}\int_{-\infty}^{\infty} \hat{\bar{\delta}}_L^{-1/2}(\tau,s)\frac{d\tau}{(\tau-p)}\right]$$

in which $\hat{\bar{\delta}}_L^{-1/2}(p,s)$ is given by

$$\hat{\bar{\delta}}_L^{-1/2}(p,s) \equiv \sqrt{s+1/\tau+ipc}\,\hat{\bar{d}}_R(p,s)$$

in place of (4.29). Corresponding to (4.39) one then shows that

$$d_L(x,t) = d_0(x,t) - \frac{1}{\pi}\int_{\max\{0,t+x/c\}}^{t} \frac{dr}{\sqrt{t-r}} \times \frac{\partial}{\partial r}\int_{\max\{0,b_1^{-1}(t+x/c)\}}^{\eta} e^{-(t-q)/\tau} d_R(c(t+x/c-q),q)\frac{dq}{\sqrt{r-q}}. \tag{5.16}$$

The principal modification of the elastic calculations occurs in deriving the stress/displacement boundary relation. Specifically, it is useful to define a 'pseudo-displacement' $D_R(x,t)$ through

$$\hat{\bar{D}}_R(p,s) = \bar{\mu}(s)\hat{\bar{d}}_R(p,s). \tag{5.18}$$

The key observation from (5.18) is that both $D_R(x,t)$ and $d_R(x,t)$ vanish off of the (x,t)-set $\{(x,t)\ :\ 0<x<a(t)\}$. The counterpart to (4.40) now becomes

$$f_R(x,t) = t_0(x,t) - \frac{2e^{-t/\tau}}{c\sqrt{\pi}}\frac{\partial}{\partial\eta}\int_{\max\{0,\xi\}}^{t} \Delta_R^{-1/2}(c(r-\xi),r)\frac{e^{r/\tau}dr}{\sqrt{t-r}} \tag{5.19}$$

with

$$\Delta_R^{-1/2}(x,t) = 2\frac{e^{-t/\tau}}{\sqrt{\pi}}\frac{\partial}{\partial\xi}\int_{\max\{0,b_1^{-1}(\eta)\}}^{\min\{t,\eta\}} e^{q/\tau} D_R(c(\eta-q),q)\frac{dq}{\sqrt{t-q}} \tag{5.20}$$

where $\xi = t - x/c$ and $\eta = t + x/c$ are the (elastic) characteristic coordinates introduced in chapter 4. Restricting (x,t) in (5.20) to the set $\{(x,t)\ :\ 0<x<a(t)\}$ one obtains the integral equation

$$t_0(x,t) = \frac{2e^{-t/\tau}}{c\sqrt{\pi}}\frac{\partial}{\partial\eta}\int_{\xi}^{t} \Delta_R^{-1/2}(c(r-\xi),r)\frac{e^{r/\tau}dr}{\sqrt{t-r}}. \tag{5.21}$$

Solving the Abel equation (5.21) for $\Delta_R^{-1/2}(x,t)$ subsequently followed by solving the Abel equation (5.20) for $D_R(x,t)$, one can prove that

$$\begin{aligned} D_R(x,t) = \frac{-c}{\pi} \int_{b_1^{-1}(t+x/c)}^{t} \frac{dq}{\sqrt{t-q}} \\ \int_0^q e^{-(t-r)/\tau} \sigma^-(x + c(-2q + t + b_0(r)), 0, r) \frac{dr}{\sqrt{q-r}}. \end{aligned} \tag{5.22}$$

From (5.18) and (5.22) one may then construct the crack-face displacement

$$\begin{aligned} w(x,0,t) &= d_R(x + a(t), t) \\ &= [K_1 + I]_r^2 (\frac{1}{\mu_0} D_R(x + a(t), r)) \end{aligned} \tag{5.23}$$

in which $[K_1 + I]_r(g(x,r))$ denotes the operator

$$[K_1 + I]_r(g(x,r)) = \frac{1}{\tau} \int_{a^{-1}(x)}^{t} g(x,r)dr + g(x,t). \tag{5.24}$$

Returning to (5.19) and restricting (x,t) to the set $\{(x,t) \ : \ a(t) < x\}$ one can construct an expression for the stress ahead of the crack-tip and extract the stress intensity factor

$$K(t) = -\sqrt{c - \dot{a}(t)} \, \frac{1}{\pi} \int_0^t \sigma^-(c(b_0(q) - b_0(t)), 0, q) e^{-(t-q)/\tau} \frac{dq}{\sqrt{t-q}} \tag{5.25}$$

as the viscoelastic counterpart to the elastic result (4.58).

It should be emphasized that (4.58) and (5.25) are not related through a simple elastic/viscoelastic correspondence principle. Thus, unfortunately, the form (5.25) which is valid for the special Achenbach-Chao model (5.13) gives no guidance as to the corresponding form for the stress intensify factor for more general material models. Indeed, it remains an open question as to whether or not closed form expressions corresponding to (5.23) and (5.25) can be constructed for the simple standard linear solid and Maxwell models or even for the Achenbach-Chao model with $\delta > 0$. Nevertheless, it is reasonable to conjecture that for general viscoelastic models, one can derive from (5.8)–(5.11) integral equations which can be solved numerically for the crack-face displacement and stress intensity factor. However, carrying out this program remains an open problem to date.

5.3 The Transition to Steady Crack Motion

One of the remarkable results of the general dynamic fracture theory presented by Freund [4] for elastic material is the limited dependence of the stress intensity factor and crack-face displacement upon crack growth history. In [30] this issue was explored for the special Achenbach-Chao model (5.13) through the consideration of a simple thought experiment in which the crack motion undergoes an initial general acceleration phase followed by a sudden transition to steady, constant crack growth. Specifically, on the interval $0 < t < t_s$, the crack-tip motion $a(t)$ is subject only to the restriction $0 \leq \dot{a}(t) < c$, but for $t > t_s$ $\dot{a}(t) = V$. Thus, $t = t_s$ is the time of transition to steady crack growth. The question to be addressed is: For how long after the crack motion becomes steady do the stress intensity factor and crack-face displacement 'remember' the initial acceleration phase? As shown in [30], the answer depends upon the nature of the loading.

This point can be effectively illustrated by considerating two special loading cases:

Case 1.

$$\text{supp}[\sigma^-(x,t)] = (-L, 0),$$

i.e. the crack-face traction vanishes off of the stationary interval $(-L, 0)$ located *behind and travelling with the crack-tip*;

Case 2.

$$\text{supp}[\sigma^-(x,t)] = (-L - a(t), -a(t)),$$

i.e. the crack-face traction vanishes off of the interval $(-L, 0)$ situated *behind the initial crack-tip location*. In order to study the behavior of (5.13) and (5.15) for $t > t_s$, one must make use of the obvious identity

$$a(t) = V(t - t_s) + a(t_s), \qquad \text{for} \quad t \geq t_s$$

which combined with (4.28) yields

$$b_0(t) = (1 - V/c)(t - t_s) + b_0(t_s), \qquad \text{for} \quad t \geq t_s$$

and

$$b_0^{-1}(t) = \frac{t - b_0(t_s)}{1 - V/c} + t_s, \qquad \text{for} \quad t \geq b_0(t_s). \tag{5.26}$$

Investigating first the stress intensity factor for Case 1 loading, one obtains from (5.15)

$$K(t) = -\sqrt{c - \dot{a}(t)}\,\frac{1}{\pi}\int_{b_0^{-1}(\max\{0, b_0(t) - L/c\})}^{t} \sigma^-(c(b_0(q) - b_0(t)), q)e^{-(t-q)/\tau}\frac{dq}{\sqrt{t-q}}. \tag{5.27}$$

For

$$t > t_s + \frac{L}{c - V} \tag{5.28}$$

one has from (5.26) that

$$b_0^{-1}(b_0(t) - L/c) = t - \frac{L}{c - V}$$

which upon substitution into (5.27) yields the expression

$$K(t) = -\sqrt{c - V}\frac{1}{\pi}\int_0^{\frac{L}{c-V}} \sigma^-(-(c - V)q, t - q)e^{-q/\tau}\frac{dq}{\sqrt{q}}. \tag{5.29}$$

The right side of (5.29) is easily seen to depend only upon the instantaneous amount of crack extension and is independent of the initial acceleration phase on $0 \le t \le t_s$. Moreover, it the applied traction also becomes steady for $t > t_s$, i.e. $\sigma^-(x,t) = \sigma^-(x)$ for $t > t_s$, then for t satisfying (5.28),

$$K(t) = -\sqrt{c - V}\frac{1}{\pi}\int_0^{\frac{L}{c-V}} \sigma^-(-(c - V)q)e^{-q/\tau}\frac{dq}{\sqrt{q}}, \tag{5.30}$$

which is the steady-state value. Thus, for Case 1 loading, the initial acceleration phase is 'remembered' only for a time interval of length $L/(c - V)$ after steady crack growth begins. This result is the same as for elastic material which one can obtain by taking the limit $\tau \to \infty$ in (5.30) .

For Case 2 loading, it is convenient to take $\sigma^-(x,t) = L_e f(x + a(t), t)$ with $\text{supp}[f(x,t)] = (-L, 0)$. The stress intensity factor (5.15) then becomes (after an obvious change of variable in the integral)

$$K(t) = \sqrt{1 - \dot{a}(t)/c}\,\frac{L_e}{\pi}\int_0^{\min\{L, cb_0(t)\}} f(-r, b_0(t) - r/c)e^{-(r+a(t))/(c\tau)}\frac{dr}{\sqrt{r + a(t)}}. \tag{5.31}$$

For $t > t_s$, (5.31) depends only upon the current amount of crack extension and the current crack speed; in particular, there is no dependence upon the initial acceleration phase $0 \le t \le t_s$ for either elastic or viscoelastic material. Thus, in contrast to Case 1 in which the loading travels with the crack-tip, when the load does not travel with the crack-tip the stress intensity factor assumes its constant crack speed value the instant the crack motion becomes steady ($\dot{a}(t) = V$).

Consideration is given next to the crack-face opening displacement at a location $x < 0$ behind the crack-tip. For elastic material one derives from (4.9), (4.16) and (4.49) that

$$\begin{aligned} w(x,0,t) &= d_R(x + a(t), t) \\ &= \frac{-c}{\pi}\int_{b_1^{-1}(b_1(t)+x/c)}^{t} \frac{dq}{\sqrt{t - q}} \\ &\quad \times \int_0^q \sigma^-(x + c(-2q + b_1(t) + b_0(r)), 0, r)\frac{dr}{\sqrt{q - r}}. \end{aligned} \tag{5.32}$$

For Case 1 loading (travelling with the crack-tip), one sees from (5.26) and (5.32) that for

$$t \geq t_s + \frac{L}{c-V} + \frac{|x|}{c+V}, \tag{5.33}$$

the crack-face separation becomes

$$w(x,0,t) = -\frac{c}{\pi\sqrt{c-V}} \int_0^{|x|/(c+V)} \frac{dq}{\sqrt{q}} \times \int_{|x|-q(c+V)}^{L-qV} \sigma^-(-r, t + (|x| - r - 2qc)/(c-V)) \frac{dr}{\sqrt{r - |x| + q(c+V)}}. \tag{5.34}$$

If as in (5.30) one assumes further that the applied traction also becomes steady for $t > t_s$, then (5.34) reduces to the steady-state expression

$$w(x,0,t) = -\frac{c}{\pi\sqrt{c-V}} \int_0^{|x|/(c+V)} \frac{dq}{\sqrt{q}} \times \int_{|x|-q(c+V)}^{L-qV} \sigma^-(-r) \frac{dr}{\sqrt{r - |x| + q(c+V)}}. \tag{5.35}$$

Thus for Case 1 loading in elastic material, steady-state conditions for the crack-face displacement at a location $x < 0$ behind the crack-tip set in at a time $|x|/(c+V)$ greater than that required for the stress intensity factor.

In contrast, for Case 2 loading with $\sigma^-(x,t) = L_e f(x+a(t),t)$ and $\text{supp}[f(x,t)] = (-L, 0)$, one has for elastic material

$$w(x,0,t) = \frac{-cL_e}{\pi} \int_{b_1^{-1}(b_1(t)+x/c)}^{t} \frac{dq}{\sqrt{t-q}} \times \int_0^q f(x + c(-2q + b_1(t) + r), r) \frac{dr}{\sqrt{q-r}}. \tag{5.36}$$

One sees now from (5.36), that for

$$t \geq t_s + \frac{|x|}{c+V}, \tag{5.37}$$

the crack-face displacement becomes

$$w(x,0,t) = \frac{-cL_e}{\pi} \int_{t-|x|/(c+V)}^{t} \frac{dq}{\sqrt{t-q}} \times \int_0^q f(x + c(-2q + b_1(t) + r), r) \frac{dr}{\sqrt{q-r}} \tag{5.38}$$

which exhibits no memory of the initial acceleration phase $0 \leq t \leq t_s$.

For the crack-face displacement in viscoelastic material, one sees from (5.23) that a fading memory of the initial acceleration phase persists at every location $x < 0$ behind the crack-tip and for all time. Thus, the elastic and viscoelastic stress fields exhibit similar memory behavior of initial acceleration transients, even though the two stress fields are not the same or even related through an elastic/viscoelastic correspondence principle. In contrast, the elastic and viscoelastic crack-face displacements behave quite differently with the former having a time limited dependence and the latter a persistent but fading dependence upon the initial acceleration phase. It must be emphasized that these results have only been proven in [30] for the special Achenbach-Chao model (5.13). Nevertheless, it seems quite likely that they reflect, at least qualitatively, the behavior of more general viscoelastic material.

REFERENCES

1. Kaninnen, M.F. and C.H. Popelar: Advanced Fracture Mechanics, Oxford Univ. Press, New York 1985.
2. Ferry, J.D.: Viscoelastic Behaviour and Analysis of Composite Materials, 2nd, Wiley, New York 1970.
3. Lockett, F.J.: Nonlinear Viscoelastic Solids, Academic Press, New York 1972.
4. Freund, L.B.: Dynamic Fracture Mechanics, Cambridge University Press, Cambridge 1990.
5. Willis, J.R.: Accelerating Cracks and Related Problems, in: Elasticity, Mathematical Methods and Applications (Ed. G. Eason and R.W. Ogden), Ellis Horwood Ltd., Chichester 1990, 397-409.
6. Graham, G.A.C.: The Correspondence Principle of Linear Viscoelasticity Theory for Mixed Boundary Value Problems Involving Time-dependent Boundary Regions, Quart. Appl. Math., 26 (1968), 167-174.
7. Schapery, R.A.: Correspondence Principles and a Generalized J Integral for Large Deformation and Fracture Analysis of Viscoelastic Media, Int. J. Frac., 25 (1984), 195-223.
8. Golden, J.M. and G.A.C. Graham: Boundary Value Problems in Linear Viscoelasticity, Springer-Verlag, Berlin 1988.
9. Willis, J.R.: Crack Propagation in Viscoelastic Media, J. Mech. Phys. Solids, 15 (1967), 229-240.
10. Noble, B.: Methods Based on the Wiener-Hopf Technique for the Solution of Partial Differential Equations, Pergamon Press, New York 1958.
11. Walton, J.R.: On the Steady-State Propagation of an Anti-Plane Shear Crack in an Infinite General Linearly Viscoelastic Body, Quart. Appl. Math., 40 No. 1 (1982), 37-52.
12. Fabrizio M. and A. Morro: Mathematical Problems in Linear Viscoelasticity, SIAM, Philadelphia 1992.

13. Atkinson, C. and R.D. List: A Moving Crack Problem in a Viscoelastic Solid, Int. J. Engng. Sci., 10 (1972), 309-322.
14. Atkinson, C. and C.J. Coleman: On Some Steady-State Moving Boundary Problems in the Linear Theory of Viscoelasticity, J. Inst. Maths. Applics., 20 (1977), 85-106.
15. Atkinson, C.: A Note on Some Dynamic Crack Problems in Linear Viscoelasticity, Arch. Mech. Stos., 31 (1979), 829-849.
16. Atkinson, C. and C.H. Popelar: Antiplane Dynamic Crack Propagation in a Viscoelastic Strip, J. Mech. Phys. Solids, 27 (1979), 431-439.
17. Popelar, C.H. and C. Atkinson: Dynamic Crack Propagation in a Viscoelastic Strip, J. Mech. Phys. Solids, 28 (1980), 79-83.
18. Walton, J.R.: The Dynamic Steady-State Propagation of an Anti-Plane Shear Crack in a General Linearly Viscoelastic Layer, J. Appl. Mech., 52 (1985), 853-856.
19. Walton, J.R.: The Dynamic Energy Release Rate for a Steadily Propagating Anti-Plane Shear Crack in a Linearly Viscoelastic Body, J. Appl. Mech., 54 (1987), 635-641.
20. Herrmann, J.M. and J.R. Walton: On the Energy Release Rate for Dynamic Transient Anti-Plane Shear Crack Propagation in a General Linear Viscoelastic Body, J. Mech. Phys. Solids, 37 (1989), 619-645.
21. Schovanec L. and J.R. Walton: The Dynamic Energy Release Rate for Two Parallel Steadily Propagating Mode III Cracks in a Viscoelastic Body, Int. J. Frac., 41 (1989), 133-155.
22. Walton, J.R.: The Dynamic Energy Release Rate for a Steadily Propagating Mode I Crack in an Infinite Linear Viscoelastic Body, J. Appl. Mech., 57 (1990), 343-353.
23. Gakov, F.D.: Boundary Value Problems, Pergamon Press, London 1966.
24. Titchmarsh, E.C.: Introduction to the Theory of Fourier Integrals, Oxford Univ. Press, London 1975.
25. Goleniewski, G.: Dynamic Crack Growth in a Viscoelastic Material, Int. J. Frac., 37 No. 3 (1988), R39-R44.
26. Achenbach, J.D. and Z.P. Bazant: Elastodynamic Near Tip Stress and Displacement Fields for Rapidly Propagating Cracks in Orthotropic Materials, J. Appl. Mech., 42 No. 1 (1975), 183-189.
27. Achenbach, J.D.: Wave Propagation in Elastic Solids, North-Holland, Amsterdam 1973.
28. Herrmann, J.M. and J.R. Walton: On the Energy Release Rate for Dynamic Transient Mode I Crack Propagation in a General Linear Viscoelastic Body, Quart. Appl. Math., (52) No. 2 (1994), 201-228.
29. Herrmann, J.M. and J.R. Walton: [1988], A Comparison of the Dynamic Transient Anti-Plane Shear Crack Energy Release Rate for Standard Linear Solid and Power-Law Type Viscoelastic Materials, in: Elastic-Plastic Failure Modelling of

Structures with Applications, Eds. D. Hui and T.J. Kozik, ASME PVP-Vol 141 (1988), 1-11.
30. Bourne, J.P. and J.R. Walton: On a Dynamically Accelerating Crack in an Achenbach-Chao Viscoelastic Solid, Int. J. Engng. Sci., 31 No. 4 (1993), 569-581.
31. Walton, J.R. and J.M. Herrmann: A New Method for Solving Dynamically Accelerating Crack Problems: Part 1. The Case of a Semi-Infinite Mode III Crack in Elastic Material Revisited, Quart. Appl. Math., 50 No.2 (1992), 373-387.
32. Ravi-Chandar, K. & Knauss, W.G.: Dynamic Crack-Tip Stresses Under Stress Wave Loading—A Comparison of Theory and Experiment, Int. J. Frac., 20 (1982), 209–222.
33. Ravi-Chandar, K. & Knauss, W.G.: An Experimental Investigation into Dynamic Fracture: I. Crack Initiation and Arrest, ibid., 25 (1984), 247–262.
34. Ravi-Chandar, K. & Knauss, W.G.: An Experimental Investigation into Dynamic Fracture: II. Microstructural Aspects, ibid., 26 (1984), 65–80.
35. Ravi-Chandar, K. & Knauss, W.G.: An Experimental Investigation into Dynamic Fracture: III. On Steady-State Crack Propagation and Crack Branching, ibid., 26 (1984), 141–154.
36. Knauss, W.G.: On the Steady Propagation of a Crack in a Viscoelastic Sheet: Experiments and Analysis, in: Deformation and Fracture of High Polymers, H.H. Kausch & R. Jaffee, eds.,Plenum Press, New York (1973), 501–541.
37. Schapery, R.A.: A Theory of Crack Initiation and Growth in Viscoelastic Media I. Theoretical Development, Int. J. Frac., 11 No.1(1975),141–159.
38. Barenblatt, G.I.: The Mathematical Theory of Equilibrium Cracks in Brittle Fracture, in: Advances in Applied Mechanics, vol. VII, Academic Press, New York (1962), 55–129.
39. Kostrov, B.V. and L.V. Nikitin: Some General Problems of Mechanics of Brittle Fracture, Arch. Mech. Stos., 22 (1970), 749-775.
40. Mueller, H.K. and W.G. Knauss: Crack Propagation in a Viscoelastic Strip, J. Appl. Mech., (93) Series E (1971), 483–488.
41. Schovanec, L. and J.R. Walton: The Energy Release Rate for a Quasi-Static Mode I Crack in a Nonhomogeneous Linearly Viscoelastic Body, Engng. Frac. Mech., (28) No. 4 (1987), 445–454.
42. Kostrov, B.V.: Unsteady Propagation of Longitudinal Shear Cracks, Appl. Math. and Mech. (PMM), (30) (1966), 1241-1248.
43. Burridge, R.: An Influence Function for the Intensity Factor in Tensile Fracture, Int. J. Engng. Sci., (14) (1976), 725-734.
44. Willis, J.R.: Accelerating Cracks and Related Problems, in: Elasticity, Mathematical Methods and Applications, G. Eason and R.W. Ogden, Ed.'s, Ellis Horwood Ltd., Chichester (1990), 397-409.

45. Achenbach, J.D. and C.C. Chao: A Three-Parameter Viscoelastic Model Particularly Suited for Dynamic Problems, J. Mech. Phys. Solids, (10) (1962), 245-252.

MIX
Papier aus verantwortungsvollen Quellen
Paper from responsible sources
FSC® C105338

If you have any concerns about our products,
you can contact us on
ProductSafety@springernature.com

In case Publisher is established outside the EU,
the EU authorized representative is:
Springer Nature Customer Service Center GmbH
Europaplatz 3, 69115 Heidelberg, Germany

Printed by Libri Plureos GmbH
in Hamburg, Germany